DC/AC Circuits:
Concepts and Applications

Other books for basic electronics courses

DIGITAL ELECTRONICS: CONCEPTS AND APPLICATIONS
  by Terry L. M. Bartelt

ELECTRONIC DEVICES: CONCEPTS AND APPLICATIONS
  by John Henderson

DC/AC Circuits:
Concepts and Applications

RICHARD PARRETT

Prentice Hall, Englewood Cliffs, New Jersey 07632

Library of Congress Cataloging-in-Publication Data

Parrett, Richard
   DC /AC circuits: concepts and applications / Richard Parrett.
     p   cm.
   Includes index.
   ISBN 0-13-200858-0
   1. Electric circuits—Direct current. 2. Electric circuits—Alternating current. I. Title.
   TK454.15.D57P37 1991   90-22247
   621.349'15—dc20   CIP

Editorial/production supervision and
  interior design: Eileen M. O'Sullivan
Cover design: Computer Graphic Resources
Manufacturing buyers: Mary McCartney and Ed O'Dougherty

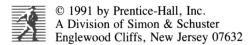 © 1991 by Prentice-Hall, Inc.
A Division of Simon & Schuster
Englewood Cliffs, New Jersey 07632

All rights reserved. No part of this book may be
reproduced, in any form or by any means,
without permission in writing from the publisher.

Printed in the United States of America
10 9 8 7 6 5 4 3 2 1

ISBN 0-13-200858-0

PRENTICE-HALL INTERNATIONAL (UK) LIMITED, *London*
PRENTICE-HALL OF AUSTRALIA PTY. LIMITED, *Sydney*
PRENTICE-HALL CANADA INC., *Toronto*
PRENTICE-HALL HISPANOAMERICANA, S.A., *Mexico*
PRENTICE-HALL OF INDIA PRIVATE LIMITED, *New Delhi*
PRENTICE-HALL OF JAPAN, INC., *Tokyo*
SIMON & SCHUSTER ASIA PTE. LTD., *Singapore*
EDITORA PRENTICE-HALL DO BRASIL, LTDA., *Rio de Janeiro*

Contents

Preface ix

xi

1

Overview 1
EMPLOYMENT OPPORTUNITIES 2
USE OF THE MIND'S EYE 3

2

Physical Definitions 5
PROPERTIES OF CHARGED PARTICLES 6
ATOMIC STRUCTURE 6
ELECTRON MOVEMENT 14
CURRENT 15
ENERGY 17
ELECTROMOTIVE FORCE 20
RESISTANCE 21
CONDUCTANCE 22
SUMMARY 23
QUESTIONS 24
PRACTICE PROBLEMS 25

3

Ohm's Law 27
GETTING STARTED: BASIC ALGEBRA 28
OHM'S FIRST LAW ($E = IR$) 32
OHM'S SECOND LAW ($I = E/R$) 34
OHM'S THIRD LAW ($R = E/I$) 35
POWER (HEAT DISSIPATION) 37
POWER AND OHM'S LAW 37
USING OHM'S LAW 41
CIRCUIT REQUIREMENTS 43
SCHEMATIC SYMBOLS 47
SUMMARY 47
QUESTIONS 47
PRACTICE PROBLEMS 48

4

Series Circuits 50
DIRECTION OF CURRENT FLOW 51
CIRCUIT IDENTIFICATION 51
SERIES-CIRCUIT CONCEPTS 52
SERIES-CIRCUIT ANALYSIS 56
USING THE CIRCUIT FOR THE MATRIX 60
SUMMARY 60
QUESTIONS 61
PRACTICE PROBLEMS 61

5
Parallel Circuits 63
- CIRCUIT IDENTIFICATION 64
- PARALLEL-CIRCUIT CONCEPTS 65
- PARALLEL-CIRCUIT ANALYSIS 71
- FABRICATING RESISTOR VALUES 76
- SUMMARY 77
- QUESTIONS 78
- PRACTICE PROBLEMS 78

6
Series and Parallel Combination Circuits 81
- KIRCHHOFF'S LAWS 82
- FINDING TOTAL RESISTANCE 85
- SERIES–PARALLEL CIRCUIT ANALYSIS 89
- A PRACTICAL CIRCUIT 104
- SUMMARY 106
- QUESTIONS 107
- PRACTICE PROBLEMS 108

7
Troubleshooting DC Circuits 112
- METER USE 113
- GROUND 120
- STEPS IN TROUBLESHOOTING 125
- SPECIAL TESTS 129
- SUMMARY 132
- QUESTIONS 133
- PRACTICE PROBLEMS 133

8
Voltage and Current Dividers 135
- GETTING STARTED: UNDERSTANDING RATIOS 136
- UNLOADED VOLTAGE DIVISION 139
- CURRENT DIVISION 143
- LOADED VOLTAGE DIVISION 145
- SUMMARY 147
- QUESTIONS 147
- PRACTICE PROBLEMS 148

9
Network Theorems 150
- SUPERPOSITION 151
- THÉVENIN'S RULE 154
- NORTON'S RULE 159
- SOURCE IMPEDANCE AND LOAD MATCHING 163
- MILLMAN'S THOREM 165
- DELTA AND WYE CIRCUITS 169
- SUMMARY 172
- QUESTIONS 172
- PRACTICE PROBLEMS 173

10
Additional Methods of Analysis 175
- GETTING STARTED: UNDERSTANDING LINEAR ALGEBRA 176
- KIRCHHOFF'S LAWS 182
- MESH CURRENT ANALYSIS 183
- SUMMARY 193
- QUESTIONS 193
- PRACTICE PROBLEMS 194

11
Circuit Conductors and Controlling Devices 198
- CONDUCTORS 199
- RESISTIVITY 200
- RESISTIVITY OF WIRE 201
- WIRE RESISTANCE 205
- TEMPERATURE COEFFICIENT OF RESISTANCE 208
- WIRE CONSTRUCTION 210
- SWITCHES 211
- CIRCUIT PROTECTION 212
- SUMMARY 214
- QUESTIONS 214
- PRACTICE PROBLEMS 215

12
VOM Meter Design and Considerations 217
- METER LOADING 218
- METER MOVEMENTS AND THEIR CONSTRUCTION 219

CONSTRUCTION AND DESIGN OF VOLTMETERS   221
CONSTRUCTION AND DESIGN OF CURRENT METERS   225
DESIGN AND CONSTRUCTION OF RESISTANCE METERS   229
EXTENDING METER RANGES   235
SUMMARY   236
QUESTONS   236
PRACTICE PROBLEMS   237

13

Magnetism   240
MAGNETIC CONCEPTS   241
MAGNETIC UNITS AND SYSTEMS   247
ELECTROMAGNETIC CONCEPTS   250
B H CURVE: FLUX DENSITY VERSUS FIELD INTENSITY   252
SUMMARY   254
QUESTIONS   254
PRACTICE PROBLEMS   255

14

AC Voltage   258
GETTING STARTED: SINE WAVES   259
CALCULATOR USE   263
AC THEORY   267
VOLTAGE MEASUREMENT   268
PERIOD AND FREQUENCY   273
NONSINUSOIDAL WAVEFORMS   277
SUMMARY   281
QUESTIONS   282
PRACTICE PROBLEMS   283

15

Inductors   286
GETTING STARTED: RIGHT TRIANGLES   287
CONSTRUCTION AND DESIGN   292
INDUCTANCE   295
MUTUAL INDUCTANCE   297
TRANSFORMERS   301
INDUCTIVE REACTANCE   306
SOLUTION OF LR CIRCUITS   311
TROUBLESHOOTING INDUCTORS   317
SUMMARY   318
QUESTIONS   319
PRACTICE PROBLEMS   321

16

Capacitors   324
CONSTRUCTION AND DESIGN   325
CAPACITANCE   329
CAPACITIVE REACTANCE   330
CAPACITIVE VOLTAGE DIVIDERS   335
SOLUTION OF RC CIRCUITS   337
TROUBLESHOOTING CAPACITORS   342
SUMMARY   343
QUESTIONS   344
PRACTICE PROBLEMS   346

17

Transients in Current and Voltage   348
GETTING STARTED: LOGARITHMS   349
VOLTAGE LEVELS AND CURRENT LEVELS   351
METHODS FOR FINDING VOLTAGE AND CURRENT LEVELS   353
PULSE AND WAVE SHAPING   362
PRACTICAL APPLICATIONS   363
SUMMARY   365
QUESTIONS   366
PRACTICE PROBLEMS   367

18

Batteries   370
BATTERY CONSTRUCTION   371
BATTERY TYPES   372
BATTERY CAPACITY   377
BATTERY CHARGING   378
INTERNAL RESISTANCE   380
SUMMARY   382
QUESTIONS   382
PRACTICE PROBLEMS   384

19

RCL Circuit Analysis   387
CAPACITIVE REACTANCE VERSUS INDUCTIVE REACTANCE   388
POWER FACTOR   392
APPLIED VOLTAGE AS A REFERENCE   394
RECTANGULAR NOTATION   396

POLAR NOTATION 398
IMPEDANCE BLOCKS 399
RESONANCE 402
SUMMARY 403
QUESTIONS 404
PRACTICE PROBLEMS 405

20

Filters 408
DECIBELS, POWER, AND VOLTAGE 409
FILTER TYPES 412
FILTER FAMILIES 416
FILTER CONSTRUCTION 420
FLUCTUATING DC VOLTAGES 422
SUMMARY 426
QUESTIONS 427
PRACTICE PROBLEMS 428

21

Resonance 431
SERIES RESONANT CONCEPTS 432
THE TANK CIRCUIT 432
PARALLEL RESONANT CONCEPTS 433

THE QUALITY FACTOR 435
RESONANT FREQUENCY AND BANDWIDTH 435
CIRCUIT ANALYSIS 436
DAMPING RESISTANCE 447
TANK DESIGN 452
TROUBLESHOOTING 455
SUMMARY 456
QUESTIONS 457
PRACTICE PROBLEMS 458

Appendix 1: Calculators 461
TYPES OF CALCULATORS 461
CALCULATOR FUNCTIONS 462
ACCURACY 464

Appendix 2: Engineering and Scientific Notation 465
SCIENTIFIC NOTATION 465
ENGINEERING NOTATION 467

Glossary 470

Answers to Selected Practice Problems 475

Preface

After teaching dc/ac theory for a number of years, it became apparent that there was something lacking in the textbooks offered. It seems that some students had difficulty with the mathematics involved in electronics and others had difficulty in understanding the reading terminology. Textbooks—and I've looked at quite a few—either rely heavily on mathematic relations or on verbal descriptions. My intent in writing this book is to provide approximatly an equal distribution between the two.

Mathematic relations are explained using simple analogies and start at the beginning of algebra. The process is simple at the onset but becomes more rigorous in the latter chapters. The intent is to provide meaningful descriptions, which can be built upon, to raise the math level of the reader from a low level to a fairly high level as painlessly as possible.

Verbal descriptions are provided in an easy-reading, conversational style. The result, I think, is an understandable text, which provides maximum knowledge with the first reading. Simple analogies are used to relate complicated theory to everyday life and general knowledge. Again, the reading level in the earlier chapters is low, but becomes more complex as the book progresses. The intent is to raise the comprehension level of the reader to an adequate point for subsequent study in electronics.

Richard Parrett

ACKNOWLEDGEMENTS

I wish to express my appreciation to the members of my review team. Their comments and suggestions were more than helpful.
    Randall G. Epstein, Total Technical Institute
    James D. Everett, Platte Co. Tech School

Albert J. Gabryash, St. Phillips College
Edward Lee Hoffman, National Education Center
Rick Hoover, Owens Technical College
T. K. Ishii, Marquette University
Samuel James Kirk, Bryan Institute
John Redden
John Steven Richards, ITT Technical Institute

I also wish to thank:

Prentice-Hall for giving me the opportunity to write this book.

Paul H. Smith who provided the photographs.

Terry L. Stivers who provided continual technical support.

The staff of ITT Technical Institute, St. Louis, for providing encouragement, AND my family for understanding my absence.

Safety and Electronics

It never seems to fail!! A little knowledge is dangerous, especially where electronics is concerned. The purpose of this section is to give you an idea of the safe way to work with electricity. First let's examine some common things that do not conduct electricity. Wood, rubber, plastic, and glass all fall in this category. All types of metals, concrete, water, and carbon (pencil lead) are common conductors of electricity.

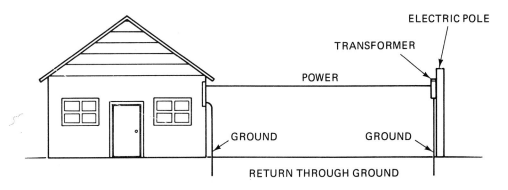

Now let's examine the electrical system that provides electricity to your home and school. It consists of a power wire and a return. The return is always a path to ground. Ground is exactly what it sounds like—the earth or any conductor that is in contact with the earth. Most electrical systems have two return paths, one called the return and the other called ground. Safety necessitates these two paths. To work safely with electricity it is necessary *not* to provide a path for electricity through your body. This is the number 1 rule for safety. One way to do this is to work with only one wire at a time and always to have a dry nonconductor between yourself and ground. Also, always treat every electrical circuit as if it had power to it. Of course, you should never work on an electrical circuit that has not been disconnected; however, like a loaded gun, a live circuit seldom accidentally hurts anyone. You should treat all weapons as if they are loaded, and, similarly, you should treat all circuits as if they are live.

Two good habits to develop are: (1) work with one hand in your pocket, because this reduces the chances of accidentally touching power with one hand while the other is touching ground; and (2) never wear jewelry when working on electrical circuits, because most jewelry is made from metals that are good conductors of electricity and can inadvertently cause a serious shock.

It is the combination of voltage and current that kills. The critical factor is the amount of current. Under the right circumstances as little as 40 V can kill you. Think about it. Your pocket radio is probably powered by a 9-V battery (one-fourth of the critical voltage). Your household voltage is 120 V (three times the critical voltage). The spark plugs in your car are powered by 10,000 V (250 times the critical voltage). Under the right circumstances as little as 0.001 A of current can kill you. Think about it. When you start your car's engine the battery is producing 100 A of current (100,000 times the critical current). A 100-W light bulb in your home requires 0.833 A (833 times the critical current) of current to burn.

When safely working with electricity, nothing can replace common sense. It is important that you avoid working when you are overtired, when your judgment is impaired (due to drugs or alcohol), or when you are in an excessive hurry to get done. Any of these things can cause you to make a life-threatening (yours or another's) mistake.

Most people begin any study in electronics by following all the safety rules. It is understandable that there is a large amount of material to remember and at times you will forget one or more of the safety requirements. So check and recheck all variables. And if you do forget and you are lucky and nothing unfortunate happens, remember that these practices are important to your safety. Count yourself lucky. Try hard not to forget, and above all, do not allow yourself to develop unsafe work habits. **Your life is at stake!**

Overview

1

EMPLOYMENT OPPORTUNITIES

Electronic Fields There are three major fields in electronics: communications, digital, and industrial. A description of each follows.

Communications is the study of television, radio, and telephone systems. Each of these areas has changed greatly in the last few years. Camcorders have made it possible to make home movies for your TV set. AM radio stations are now playing stereo. Cellular phones are available for your car. Communications includes radar, lasers and fiber optics, and microwave and satellite systems.

Digital is the study of the logic gates that make up computers and microprocessors. There have been more advances in this area than in almost any other area of electronics and there will continue to be advances. In 1970 a computer having 256 kilobytes (kB) of memory would have been thought to be extravagant if not impossible. In 1980 a standard computer had 32 kB. Today many programs require that your computer have 640 kB and some systems are configured for 4000 kB. Microprocessors of the past have been capable of performing tasks one at a time. Future (near future) microprocessors will be multitasking and have 1000 or 2000 kB of memory on one chip. Almost every home has a personal computer. If the home does not have a personal computer, it will have a VCR, microwave oven, or television set, all of which have a microprocessor. Even your car (1980 or later) has a microprocessor.

Industrial is the study of motor controls (SCRs, diacs and triacs, relays, and solenoids) and process controls (speed sensors, pressure sensors, and temperature sensors). In the last few years, automated manufacturing techniques (robots) have changed industrial electronics. Here again, the microprocessor and computer have had a great impact. Programmable controllers are used to determine when and how a particular robot will function. The electronics on such systems is simple but must be interfaced with hydraulic (fluid pressure) or pneumatics (air pressure) and require that the technician have good mechanical skills.

Biomedical is a specialized field of electronics that is a blend of all three groups. Electrical impulses are used to speed healing. Lasers are used for surgery. Mechanical prostheses are used to aid the handicapped. Computers are used to speed diagnosis. Where x-rays were once used, CAT scans are now substituted. All of these require that the technician have a much larger base of knowledge than ever before.

Employment Areas There are four areas of employment in each of these fields: sales, operation, service and repair, and design. We will concentrate on television service and repair as an example; however, there are similar examples of other appliances and for each of the other three areas.

In 1960, it was possible to go into the TV repair business with a small investment in tubes and a little knowledge of the system. Most TV sets could be repaired by changing a tube. In 1970, the same was true, except that most sets had some transistors. The required investment was more and you had to have a greater knowledge of the system. In 1980, systems began to have more and more extras: remote control, electronic tuning, and so on, each requiring that the technician have a more extensive knowledge of systems. In 1990, everything has changed.

First, a 19-inch color television set with remote control (in 1960, people would have died for such a set) costs about $270. Labor cost (varies by locale) is about $50 an hour. When presented with a choice of paying $100 to fix a three-year-old set or buying a new one, most people buy a new one. This would make it seem that the repair business is not a lucrative area to work in, but that is not the case. True—people don't repair $270 sets, but large-screen systems with stereo and other options cost $2000+ and do need repair. People want their sets back quickly. The technician must have a good knowledge of many more systems. Also, to provide quick repair, he or she must be able to narrow down the problem to a specific area in a short time.

There are similar examples for other items: computers, telephone systems, alarm systems, robots, and industrial machinery. With these, downtime costs the operator money. Lots of money! So quick repair is mandatory. In most cases, the time needed to find the component at fault costs more than replacement of the faulty section. So board troubleshooting, followed by board replacement, are done to reduce downtime. Later, a bench technician will identify and replace the faulty component allowing the board to be used for future repairs.

You and Electronics At this time it may not be possible for you to know exactly what your future employment will be. But no matter what field or area of electronics you specialize in, the information provided in the remainder of this book is necessary to your future. Volts, ohms, amperes, and watts are the basic units of electricity and form the foundation for all fields of electronics. Communication technicians, digital technicians, and industrial technicians work with these units every day. It is necessary for technicians to understand how the units relate to each other. You will study each of these in detail in the following chapters.

USE OF THE MIND'S EYE

Psychologists have determined that verbal and written material is stored in one side of the brain; pictures and ideas are stored in the other. It is important to develop both sides of the brain. Learning to think in terms of pictures will help to strengthen your mind's eye. It is important in electronics to think of abstract ideas in terms of the concrete. Chances are you will never see an electron flow through a piece of wire; however, you must be able to picture this movement in your mind in order to trace the path of electric flow through a circuit. This will be helpful when you are analyzing circuits.

One way to exercise both sides of your brain is to use your calculator as a tool—not as a replacement for paper and pencil. Doing your calculations one step at a time and neatly writing the results for each step provides equal stimulation to both sides of your brain. Analyzing the problem and deciding what function to perform exercises one side (pictures and ideas); writing the values down exercises, the other (verbal and written). Think of it this way. Picturing how a thing works is not easy until you have seen or made a sketch of it. Using a calculator is a lot like picturing how a thing works; Writing the steps down gives you a sketch of its operation. Although writing each step may seem time consuming, it will save you time (and confusion) in the long run.

All too often students get caught up in a trap! They begin to memorize formulas blindly. Memorizing any formula exercises only one side of the brain. This section is an attempt to discourage the practice of memorizing every form of equation or formula that you contact.

There are nine different forms of Ohm's law. Most students begin electronics by memorizing all of these forms. They are useful, but using a little algebra can cut by two-thirds the amount you need to memorize. When you come in contact with a new equation, compare it to one with which you are already familiar. Surprisingly, in electronics the formulas are not very different from each other. Develop the habit of thinking about what a formula is indicating, and form a picture of this relationship in your mind. For example, the equal sign is a lot like a balance scale and can be thought of as such. When looking at a formula, picture each side of the equation as a weight placed on a balance scale. Removing something from either side will tip the scale.

Physical Definitions

Chapter objectives

After reading this chapter and answering the questions and problems, you should be able to:

- Describe the properties of like and unlike charges, including the relation between electrons and positive, negative, and neutral charges.
- Describe the structure of an atom.
- List each of the orbital shells and state the maximum number of electrons that each can contain.
- Use the valance (number of valance electrons) to define a conductor, semiconductor, and insulator.
- Use the periodic table to find metal, nonmetal, and transitional (semiconductor) elements.
- Define subshells.
- Describe electron movement (current) in terms of electron theory and hole theory.
- Define, and state the relation between, current, coulomb, energy, calorie, joule, electromotive force, voltage, potential difference, resistance, conductance, ohm, and mho.
- Given the number of ohms of resistance, calculate the amount of conductance, and vice versa.
- Given any two of force, distance, or work, find the missing quantity.
- Given any two of amperage, charge (in coulombs), and time, find the missing quantity.
- Given any two of voltage, current, or resistance, find the missing quantity.
- List the symbol, unit, and unit symbol for each of the following: electromotive force, resistance, charge, current, power, work, and conductance.
- Describe the name, use, and function of the 1/x, EE, and = keys on your calculator.

The first step in learning any subject is to familiarize yourself with the language of that subject. The study of electronics entails the use of physical terms. Scientists and engineers try to relate everything to physics, but most people do not have a great knowledge of physics. For that reason, in this chapter we define and explain the physical terms normally used to describe electrical quantities in words that are meaningful to you. Any in-depth study of electronics requires that you develop a working knowledge of physics. This chapter is intended to help you gain that knowledge.

PROPERTIES OF CHARGED PARTICLES

There are two main things to remember about charged particles: (1) *like charges repel*, and (2) *unlike charges attract* each other (Figure 2-1). Everything in the universe has either a positive, a negative, or a neutral charge. This charge is determined by the number of electrons that are present or missing from the material in question. An excess of electrons leaves the material with a *negative charge*, and a deficiency of electrons provides a *positive charge*. When there are an equal number of positive and negative charges in a material, neither an excess nor a deficiency is present, and that material is said to be *balanced* (*neutral*). In the sections that follow we explain these concepts in more detail.

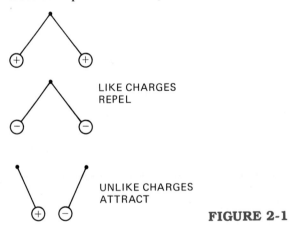

FIGURE 2-1

ATOMIC STRUCTURE

All materials are made up of *atoms*. The atom is the basic building block for all things. Iron is different from silver because of the difference in the structure of their atoms. The atom is the smallest particle of iron that we can examine and still be looking at iron. The atom can be broken down further, but when it is, no difference can be seen between iron or silver except for the quantities of particles that remain.

Atoms are divided up into two sections: the center (*nucleus*) and the outer section (*electron orbitals*). The center consists of two differing particles: (1) the *proton*, which is the positively charged portion of the atom, and (2) the *neutron*, which is neutral in charge (Figure 2-2). Around the nucleus in several distinct patterns are the orbits traveled by the electrons. Remember that the electrons are the negatively charged portion of the atom.

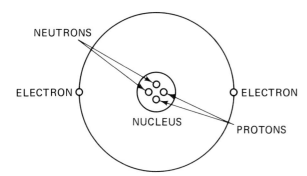

FIGURE 2-2 Helium molecule.

In its normal state, the atom of any material is neutral in charge. This means that there are an equal number of electrons and protons in a molecule of iron, silver, or any other element in its normal state. *Isotopes* are molecules of iron (or another element) which are heavier than the normal iron molecule. This extra weight is caused by a larger than normal number of neutrons present in the nucleus, or center of the atom. The electrons are light and contribute very little to the weight of an atom. The atomic weight is typically the sum of the neutrons and protons of an atom (see Figure 2-3).

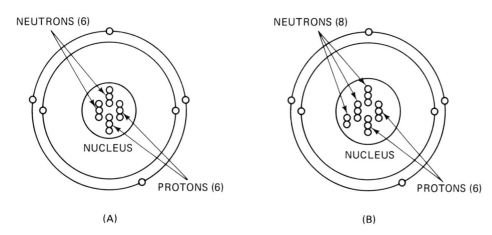

FIGURE 2-3 (A) Carbon-12 molecule; (B) Carbon-14 molecule.

The electrons travel around the nucleus at a very high rate of speed. If they did not, there would be enough attraction from the protons to cause the electrons to enter the center of the atom (unlike charges attract). Not only do the electrons travel around the nucleus at a high rate of speed, but they also travel in predetermined paths. These paths are called *orbits*. The orbits are lettered for identification. They begin with the K shell or orbital and follow an alphabetic pattern (K shell, L shell, M shell, N shell, etc.; see Figure 2-4). Each orbital can hold only a specified number of

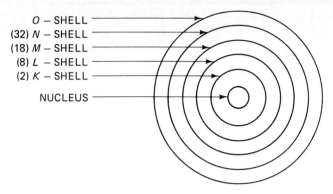

FIGURE 2-4 Orbital shells.

electrons (the *K* shell can hold only two electrons, and the *L* shell can hold only eight electrons). The number of electrons possible for the first few shells can be calculated by the formula $2n^2$, where *n* is the number of the shell's location (*n* for the *K* shell is 1, *n* for the *L* shell is 2, etc.). This formula gives the maximum number of electrons allowed in each shell.

The outermost orbital is called the *valence shell* (Figure 2-5). It is the valence shell that determines the electrical properties of the material. The ideal number of electrons in the valence shell is eight, except for the *K* shell, which contains only two. Materials that contain seven or eight electrons in their valence shell are not willing to give up their electrons. They are at a stable state. These materials make good insulators (nonconductors). Materials that contain one or two electrons in their valence shell are willing to give up their electrons to reach a stable state. Understand

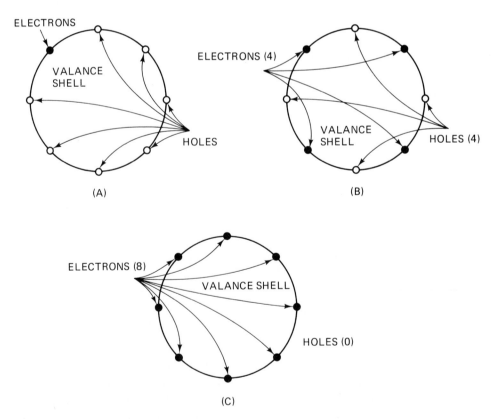

FIGURE 2-5 (A) Outer shell of a conductor; (B) Shell of a semiconductor; (C) Shell of an insulator.

CHAP. 2 / PHYSICAL DEFINITIONS

that such a material will have a full outer shell containing eight electrons after giving up these one or two outer electrons. After such an exchange, the next-lower orbital is completely filled. Materials with one or two outer electrons make good conductors. Materials containing four electrons in their valence shell are equally willing to give up electrons or to pick up electrons from another atom. They are conductors at times and insulators at others. They are called *semiconductors* and are used to make transistors and diodes.

There are some exceptions to these rules, which stem from the fact that the shells around the nucleus are broken down into suborbitals (s, p, d, f). These suborbitals do not fill in a one-layer-at-a-time method, but instead, fill in a piecemeal manner. So elements that have an atomic number of 30 or more may not follow this pattern. It is not my intention for you to have a total understanding of the inner workings of the atoms. At this time, it is my intention for you to understand the following.

1. Any material that contains free electrons (valence shell = one or two electrons) will easily conduct electricity.
2. Materials that have few or no free electrons (valence shell = seven or eight electrons) will not easily conduct electricity.
3. Materials with a valence of 4 are classified as semiconductors.

Let's think about it another way: For an electron to move (electricity), there must be a hole for it to move into. Conductors have many holes, so movement is easy (see Figure 2-6).

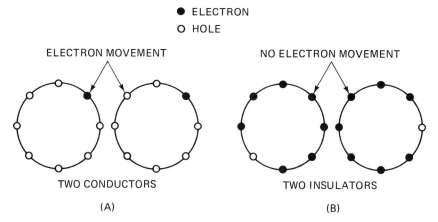

FIGURE 2-6 (A) The many holes in conductors allow easy movement of electrons; (B) The absence of holes inhibits the movement of electrons.

In examining the following charts, it is necessary to know the symbol for each element. Table 2-1 lists each element and its symbol. The atomic number has also been included, for easier reference.

Figure 2-7 is a periodic table of elements. It contains the atomic number, element symbol, and possible valence for each element. The elements are grouped in columns according to similar properties. You will also notice a stair-step pattern; this separates the metals from the nonmetals. Any element on the left of this stair step is classified as a metal. These metals are all malleable, conductive, and have a bright luster. Also, any element that has a valence of +4 and/or −4 is classified as a semiconductor. Carbon, silicon, and germanium are common semiconductors. The

ATOMIC STRUCTURE

TABLE 2-1 The Elements and Their Symbols

Element	Symbol	Atomic Number	Element	Symbol	Atomic Number
Actinium	Ac	89	Molybdenum	Mo	42
Aluminum	Al	13	Neodymium	Nd	60
Americium	Am	95	Neon	Ne	10
Antimony	Sb	51	Neptunium	Np	93
Argon	Ar	18	Nickle	Ni	28
Arsenic	As	33	Niobium	Nb	41
Asatine	At	85	Nitrogen	N	7
Barium	Ba	56	Nobelium	No	102
Berkelium	Bk	97	Osmium	Os	76
Beryllium	Be	4	Oxygen	O	8
Bismuth	Bi	83	Palladium	Pd	46
Boron	B	5	Phosphorus	P	15
Bromine	Br	35	Platinum	Pt	78
Cadmium	Cd	48	Plutonium	Pu	94
Calcium	Ca	20	Polonium	Po	84
Californium	Cf	98	Potassium	K	19
Carbon	C	6	Praseodymium	Pr	59
Cerium	Ce	58	Promethium	Pm	61
Cesium	Cs	55	Protactinium	Pa	91
Chlorine	Cl	17	Radium	Ra	88
Chromium	Cr	24	Radon	Rn	86
Cobalt	Co	27	Rhenium	Re	75
Copper	Cu	29	Rhodium	Rh	45
Curium	Cm	96	Rubidium	Rb	37
Dysprosium	Dy	66	Ruthinium	Ru	44
Einsteinium	Es	99	Samarium	Sm	62
Erbium	Er	68	Scandium	Sc	21
Europium	Eu	63	Selenium	Se	34
Fermium	Fm	100	Silicon	Si	14
Fluorine	F	9	Silver	Ag	47
Francium	Fr	87	Sodium	Na	11
Gadolinium	Gd	64	Strontium	Sr	38
Gallium	Ga	31	Sulfur	S	16
Germanium	Ge	32	Tantalum	Ta	73
Gold	Au	79	Technetium	Tc	43
Hafnium	Hf	72	Tellurium	Te	52
Helium	He	2	Terbium	Tb	65
Holmium	Ho	67	Thallium	Tl	81
Hydrogen	H	1	Thorium	Th	90
Indium	In	49	Thulium	Tm	69
Iodine	I	53	Tin	Sn	50
Iridium	Ir	77	Titanium	Ti	22
Iron	Fe	26	Tungsten	W	74
Krypton	Kr	36	Uranium	U	92
Landthanum	La	57	Vanadium	V	23
Lead	Pb	82	Xenon	Xe	54
Lithium	Li	3	Ytterbium	Yb	70
Lutetium	Lu	71	Yttrium	Y	39
Magnesium	Mg	12	Zinc	Zn	30
Mendelebium	Md	101	Zirconium	Zr	40
Mercury	Hg	80			

Group Number	IA	IIA	IIIB	IVB	VB	VIB	VIIB	VIII			IB	IIB	IIIA	IVA	VA	VIA	VIIA	0
Atomic number	1																	2
Element	H																	He
Valence	+1, −1																	0
Atomic number	3	4											5	6	7	8	9	10
Element	Li	Be											B	C	N	O	F	Ne
Valence	+1	+2											+3	+2, +4, −4	+1, +2, +3, +4, +5, −1, −2, −3	−2	−1	0
Atomic number	11	12											13	14	15	16	17	18
Element	Na	Mg											Al	Si	P	S	Cl	Ar
Valence	+1	+2											+3	+2, +4, −4	+3, +5, −3	+4, +6, −2	+1, +5, +7, −1	0
Atomic number	19	20	21	22	23	24	25	26	27	28	29	30	31	32	33	34	35	36
Element	K	Ca	Sc	Ti	V	Cr	Mn	Fe	Co	Ni	Cu	Zn	Ga	Ge	As	Se	Br	Kr
Valence	+1	+2	+3	+2, +3, +4	+2, +3, +4, +5	+2, +3, +6	+2, +3, +4, +7	+2, +3	+2, +3	+2, +3	+1, +2	+2	+3	+2, +4	+3, +5, −3	+4, +6, −2	+1, +5, −1	0
Atomic number	37	38	39	40	41	42	43	44	45	46	47	48	49	50	51	52	53	54
Element	Rb	Sr	Y	Zr	Nb	Mo	Tc	Ru	Rh	Pd	Ag	Cd	In	Sn	Sb	Te	I	Xe
Valence	+1	+2	+3	+4	+3, +5	+6	+4, +6, +7	+3	+3	+2, +4	+1	+2	+3	+2, +4	+3, +5, −3	+4, +6, −2	+1, +5, −1	0
Atomic number	55	56	57	72	73	74	75	76	77	78	79	80	81	82	83	84	85	86
Element	Cs	Ba	La	Hf	Ta	W	Re	Os	Ir	Pt	Au	Hg	Tl	Pb	Bi	Po	At	Rn
Valence	+1	+2	+3	+4	+5	+6	+4, +6, +7	+3, +4	+3, +4	+2, +4	+1, +3	+1, +2	+1, +3	+2, +4	+3, +5	+2, +4		0
Atomic number	87	88	89	104	105	106	107	108										
Element	Fr	Ra	Ac															
Valence	+1	+2	+3															

Lanthanides

Atomic number	58	59	60	61	62	63	64	65	66	67	68	69	70	71
Element	Ce	Pr	Nd	Pm	Sm	Eu	Gd	Tb	Dy	Ho	Er	Tm	Yb	Lu
Valence	+3, +4	+3	+3	+3	+2, +3	+2, +3	+3	+3	+3	+3	+3	+3	+2, +3	+3

Actinides

Atomic number	90	91	92	93	94	95	96	97	98	99	100	101	102	103
Element	Th	Pa	U	Np	Pu	Am	Cm	Bk	Cf	Es	Fm	Md		Lw
Valence	+4	+5, +4	+3, +4, +5, +6	+3, +4, +5, +6	+3, +4, +5, +6	+3, +4, +5, +6	+3	+3, +4	+3					

FIGURE 2-7

elements to the right of the stair step are classified as nonmetals. They are generally nonconductive.

Table 2-2 contains a list of the basic elements, the number of electrons, and the shells in which they are found. Looking at the table, you will note that there are seven shells (K, L, M, N, O, P, and Q). These shells may have from one to four

TABLE 2-2 Basic Elements: Electrons and Shells

Location Orbital Suborbital Element Number		1 K s	2 L s p	3 M s p d	4 N s p d f	5 O s p d f	6 P s p d f	7 Q s p d f
H	1	1						
He	2	2						
Li	3	2	1					
Be	4	2	2					
B	5	2	2 1					
C	6	2	2 2					
N	7	2	2 3					
O	8	2	2 4					
F	9	2	2 5					
Ne	10	2	2 6					
Na	11	2	2 6	1				
Mg	12	2	2 6	2				
Al	13	2	2 6	2 1				
Si	14	2	2 6	2 2				
P	15	2	2 6	2 3				
S	16	2	2 6	2 4				
Cl	17	2	2 6	2 5				
Ar	18	2	2 6	2 6				
K	19	2	2 6	2 6	1			
Ca	20	2	2 6	2 6	2			
Sc	21	2	2 6	2 6 1	2			
Ti	22	2	2 6	2 6 2	2			
V	23	2	2 6	2 6 3	2			
Cr	24	2	2 6	2 6 5	1			
Ma	25	2	2 6	2 6 5	2			
Fe	26	2	2 6	2 6 6	2			
Co	27	2	2 6	2 6 7	2			
Ni	28	2	2 6	2 6 8	2			
Cu	29	2	2 6	2 6 10	1			
Zn	30	2	2 6	2 6 10	2			
Ga	31	2	2 6	2 6 10	2 1			
Ge	32	2	2 6	2 6 10	2 2			
As	33	2	2 6	2 6 10	2 3			
Se	34	2	2 6	2 6 10	2 4			
Br	35	2	2 6	2 6 10	2 5			
Kr	36	2	2 6	2 7 10	2 6			
Rb	37	2	2 6	2 6 10	2 6	1		
Sr	38	2	2 6	2 6 10	2 6	2		
Y	39	2	2 6	2 6 10	2 6 1	2		
Zr	40	2	2 6	2 6 10	2 6 2	2		
Cb	41	2	2 6	2 6 10	2 6 4	1		
Mo	42	2	2 6	2 6 10	2 6 5	1		
Tc	43	2	2 6	2 6 10	2 6 6	1		
Ru	44	2	2 6	2 6 10	2 6 7	1		
Rh	45	2	2 6	2 6 10	2 6 8	1		
Pd	46	2	2 6	2 6 10	2 6 10			
Ag	47	2	2 6	2 6 10	2 6 10	1		
Cd	48	2	2 6	2 6 10	2 6 10	2		
In	49	2	2 6	2 6 10	2 6 10	2 1		

TABLE 2-2 Basic Elements: Electrons and Shells, Cont.

Location Orbital Suborbital Element Number		1 K s	2 L s p	3 M s p d	4 N s p d f	5 O s p d f	6 P s p d f	7 Q s p d f
Sn	50	2	2 6	2 6 10	2 6 10	2 2		
Sb	51	2	2 6	2 6 10	2 6 10	2 3		
Te	52	2	2 6	2 6 10	2 6 10	2 4		
I	53	2	2 6	2 6 10	2 6 10	2 5		
Xe	54	2	2 6	2 6 10	2 6 10	2 6		
Cs	55	2	2 6	2 6 10	2 6 10	2 6	1	
Ba	56	2	2 6	2 6 10	2 6 10	2 6	2	
La	57	2	2 6	2 6 10	2 6 10	2 6 1	2	
Ce	58	2	2 6	2 6 10	2 6 10 2	2 6	2	
Pr	59	2	2 6	2 6 10	2 6 10 3	2 6	2	
Nd	60	2	2 6	2 6 10	2 6 10 4	2 6	2	
Pm	61	2	2 6	2 6 10	2 6 10 5	2 6	2	
Sm	62	2	2 6	2 6 10	2 6 10 6	2 6	2	
Eu	63	2	2 6	2 6 10	2 6 10 7	2 6	2	
Gd	64	2	2 6	2 6 10	2 6 10 7	2 6 1	2	
Tb	65	2	2 6	2 6 10	2 6 10 9	2 6	2	
Dy	66	2	2 6	2 6 10	2 6 10 10	2 6	2	
Ho	67	2	2 6	2 6 10	2 6 10 11	2 6	2	
Er	68	2	2 6	2 6 10	2 6 10 12	2 6	2	
Tm	69	2	2 6	2 6 10	2 6 10 13	2 6	2	
Yb	70	2	2 6	2 6 10	2 6 10 14	2 6	2	
Lu	71	2	2 6	2 6 10	2 6 10 14	2 6 1	2	
Hf	72	2	2 6	2 6 10	2 6 10 14	2 6 2	2	
Ta	73	2	2 6	2 6 10	2 6 10 14	2 6 3	2	
W	74	2	2 6	2 6 10	2 6 10 14	2 6 4	2	
Re	75	2	2 6	2 6 10	2 6 10 14	2 6 5	2	
Os	76	2	2 6	2 6 10	2 6 10 14	2 6 6	2	
Ir	77	2	2 6	2 6 10	2 6 10 14	2 6 9		
Pt	78	2	2 6	2 6 10	2 6 10 14	2 6 9	1	
Au	79	2	2 6	2 6 10	2 6 10 14	2 6 10	1	
Hg	80	2	2 6	2 6 10	2 6 10 14	2 6 10	2	
Tl	81	2	2 6	2 6 10	2 6 10 14	2 6 10	2 1	
Pb	82	2	2 6	2 6 10	2 6 10 14	2 6 10	2 2	
Bi	83	2	2 6	2 6 10	2 6 10 14	2 6 10	2 3	
Po	84	2	2 6	2 6 10	2 6 10 14	2 6 10	2 4	
At	85	2	2 6	2 6 10	2 6 10 14	2 6 10	2 5	
Rn	86	2	2 6	2 6 10	2 6 10 14	2 6 10	2 6	
Fr	87	2	2 6	2 6 10	2 6 10 14	2 6 10	2 6	1
Ra	88	2	2 6	2 6 10	2 6 10 14	2 6 10	2 6	2
Ac	89	2	2 6	2 6 10	2 6 10 14	2 6 10	2 6 1	2
Th	90	2	2 6	2 6 10	2 6 10 14	2 6 10	2 6 2	2
Pa	91	2	2 6	2 6 10	2 6 10 14	2 6 10 2	2 6 1	2
U	92	2	2 6	2 6 10	2 6 10 14	2 6 10 3	2 6 1	2
Np	93	2	2 6	2 6 10	2 6 10 14	2 6 10 4	2 6 1	2
Pu	94	2	2 6	2 6 10	2 6 10 14	2 6 10 6	2 6	2
Am	95	2	2 6	2 6 10	2 6 10 14	2 6 10 7	2 6	2
Cm	96	2	2 6	2 6 10	2 6 10 14	2 6 10 7	2 6 1	2
Bk	97	2	2 6	2 6 10	2 6 10 14	2 6 10 8	2 6 1	2
Cf	98	2	2 6	2 6 10	2 6 10 14	2 6 10 10	2 6	2
Es	99	2	2 6	2 6 10	2 6 10 14	2 6 10 11	2 6	2
Fm	100	2	2 6	2 6 10	2 6 10 14	2 6 10 12	2 6	2
Md	101	2	2 6	2 6 10	2 6 10 14	2 6 10 13	2 6	2
No	102	2	2 6	2 6 10	2 6 10 14	2 6		
Lr	103	2	2 6	2 6 10	2 6 10 14	2 6		
Rf	104	2	2 6	2 6 10	2 6 10 14	2 6		
Ha	105	2	2 6	2 6 10	2 6 10 14	2 6		
	106	2	2 6	2 6 10	2 6 10 14	2 6		
	107	2	2 6	2 6 10	2 6 10 14	2 6		
	108	2	2 6	2 6 10	2 6 10 14	2 6		

ATOMIC STRUCTURE

subshells (s, p, d, f) depending on the shell. Using the formula $2n^2$ and counting the total number of shells indicates this formula to be valid only for the shells K, L, M, and N (Table 2-3). There is not enough data to validate this formula for the last few shells. (Also, the last few elements remain unnamed at this writing.)

TABLE 2-3 Relation of Shells to Electrons

Shell	Shell Number	$2 \times n^2$	=	Total Electron
K	1	2×1^2	=	2
L	2	2×2^2	=	8
M	3	2×3^2	=	18
N	4	2×4^2	=	32
O	5	2×5^2	=	50

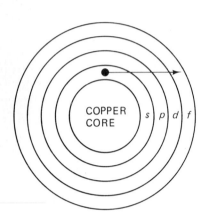

FIGURE 2-8 The electrons of an atom are free to move from one sub-shell to another as they absorb and release energy. A high energy copper atom would have its free electron in the F sub-shell, while a low energy copper atom would have its free electron in the S sub-shell.

From what scientists tell us, it is doubtful that we will have enough data to validate this formula for several more years. Notice also that each subshell can contain a maximum number of electrons ($s = 2$, $p = 6$, $d = 10$, $f = 14$) and these subshells do not necessarily follow a set pattern in filling. Also, electrons are not forced to stay within their suborbital; They are free to move to other orbitals. The only requirement for this movement is the absorption of an extra amount of energy.

Let us examine the copper atom (see Figure 2-8). It has one electron in the N shell. This electron is normally in the s subshell. As the copper atom picks up energy, usually in the form of heat, this electron can move to a higher-level subshell. Understand that the farther away the electron moves from the nucleus, the less attraction it feels from the protons in the nucleus. That is, when the electron of the copper atom is in the s subshell, it is being strongly held in place by the protons (unlike charges attract) in the nucleus. When the electron acquires energy, it moves at a higher rate of speed, which forces the electron to the d subshell through centrifugal action. At the d subshell the electron feels less attraction from the nucleus. If the electron would move to the f subshell, it (the electron) would feel even less attraction. It is this phenomenon (the orbital and suborbital shifts) that permits electron movement and electricity.

ELECTRON MOVEMENT

Electricity is the movement of electrons. To fully understand this movement, we must remember that unlike charges attract and like charges repel. A piece of wire is a conductor and is made up of atoms containing free electrons. When we connect a battery to it, the positive terminal forms a strong attraction for these free electrons, while the negative terminal forms a strong repelling force against these free electrons. The result is *electron movement* (Figure 2-9). Notice that the direction of

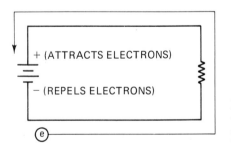

FIGURE 2-9 Simple circuit. Electrons flow from negative to positive to light the bulb.

FIGURE 2-9 (continued)

movement is from negative to positive. This is the *electron theory of current flow* (electron movement).

Unfortunately, in electronics we are recurrently subjected to alternative theories. The alternative for the electron theory is the *conventional* or *hole theory of electron movement*. The conventional theory states that for an electron to move, it must first have a hole (an unfilled valence shell) to move into. This theory deals with the movement of these holes, which is from positive to negative (Figure 2-10). It really doesn't matter which theory is used; they both work.

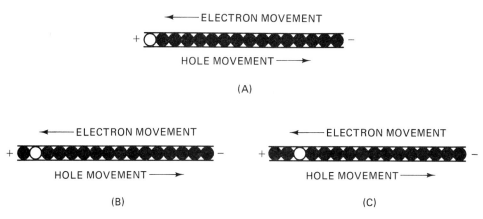

FIGURE 2-10 (A–C) Conventional flow and electron flow for current movement.

CURRENT

On a large scale, the rate of electron movement is called *current flow* (Figure 2-11). Electrons are negatively charged particles and as such, form the basis for the unit of charge, the coulomb. The *coulomb* is defined as a concentration of 6.25×10^{18} electrons. Understand that this would be termed -1 coulomb, because 6.25×10^{18} electrons are needed for a charge of 1 coulomb (C), and electrons have a negative charge. $+1$ C would be the result of a deficiency of 6.25×10^{18} electrons. An average bolt of lightning has a charge of 20 C. It would contain

$$20 \times 6.25 \times 10^{18} = 125.0 \times 10^{18} \quad \text{electrons}$$

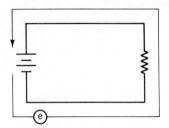

CURRENT IS THE MOVEMENT OF ELECTRONS

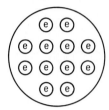

GROUPS OF ELECTRONS ARE MEASURED
IN TERMS OF ELECTRICAL CHARGE

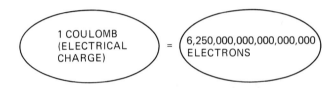

FIGURE 2-11

Current is the rate of electron movement and is based on the coulomb unit of charge. The unit for current is the *ampere* (A), defined as the movement of 1 coulomb of charge past a given point in 1 second (s). Sometimes this is expressed as 1 A = 1 C/s. Pay close attention to the way that compound units are handled. They will show you what mathematic operation was used to get the answer. Here C/s indicates that the total number of coulombs was divided by the total length of time (measured in seconds).

EXAMPLE 2-1

Suppose that 6 C passes a given point in 2 s. How much is the resultant current?

Solution:

$$1 \text{ A} = 1 \text{ C/s}$$

So the number of coulombs must be divided by the number of seconds:

$$\frac{6 \text{ C}}{2 \text{ s}} = 3 \text{ A of current}$$

Calculator Use: Press the following sequence of keys:

$$[6] \, [\div] \, [2] \, [=]$$

The calculator should display

$$3.000 \qquad 00$$

■

If you memorize relationships like

$$1 \text{ ampere} = 1 \text{ coulomb/second}$$

it will reduce the number of formulas you must remember, and at the same time it will make the units more meaningful. Again an ampere is a unit of current and is equivalent to 1 coulomb of charge passing through a point in 1 second.

This is a physical relationship that you might have learned by taking a course in physics. More such relationships follow. These relationships are important from a definition standpoint. They are not terms in which electronic technicians or engineers converse; however, they do show the relation of electrical terms to classical physics.

ENERGY

Energy or *work* is defined as *force times distance*. Force is generally measured in pounds or kilograms, depending on the system in which you are working. Distance can be measured in feet or meters. This leads to several different units of energy. First let's examine the process of calculating work.

EXAMPLE 2-2

Calculate the amount of work that a 150 lb person does as he or she walks down a 20-ft hall.

Solution: The force involved is the weight of the person, 150 lb. The distance is 20 ft. Work is equal to force times distance:

$$20 \text{ ft} \times 150 \text{ lb} = 3000 \text{ ft-lb}$$

The person would be doing 3000 ft-lb of work.

Calculator Use: This is a simple exercise in calculator usage. You would enter the numbers and operations just as the formula indicates. Use the following sequence of keys:

$$[2]\ [0]\ [\times]\ [1]\ [5]\ [0]\ [=]$$

The calculator should display

$$3.000 \qquad 03 \qquad \blacksquare$$

It is just that easy to calculate work (force × distance). The thing that makes these calculations difficult is determining which unit of work to use. Table 2-4 contains several units in which work is expressed, together with some common conver-

TABLE 2-4 Selected Energy Units

Unit	ft-lb	joule	kg-meter	watt-second	calorie	Btu
ft-lb	1	1.35582	0.138255	1.35582	0.323582	0.00128408
joule	0.737562	1	0.1019716	1	0.238662	0.000947831
kg-meter	7.23301	9.80665	1	9.80665	2.34048	0.00928776
watt-second	0.737562	1	0.1019716	1	0.238662	0.000947831
calorie	3.08596	4.18331	0.426649	4.18331	1	0.00396262
Btu	777.649	1055.87	107.514	1055.87	251.634	1

sion factors. Several of the energy units listed in Table 2-4 should be familiar to you. The calorie listed is the gram calorie. It is 1/1000 of the Calorie with which dieters are familiar. The Btu is the standard way of rating heaters, furnaces, hot-water heaters, and the like. The watt-second is a smaller version of the kilowatt-hour unit that the electric company uses in figuring your electric bill.

EXAMPLE 2-3

Convert 1 kilowatt-hour (kWh) into its equivalent number of watt-seconds (W-s).

Solution: Remember that *kilo* means 10^3 and is equal to 1000.

$$1 \text{ kilowatt-hour} = 1000 \text{ watt-hours}$$

There are 60 minutes in an hour.

$$1 \text{ kilowatt-hour} = 1000 \text{ watt-hours} \times 60 \text{ minutes/hour}$$

$$1 \text{ kilowatt-hour} = 60{,}000 \text{ watt-minutes}$$

Note: The hour units cancel.
There are 60 seconds in 1 minute.

$$1 \text{ kilowatt-hour} = 60{,}000 \text{ watt-minutes} \times 60 \text{ seconds/minute}$$

$$1 \text{ kilowatt-hour} = 3{,}600{,}000 \text{ watt-seconds}$$

Note: The minute units cancel.
   One kilowatt-hour is equivalent to 3,600,000 watt-seconds.

You would normally solve for the units first. Begin by setting up a chart like this:

kilowatt-hours		minutes	seconds
	kilo	hours	minutes

Notice that the units on top are canceled by the units on the bottom. After all cancellations, the only units left are watt-seconds. Now insert the values for each of these units:

1 kilowatt-hours		60 minutes	60 seconds
	1 kilo	1 hour	1 minute

Calculator Use: Calculations of this type can be entered on the calculator in sequence all at one time. Enter the following sequence of keystrokes:

$$[1] \; [EE] \; [3] \; [\times] \; [6] \; [0] \; [\times]$$

The calculator should display

$$6.000 \quad 04$$

Notice that the equal sign was not pressed at this time. Whenever the next operation is known, press it instead of equals. You will find this a time-saving habit.

Then press

[6] [0] [=]

The calculator should display

3.600   06   ■

A close examination of Table 2-4 will show that the joule unit and the watt-second unit are equivalent. The physical aspects of electronics are based on the joule unit of work. It is this unit that is most important to our discussion. First, let's try to get an idea of the size of a joule and of how much energy it consists. A 100-watt (W) light bulb uses 100 joules (J) every second it is in operation.

EXAMPLE 2-4

In Example 2-2 a person did 3000 ft-lb of work simply by walking down a hall. Convert this amount of work into units of joules.

Solution: Table 2-4 indicates that the conversion factor is 1.35582. Multiply the number of foot-pounds by this factor to find the equivalent number of joules.

$$3000 \text{ ft-lb} \times 1.35582 = 4067.46 \text{ J}$$

There are 4067 J in 3000 ft-lb.   ■

EXAMPLE 2-5

One can of cola contains 115 Calories. How many joules would this soda contain?

Solution: The unit Calorie is sometimes referred to as the kilocalorie. There are 1000 calories in 1 Calorie. Multiplying 115 by 1000 will convert the Calorie unit to the calorie unit.

$$115 \text{ Calories} \times \frac{1000 \text{ calories}}{\text{Calorie}} = 115{,}000 \text{ calories}$$

Note: The uppercase letter C is commonly used to differentiate between the gram-calorie and the kilogram-calorie. The Calorie units cancel.

Table 2-4 indicates that the conversion factor from calories to joules is 4.18331. Multiplying by this number will convert this unit to joules.

$$115{,}000 \times 4.18331 = 481{,}080.65 \text{ J}$$

There are 481,100 J in the average can of cola-flavored soda.
Note: This is not the work required to drink the soda but the amount of energy contained in the soda.
Calculator Use: This example would be performed by entering the following sequence of keys:

[1] [1] [5] [EE] [3] [×] [4] [.] [1] [8] [3] [3] [1] [=]

Notice that the EE key was used instead of multiplying by 1000.
The display would be

4.811   05   ■

ENERGY

Another consideration is the amount of work something can do. In electronics this is referred to as *potential*.

ELECTROMOTIVE FORCE

Any time 1 coulomb of charge has the potential to do 1 joule of work, we say that the electromotive force (emf) is equal to 1 volt. The unit *volt* is equal to 1 joule per coulomb. You are probably already familiar with the term *voltage* and realize that this is how we measure batteries (Figure 2-12). The concept of emf and potential difference are two concepts that always tend to confuse students.

FIGURE 2-12 Typical power sources. Function generator–top right, DC Power supply–bottom left, solar cell–center, batteries–left.

In general terms, we can say that voltage is the force (push) in electrical circuits. It is possible to use more than one battery at a time, and as when two people work together, they either help each other or, at times, work against each other. When batteries work together, their voltages are combined for more pushing power, emf. When batteries work against each other, their voltages are subtracted. We use signs (+ and −) to indicate the direction in which the batteries are pushing. Contrary to what you would think, if a + (positive) voltage terminal pushes against a − (negative) voltage terminal, the voltages are additive. When "−" pushes against "−," they are subtracted. Algebraically, this is the same as subtracting signed numbers. For this reason, emf is also referred to as a *potential difference* (because the answer to a subtraction problem is called a *difference*).

EXAMPLE 2-6

Calculate the potential difference between a −4-V power supply and a −5-V power supply.

Solution: To find the potential difference, simply subtract one supply voltage from the other:

$$-4\text{ V} - (-5\text{ V}) = +1\text{ V}$$

The potential difference would be 1 V.

Calculator Use: This calculation requires the use of the sign-change key. This key is labeled with a $+/-$. On some calculators it will be necessary to press a second function key first. Enter the following sequence of keys:

$$[4]\ [+/-]\ [-]\ [5]\ [+/-]\ [=]$$

The calculator should display

$$1.000 \qquad 00 \qquad\qquad \blacksquare$$

EXAMPLE 2-7

Calculate the potential difference between a +3-V power supply and a +2-V power supply.

Solution: To find the potential difference, simply subtract one supply voltage from the other:

$$+3\ V - (+2\ V) = +1\ V$$

The potential difference would be 1 V.

Calculator Use: This is a straightforward use of the calculator. Enter the following sequence of keys:

$$[3]\ [-]\ [2]\ [=]$$

The display would be

$$1.000 \qquad 00 \qquad\qquad \blacksquare$$

In an electrical circuit, the voltage pushes electrons through the wires. When we check the number of electrons pushed, by a certain voltage, through a specific wire, we are checking the resistance of the wire.

RESISTANCE

Resistance, simply defined, is the opposition to current flow. The unit of resistance is the ohm (Ω). One ohm is equal to 1 volt per ampere. Stated another way, it takes 1 volt to push 1 ampere through 1 ohm of resistance. One major concern in electronics is either reducing the amount of resistance in a circuit for more efficient operation, or increasing the amount of resistance for better control of circuit operation (Figure 2-13). Looking at these two concerns, you can see the conflicting nature of resis-

FIGURE 2-13 Typical Loads. Neon Bulb–upper left, Resistance Decade Box–lower left, Resistors–center, Potentiometer–upper left, Slide Potentiometer–upper right, Trim Potentiometer–center right, Speaker–lower center, Light Bulb–lower right.

tance. It is one of the first relationships we examine in Chapter 3. The inverse (reciprocal) of resistance is also important. This inverse is called *conductance*.

CONDUCTANCE

Conductance is defined as the lack of opposition to current flow. The unit for conductance is the mho ("ohm" spelled backwards) or siemens. The mho (℧) is the traditional unit and is used more in the field. Siemens (S) is a fairly new term and is used more in the classroom. They are equivalent and interchangeable. One mho is equal to 1/ohm, or 1 ampere per volt. Conductance is used primarily in working with parallel circuits. This type of circuit has more than one path for current and is discussed in full in a later chapter.

EXAMPLE 2-8

A particular component has a resistance of 100 Ω. What is this component's conductance?

Solution: The reciprocal of the resistance is the conductance. So if you divide 1 by the component's resistance, you will find that component's conductance.

$$\frac{1}{100} = 0.01 \text{ mho}$$

The conductance is 0.01 mho.

Calculator Use: First find the 1/x key on your calculator. On some models it may be necessary to press the second function key before pressing the 1/x key. Then enter the following sequence of keys.

$$[1]\ [0]\ [0]\ [1/x]$$

The calculator should display

$$1.000 \qquad -02$$

■

EXAMPLE 2-9

The conductance of a certain component is known to be 20 millimhos (m℧). What is the resistance of this component?

Solution: Again, conductance is the reciprocal of resistance. So if you divide 1 by the conductance, you will find the resistance.

$$\frac{1}{20 \times 10^{-3}} = 50 \text{ Ω}$$

The resistance is 50 Ω.

Calculator Use: Find the 1/x key. It may be necessary to press the second function key before pressing the 1/x key on some calculators. Then press the following sequence of keys:

$$[2]\ [0]\ [EE]\ [3]\ [+/-]\ [1/x]$$

The calculator should display

$$5.000 \qquad 01$$

■

Table 2-5 shows the relationships between each of the quantities mentioned previously. It can be used as a guide to help you understand these relations more fully. It shows the measurement, the symbol for the measurement, the unit in which the measurement is taken, the symbol for the unit, and a brief description of the quantity.

TABLE 2-5 Fundamental Units of Electrical Measurement

Measurement	Symbol	Unit	Symbol	Definitions, Formulas, Comments
Charge	Q_e	electron	e	Smallest unit of charge is 1e
	Q	coulomb	C	Practical unit of charge; $1\text{ C} = 6.25 \times 10^{18}\text{ e}$
Work or energy	W	joule	J	Units of energy in electronics relate to energy in other branches of physics: for example, heat energy 1 joule = 0.2388 calorie 1 calorie = 4.187 joules 1 calorie can raise 1 gram of water 1° C
		watt-second	W-s	1 joule = 1 watt × 1 second
Potential difference	PD	volt	V	PD = work/charge 1 volt = 1 joule/1 coulomb
Electromotive force	emf			
Electrical pressure	E			
Current	I	ampere	A	Charge in motion Flow of charge/unit time; 1 ampere = 1 coulomb/second
Power	P	watt	W	$P = I \times E$ 1 watt = 1 volt × 1 ampere 1 watt = 1 joule/second Power = work/time
Resistance	R	ohm	Ω	$R = E/I$ (Ohm's law) 1 ohm = 1 volt/ampere
Conductance	G	mho/siemens	℧/S	Reciprocal of resistance $G = 1/R$ $G = I/E$ 1 mho = 1 ampere/volt

SUMMARY

It is important that you have a knowledge of physics, because electronics is built around physics. Although it is not necessary to have a complete knowledge of physics to work on electrical circuits, it is desirable that you have at least a working vocabulary of physical terms. Concepts such as "opposite charges attract" and "like charges repel" must be memorized.

The structure of the atom as it relates to electrical properties must be understood. The valence shell determines the electrical properties of a material. A material that has only one valence electron will conduct electricity. A material that has four valence electrons is a semiconductor. Materials that have seven or eight electrons are normally insulators.

Electronics is heavily dependent on terms such as *voltage, current, resistance,* and *conductance*. The definition of each must be understood. Voltage is electrical pressure. Current is electrical flow or electron movement. Resistance is the opposition to current flow. Conductance is the reciprocal of resistance and is defined as the lack of resistance.

The kilowatt-hour is the standard unit of electrical work. Depending on circumstances it may be necessary to convert Btu and other units of work into kilowatt-hours. Electronics engineers and technicians seldom work in units like coulombs or joules, but it is important that they know that a coulomb is a unit of charge and a joule is a unit of work.

QUESTIONS

2-1. What determines the charge of a material?

2-2. What happens when like charges are placed close together?

2-3. What is centrifugal force, and how does it affect the electrons of an atom?

2-4. Describe the effect that energy absorption has on an electron.

2-5. In terms of the number of electrons contained in the valence shell, discuss the differences between insulators, conductors, and semiconductors.

2-6. Discuss the meaning of the word *semiconductor*. Include several electronic components considered to be semiconductors. (Some research may be necessary.)

2-7. Discuss the physical concept of work. List several units in which work could be measured.

2-8. Define *resistance* and describe its effect on electricity.

2-9. Define *voltage* and describe its relation to electricity.

2-10. What is the common electrical unit used to measure work?

2-11. What is conductance? When is this quantity used?

2-12. Explain the relationship between a watt-second and a joule.

2-13. A joule per second is equivalent to what other unit?

2-14. A volt per ampere is equivalent to what other unit?

2-15. A coulomb per second is equivalent to what other unit?

2-16. How many electrons make up 1 coulomb of charge?

2-17. How much charge is present on one electron? (Give your answer in coulombs.)

2-18. Analyze the unit joule-second per square coulomb:

$$\frac{\text{joule-second}}{\text{coulomb}^2}$$

to find the equivalent unit. (*Hint:* volts = joules/coulomb, amperes = coulombs/second.)

PRACTICE PROBLEMS

2-1. Calculate the work done in moving 50 lb 100 ft. Express the answer in foot-pounds, Btu, and joules.

2-2. A material has acquired 53.5×10^{20} electrons. Calculate its charge in coulombs. Indicate whether the charge is positive or negative.

2-3. As 25×10^{18} electrons move through a circuit, they do 12 J of work. Find the amount of voltage (pressure) needed to force these electrons to move.

2-4. It is found that 14 C of charge passes a point in a circuit every 5 s. How much current is being produced?

2-5. A battery is being charged with 4 A of current. How much charge has been added to the battery after 20 s?

2-6. A 12-V battery can force only 4.5 A of current to flow through a circuit. Find the resistance of this circuit.

2-7. A circuit has 50 Ω of resistance. What conductance does this circuit have?

2-8. How much resistance would a circuit have if its conductance were 10 μ℧?

2-9. Every 12 s 72 C enters a battery? What is the charge rate (current)?

2-10. A 20-V battery powers a circuit that has 1200 Ω of resistance. How much current can flow through this circuit?

2-11. Calculate the amount of voltage needed to push 50 A through 25 kΩ of resistance.

2-12. A conductance of 250 m℧ is equivalent to _____ ohms.

2-13. How much current can flow through a circuit having 20 μ℧ of conductance and a power source of 10 V?

2-14. A circuit does 12 J of work every 3 s. How much power does this circuit use? (The unit for power is the watt. *Hint:* 1 joule = 1 watt-second.)

2-15. A current of 5 A is equivalent to _____ coulombs every 2 s.

Match the following quantities with the correct unit.

2-16.	Electromotive force	A.	Ampere
2-17.	Resistance	B.	Watt
2-18.	Charge	C.	Joule
2-19.	Current	D.	Ohm
2-20.	Power	E.	Mho
2-21.	Work	F.	Coulomb
2-22.	Conductance	G.	Volt

Match the following quantities with the correct symbol.

2-23.	Conductance	A.	*Q*
2-24.	Charge	B.	*E*
2-25.	Current	C.	*G*
2-26.	Power	D.	*I*
2-27.	Electromotive force	E.	*W*
2-28.	Resistance	F.	*P*
2-29.	Work	G.	*R*

Match the following units with the correct symbol.

2-30.	Ampere	**A.**	W
2-31.	Watt	**B.**	℧
2-32.	Joule	**C.**	Ω
2-33.	Ohm	**D.**	J
2-34.	Mho	**E.**	A
2-35.	Coulomb	**F.**	V
2-36.	Volt	**G.**	C

Ohm's Law

3

Chapter objectives

After reading this chapter and answering the questions and problems, you should be able to:

- Define *direct and inverse proportion, load, voltage source, power dissipation, open circuit,* and *short circuit.*
- State if each of several word problems uses direct or inverse proportion and solve for the missing value.
- State the opposite function of each of the following: addition, subtraction, multiplication, division, square, and square root.
- Use opposite functions and maintain equality to solve algebra problems.
- State Ohm's law in its three forms.
- State the standard unit of power. Include at least one alternative unit.
- List the three power formulas.
- Given any two of P, V, I, and R for a component or circuit, solve for the missing values.
- Given R, I, or V, calculate minimum required power ratings.
- Given R and P ratings for that component, calculate maximum allowable component current.
- Describe the three requirements for an active circuit.
- Describe the voltage, current, and resistance contained in both a shorted circuit and an opened circuit.
- Given a graph of two quantities, state if the relationship is linear or exponential.

Ohm's law is one of the most important concepts in electronics. It was described over 150 years ago by Georg Simon Ohm. Although it was first explored in the early 1840s, it remains the single most important idea in electronics. This chapter is devoted to an explanation of Ohm's law, its conceptual implications, and the various ways to solve for circuit values using this ageless principle.

3.1 GETTING STARTED: BASIC ALGEBRA

Almost everyone has difficulties with word problems. But the simple truth is that the only problems in the real world are word problems. Two simple phrases will be used throughout this book. They will help you analyze at least one type of common word problem. The first phrase is *directly proportional*.

The higher we go in space, the thinner the atmosphere. The more you eat, the fatter you will get. The deeper you go into a cave, the darker it gets. These are all relationships that are directly proportional. *Directly proportional* refers to any quantity that can be expressed in the form of

$$A = k \times B$$

where k is the constant value.

EXAMPLE 3-1

Remember Sally and her apples? Suppose that Sally always gave half (k) her apples to Tom. If Sally was given 10 apples (B), how many apples (A) did she give to Tom?

Solution:
$$A = k \times B$$
$$= \tfrac{1}{2} \times 10 = 5$$

Tom got 5 apples. ■

This apple problem is a type of directly proportional relationship. The more apples that Sally gets, the more apples she will give to Tom. This means that if there is an increase in A, there must also be an increase in B. Or conversely, if there is a decrease in B, there must also be a decrease in A.

The other type of common word problem is an *inversely proprotional* relationship. The farther we go, the closer we get. The more you save, the less you'll spend. A hot bath gets cooler the longer it sets. These are all inversely proportional relationships. *Inversely proportional* pertains to any quantity that can be expressed in the form

$$A = \frac{k}{B}$$

where k is again a constant value.

EXAMPLE 3-2

Remember all of those pie problems? Suppose that we had two pies (k) to divide among eight (B) people. How much pie (A) would each get?

Solution:

$$A = \frac{k}{B}$$

$$= \tfrac{2}{8} = \tfrac{1}{4} \quad \text{(one-fourth)}$$

Each would receive one-fourth of a pie. ■

This pie problem is a type of inversely proportional relationship. It means just the opposite of "directly proportional." If A increases, then B must decrease and vice versa.

The majority of relations in electronics are either direct or inverse proportions. Mastering them will greatly help you master electronics. Two more examples follow.

EXAMPLE 3-3

If John always has twice as many apples as Sally, when Sally increases the number of apples she has, what happens to the number that John has?

Solution: Begin by reading carefully one section at a time. John has twice as many apples as Sally. In equation form this would be

$$J = 2S$$

When Sally increases the number she has, what happens to the number John has? Pick some numbers and see. Suppose that Sally has 2 apples. Then

$$J = 2(2) = 4$$

John has 4 apples. Now increase the number that Sally has from 2 to 3. Then

$$J = 2(3) = 6$$

John has 6 apples. So if Sally's number increases, John's increases also. ■

In Example 3-3 the value of k (the proportionality constant) was 2. It was given in the statement "John always has twice as many apples as Sally." Since both Sally's and John's number of apples increased, this is an example of direct proportionality.

EXAMPLE 3-4

A light bulb glows with an intensity of 300 candle-feet. What will be the intensity, in candles, at a distance of 10 ft? What will happen to the intensity if the distance increased?

Solution: Begin by reading the problem carefully. Notice the unit *candle-feet*. This implies that candle units and feet units have been multiplied. Dividing this unit by feet units will produce candle units. In equation form we have

$$I = \frac{300 \text{ candle-feet}}{10 \text{ ft}} = 30 \text{ candles}$$

GETTING STARTED: BASIC ALGEBRA

So at a distance of 10 ft the brightness will be 30 candles.

Now increase the distance by choosing a larger number of feet, say 20 ft, and repeat the calculation. The equation would be

$$I = \frac{300 \text{ candle-feet}}{20 \text{ ft}} = 15 \text{ candles}$$

So when the distance was increased from 10 ft to 20 ft, the brightness reduced from 30 candles to 15 candles. ∎

In Example 3-4 the proportionality constant, k, is 300 candle-feet. Notice that as the distance increased, the brightness decreased. This is an example of an inverse proportionality.

In math classes instructors always list several forms of algebra equations and expect us to understand them immediately or at least memorize them. It is my intention that you will learn to examine a problem or calculation, think about the concepts involved, and proceed as your thoughts tell you. No one is perfect, so your thoughts may not always be correct. *Being correct is not as important as thought*. If you do something wrong the first time, especially when it costs you money, you tend not to make that same mistake the second time. The only true way a person learns is by doing, and doing a task wrong the first time generally has more impact on us than doing it right.

Algebra can be summed up in two simple statements:

1. If you do something to one side of an equal sign, you must also do it to the other side to maintain the equality.
2. To remove or get rid of a portion of an equation, you must perform the opposite function.

It takes some thought and some care to maintain equality. And with long equations, it is sometimes hard to figure out exactly what the inverse function is. However, the entire idea behind basic algebra is contained in these two concepts. Maintaining equality and picking the right inverse operation involves first looking at the equation. Let's examine an equation.

EXAMPLE 3-5

Given the equation $X + 5 = 12$, solve for X.

Solution: Notice that there is an *unknown* in the equation. In all forms of algebra problems you will be looking for the value of an unknown. To find the value of an unknown, you must isolate it.

The left side of the equation, $X + 5$, has a 5 which you need to remove (get rid of). Also, this 5 is being added to your unknown. You must think to yourself: "The 5 is being added. To remove it, I must subtract, because subtract is the opposite of add." Remember to do this to both sides. Your work should look something like this:

$$X + 5 = 12$$
$$X + 5 - 5 = 12 - 5$$
$$X + 0 = 7$$
$$X = 7$$

∎

Most inverse operations should be familiar to you: add and subtract, multiply and divide, square and square root. These are primarily the ones we will work with.

EXAMPLE 3-6

Given the equation $A = 12 \times 4$, solve for A.

Solution: This equation is an example of an algebra problem that tells you what you are to do in order to find A: Multiply 12 by 4.

$$A = 48$$

because $12 \times 4 = 48$. ∎

EXAMPLE 3-7

Given the equation $3B = 48$, solve for B.

Solution: This problem is not as simple. Here you must identify 3 as not belonging. To remove the 3, you need to divide (opposite of multiply). So your work would look something like this:

$$3B = 48$$

$$\frac{3B}{3} = \frac{48}{3}$$

$$B = 16$$
∎

A more confusing version of this problem occurs when the unknown is the divisor. This is demonstrated in the following example.

EXAMPLE 3-8

Given the equation $24 = 96/k$, solve for k.

Solution: Thinking through this problem will cause some difficulties. At first glance it appears that you should divide the right side by 96 to remove it from the equation; however, doing so only results in confusion, because the right side correctly becomes $1/k$ and not k, as most assume. This form of equation requires a trick. This trick goes against your first instincts, and for some, it is hard to learn to do. The trick is to multiply both sides of the equation by k (the unknown). This effectively moves k from the right as a denominator (bottom number) to the left as a numerator (top number). Your work would look like this:

$$24 = \frac{96}{k}$$

$$24 \times k = \frac{96}{k} \times k$$

$$24k = 96$$

$$\frac{24k}{24} = \frac{96}{24}$$

$$k = 4$$

SEC. 3-1 / GETTING STARTED: BASIC ALGEBRA

Note: Experience will show you that this form of equation can be handled simply by switching places with the term on the left and the unknown on the right as:

$$24 = \frac{96}{k}$$

$$k = \frac{96}{24}$$

Working circuits using Ohm's law will lead you to one of the proportionality relations. As you read through the remainder of this chapter, notice that these Ohm's law relationships are mathematically like the ones shown here.

SELF-TEST

Solve for the unknown.

1. $D = 23.5 \times 43$
   $D = $ _____

2. $56.3 = W \times 23.9$
   $W = $ _____

3. $123 = \dfrac{23}{L}$
   $L = $ _____

4. $7.89 = \dfrac{U}{23}$
   $U = $ _____

5. $I = 45.7 \times 2.5$
   $I = $ _____

6. $8.23 = \dfrac{G}{0.0025}$
   $G = $ _____

7. $1.54 = \dfrac{0.005}{H}$
   $H = $ _____

8. $23.9 = W \times 6.5$
   $W = $ _____

9. $45.9 = 0.003 \times Y$
   $Y = $ _____

10. $0.00045 = 34 \times C$
    $C = $ _____

ANSWERS TO SELF-TEST

1. $D = 1011$
2. $W = 2.356$
3. $L = 0.1870$
4. $U = 181.5$
5. $I = 114.3$
6. $G = 0.02058$
7. $H = 3.247 \times 10^{-3}$
8. $W = 3.677$
9. $Y = 15,300$
10. $C = 1.324 \times 10^{-5}$

OHM'S FIRST LAW ($E = IR$)

Ohm's first law is an example of a directly proportional relationship. E, the voltage, is like A in Example 3-1. I, the current, is the constant, k; and R is similar to B in the same example.

Voltage across a component is directly proportional to the resistance of the component when the current is held constant. This law implies that current times resistance is equal to voltage. Ohm's law is an important factor in analyzing electrical circuits. Any time that current through a component is known and its resistance is known, the voltage can be found by multiplying the current by the resistance. Table 3-1 demonstrates how this relationship would appear on an actual circuit. Notice in the table that the current was held at 100 mA. Remember, this first law requires that current be a constant. Also, for every change in voltage there is a similar change in resistance. That is, as voltage went from 0.1 V to 0.5 V (5 × 0.1 = 0.5, a change of five times) the resistance also changed by a factor of 5 (5 × 1 = 5). This only appears to be true, as we cannot change resistance in a circuit by increasing the potential (voltage) of the supply. Nor can we actually change voltage by increasing resistance. What Ohm's law is saying is: For every increase in circuit resistance it will also be necessary to increase the voltage if we are to maintain the same current flow.

TABLE 3-1 Voltage Is Directly Proportional to Resistance

Voltage (V)	Current (A)	Resistance (Ω)
0	0.1	0
0.1	0.1	1
0.5	0.1	5
1	0.1	10
5	0.1	50
10	0.1	100

The voltage supply of the circuit provides the force (voltage) to push the electrons (current) through the circuit. If we increase the resistance of the circuit, we must also increase the voltage if we are to keep the same amount of current flow.

Graphs are sometimes used to demonstrate Ohm's first law (Figure 3-1). One simple way to construct a graph of this type is to build a circuit containing a voltage source, a resistor, and a current-measuring device (Figure 3-2). Measurements of voltage and resistance are then taken and plotted on this graph. Each of these readings is taken with the same amount of current (200 mA in Figure 3-1). After recording these measurements, a larger resistor is placed in the circuit. At this time it will be necessary to increase the voltage to maintain the same amount of current (200 mA in Figure 3-1). Figure 3-1 shows a proportional relationship between voltage and resistance. The graph indicates a linear relationship between voltage and resistance because connecting the dots will form a straight line (linear). The slope of this line represents the amount of current present in the circuit. A line that sloped upward more would indicate more current present in the circuit.

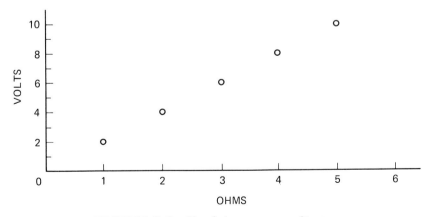

FIGURE 3-1 Resistance vs. voltage.

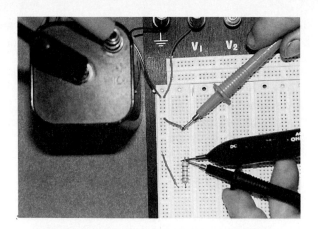

FIGURE 3-2 Volt meter and current meter used to construct graph in Figure 3-2.

OHM'S SECOND LAW ($I = E/R$)

Current is inversely proportional to resistance when voltage is held constant. As resistance changes in a circuit, the current will also change; however, it will do so in an opposite manner. That is, if the resistance increases, the current will decrease. The second law is important in calculating the electrical properties of components. When the voltage across any component is known and its resistance is known, the current can be found by dividing the voltage by the resistance. Table 3-2 demonstrates the way this would appear in an electrical circuit. Notice that the voltage is held at a constant value of 100 V. Voltage must be a constant to satisfy the second law. Notice the last entry for current. Also notice the last entry for resistance. The current is at a minimum but resistance is at a maximum. The current is gradually getting smaller and the resistance is gradually becoming a larger quantity. This indicates that an inverse relationship exists between these two values.

TABLE 3-2 Current Is Inversely Proportional to Resistance

Voltage (V)	Current (mA)	Resistance (kΩ)
100	50	2
100	25	4
100	10	10
100	5	20
100	2.5	40
100	1	100

Again the voltage provides the pressure for the circuit. It pushes the electrons through the circuit. It makes sense that the same voltage can push more current through less resistance. Current can be calculated by dividing the voltage by the resistance.

There is no way to stress the importance of your understanding these concepts. Figure 3-3 is a graphical representation of Ohm's second law. You will notice that this representation does not show a linear response. When these dots are connected they form a curve. This is referred to in mathematics as an *exponential* relationship. It will occur any time that two inversely proportional numbers are plotted against each other. There are two things to be remembered concerning this graph.

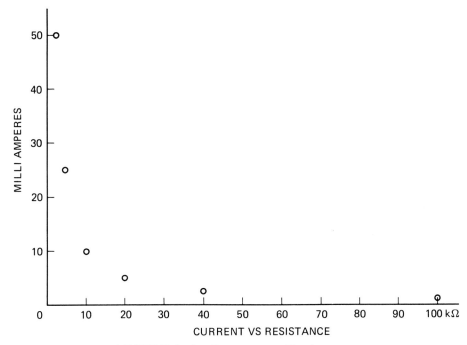

FIGURE 3-3 Current vs. Resistance

1. It implies that an infinite amount of current can flow if the resistance is zero. Such a condition is called a *short* (zero circuit resistance). See Figure 3-10 for a diagram of a shorted circuit. In real-life situations, an infinite current is impossible. Power supplies do have a limited amount of current they can provide. However, this current could be large and may damage components and/or the power supply itself.

2. This graph also indicates that zero current flows if the resistance is infinite. Such a condition is referred to as an *open*. See Figure 3-10 for a diagram of an open circuit. In real-life situations this is true and can be used to troubleshoot circuits. Any time there is zero current in a circuit there must be an open present. An open can be several complex problems or something as simple as a broken wire. Remember, there is voltage across an open. There is no voltage across a short.

OHM'S THIRD LAW ($R = E/I$)

Resistance is inversely proportional to current when voltage is held constant. This law is the converse of the second law and as such has similar implications. The only difference is that resistance can be calculated by dividing voltage by current.

The triangle in Figure 3-4 can be used as a crutch. It will help you learn Ohm's law. Place your finger over the symbol of the quantity that you are calculating. The triangle will indicate what is to be done to the other two quantities in order to solve the problem.

EXAMPLE 3-9

Describe how to use the triangle in Figure 3-4.

Solution: If you were calculating current in a problem, you would cover the I in the triangle. Doing so would leave the E over the R. This means "divide

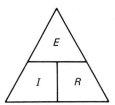
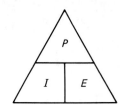
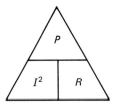
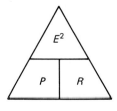

MEMORY AID
FOR OHM'S LAW

FIGURE 3-4

the voltage by the resistance." When calculating for voltage, cover the *E*. This leaves *I* and *R* on the bottom. This means "multiply the resistance by the current." ∎

SELF-TEST

Use Ohm's law to solve for the missing quantity.

1. $E = 24$ V
   $I = 2$ A
   $R = $ _____

2. $R = 1200\ \Omega$
   $I = 12$ mA
   $E = $ _____

3. $I = 1$ A
   $R = 0.5\ \Omega$
   $E = $ _____

4. $E = 12$ V
   $I = 3.5\ \mu$A
   $R = $ _____

5. $R = 2.7$ MΩ
   $I = 18\ \mu$A
   $E = $ _____

6. $R = 56$ kΩ
   $E = 15$ V
   $I = $ _____

7. $I = 150$ mA
   $E = 20$ V
   $R = $ _____

8. $E = 25$ V
   $R = 4.7$ kΩ
   $I = $ _____

9. $R = 820\ \Omega$
   $E = 9$ V
   $I = $ _____

10. $E = 18$ V
    $R = 2.7$ kΩ
    $I = $ _____

ANSWERS TO SELF-TEST

1. $R = 12\ \Omega$
2. $E = 14.4$ V
3. $E = 0.5$ V
4. $R = 35$ MΩ
5. $E = 48.6$ V

6. $I = 267.9\ \mu$A
7. $R = 133.3\ \Omega$
8. $I = 5.319$ mA
9. $I = 10.98$ mA
10. $I = 6.667$ mA

POWER (HEAT DISSIPATION)

Earlier we discussed the need for extra energy in order to cause the electron to move to a higher orbital. Remember that this move was necessary to allow the electron to move from one atom to another (current flow). The energy need for electron movement is always associated with heat energy. In other words, current flow produces a lot of heat. For a large amount of current flow, more heat is produced. Any increase in resistance causes added electron friction. The extra friction will cause an increase in the associated heat produced. The heat produced is proportional to the amount of work (number of electrons moved) being done in a particular time period. This is power!

In the mid-1700s, James Watt began working on steam engines. The first difficulty he came up against was how to compare his engines to a work horse. After some experimentation he found that the average horse could lift 33,000 lb 1 ft high in 1 minute. He defined this amount of power as 1 horsepower (hp), a unit still used today. Let's examine the unit.

$$1 \text{ horsepower} = 33,000 \text{ ft-lb/min}$$

Notice the unit foot-pounds. This is a unit of work. So what power is designating is the amount of work that can be done in a certain amount of time. A more powerful horse can, naturally, do more work in an hour than can a less powerful horse.

Since the standard unit for electrical work is the joule, the electrical unit for power is joules/second. This unit bears the name *watt*. Also notice that the product of voltage (joules/coulomb) times current (coulomb/second) is joules/second.

$$1 \text{ volt} = 1 \text{ joule/coulomb}$$
$$1 \text{ ampere} = 1 \text{ coulomb/second}$$
$$1 \text{ volt} \times 1 \text{ ampere} = 1 \text{ joule/coulomb} \times 1 \text{ coulomb/second}$$
$$1 \text{ volt} \times 1 \text{ ampere} = 1 \text{ joule/second} = 1 \text{ watt}$$

Algebraically, the coulombs in the third step cancel, leaving joules/second. Understand that volt-amperes are watts. The conversion factor from horsepower to watts is 746. To convert from horsepower to watts, simply multiply by 746. If you divide the number of watts by 746, you will change the unit to horsepower.

POWER AND OHM'S LAW

The amount of power used by a circuit is directly proportional to the amount of heat produced by the circuit. Any component that gets too hot will destroy itself. When you talk of power ratings of components, you are referring to the maximum amount of heat which that component can give off (dissipate). If we exceed this amount of heat, the component will no longer function in an acceptable manner. There are three Ohm's law–based equations that can be used to determine the amount of power being dissipated by any component:

$$P = I \times E$$
$$P = I^2 \times R$$
$$P = \frac{E^2}{R}$$

We will discuss each of these in detail. Of the three, the first two are the most used in classroom calculations; however, in laboratory environments the third is the most useful. So you must be familiar with all three.

$P = I \times E$ As shown earlier, multiplying the current times the voltage will give you the power dissipation for that component. This formula can be used in reverse to find the current (or voltage) if the power and voltage (or current) quantities are known.

EXAMPLE 3-10

Find the amount of current required by a 100-W light bulb.

Solution: We know that household voltage is 120 V. By dividing the power by the voltage, we can find the amount of current the bulb is conducting when it is burning.

$$\frac{100}{120} = 0.8333 \text{ A}$$

■

If we know the amount of current and the power, we can calculate the voltage across the component in a similar manner.

$P = I^2 \times R$ Ohm's law indicates that multiplying the current times the resistance will give the voltage for a component. As we saw earlier, voltage times current will provide the power dissipation for a given component. Further examination of this will show that current times resistance times current, a second time, will also provide power. By rearranging this product, we get current times current times resistance, which is the same as current squared times resistance. This formula is handy when we know the current through a particular component and that component's resistance.

EXAMPLE 3-11

Find the required power rating of a 3-Ω resistor that has a current of 5 A going through it.

Solution: The power dissipation would be

$$5^2 \times 3 = 75 \text{ W}$$

It would then be necessary to use a resistor that had a power rating of at least 75 W. ■

This formula can also be used in reverse. That is, if we know the power rating of a component and its resistance, we can find the maximum current through it.

EXAMPLE 3-12

Find the maximum current for a 2-W 10-Ω resistor.

Solution: This calculation is easy; however, it usually causes students a great bit of difficulty. First divide the power by the resistance. This will provide the square of the current.

$$\frac{2}{10} = 0.2$$

0.2 is the square of the current.

The opposite of the square function is the square-root function. So next, find the square root of 0.2. The square root of 0.2 is

$$\sqrt{0.2} = 0.4472 \text{ A}$$

So the maximum current such a resistor could carry would be 447.2 mA.

Calculator Use: On your calculator you should have an x^2 key and a $\sqrt{x}$ key. They should easily provide the answers for you. The keystrokes for this calculation would be

$$[2] \; [\div] \; [1] \; [0] \; [=] \; [\sqrt{x}]$$

Your calculator should display

$$4.4721 \qquad -01$$

Note: Some calculators may require that the 2nd or inv key be pressed before $\sqrt{x}$. It is necessary for you to check your operator's manual for correct operation. ∎

$P = E^2/R$ The previous power formula showed that $I^2 \times R$ was a way to calculate power dissipation for a given component. Notice that multiplying this value by resistance will give

$$I^2 \times R^2$$

which can be rearranged to provide

$$(I \times R)^2$$

and this is the equivalent of E^2. This indicates that by reversing the procedure we can arrive at an alternative formula for power. This procedure is outlined in the following proof.

$$E^2 = I^2 \times R^2$$

$$\frac{E^2}{R} = I^2 \times R = P$$

It is usually easier to measure the voltage across a component. Then square this value and divide by the component resistance. The result is the power dissipation for a given component.

EXAMPLE 3-13

Find the power dissipated by an 8-Ω resistor that drops 12 V in circuit.

Solution: To find the power, first square the voltage. Then divide this answer by the resistance, as $P = E^2/R$ indicates:

$$\frac{12^2}{8} = 18 \text{ W}$$

The resistor is dissipating 18 W of power.

Calculator Use: Find the x^2 key. Then enter the following sequence of keys:

$$[1]\ [2]\ [x^2]\ [\div]\ [8]\ [=]$$

The calculator should display

        1.8000   01   ■

Also, we can use this formula in reverse. If we know the power rating and the resistance, we can find the voltage limit for the component.

EXAMPLE 3-14

Find the maximum voltage that can be placed across a 10-Ω resistor that has a ½-W power rating.

Solution: The formula $P = E^2/R$ is used in reverse. Multiplying P by R gives us the square of the voltage. So the square root of $P \times R$ is the maximum voltage.

$$P = \frac{E^2}{R}$$

$$0.5 = \frac{E^2}{10}$$

$$0.5 \times 10 = E^2$$

$$E^2 = 5$$

$$E = \sqrt{5} = 2.236 \text{ V}$$

Calculator Use: The following keystrokes would be used:

$$[.]\ [5]\ [\times]\ [1]\ [0]\ [=]\ [\sqrt{x}]$$

The calculator should display

        2.2360   00   ■

Again, from a practical standpoint, it is difficult to measure the current in a circuit, and as a result, the first two versions of the power formulas,

$$P = I \times E$$

$$P = I^2 \times R$$

apply mostly to classroom situations. The measurement of voltage in any circuit is not as difficult a task, so the latter formula,

$$P = \frac{E^2}{R}$$

is the laboratory standard.

The triangle in Figure 3-4 can be used as a crutch. It will help you learn the power formulas. Place your finger over the symbol of the quantity that you are calculating. The triangle will indicate what is to be done to the other two quantities in order to solve the problem. For example, if you were calculating current in a problem, you would cover the *I* in the triangle. Doing so would leave the *P* over the *E*. This means "divide the power by the voltage." When calculating for power, cover the *P*. This leaves *I* and *E* on the bottom. This means "multiply the voltage by the current."

SELF-TEST

Use the power formulas to solve the following.

1. $E = 25$ V
   $I = 12$ mA
   $P = $ _____

2. $P = 2$ W
   $E = 9$ V
   $I = $ _____

3. $I = 56$ mA
   $P = 250$ mW
   $E = $ _____

4. $E = 90$ mV
   $P = 120$ μW
   $I = $ _____

5. $I = 25$ μA
   $E = 10$ V
   $P = $ _____

6. $P = 300$ mW
   $E = 24$ V
   $I = $ _____

7. $E = 18$ V
   $I = 20$ mA
   $P = $ _____

8. $P = 6.7$ W
   $I = 530$ mA
   $E = $ _____

9. $I = 0.0004$ A
   $P = 25$ mW
   $E = $ _____

10. $E = 36$ V
    $I = 2.56$ mA
    $P = $ _____

ANSWERS TO SELF-TEST

1. $P = 2.083$ kW
2. $I = 222.2$ mA
3. $E = 4.464$ V
4. $I = 1.333$ mA
5. $P = 250$ μW
6. $I = 12.5$ mA
7. $P = 360$ mW
8. $E = 12.64$ V
9. $E = 62.5$ V
10. $P = 92.16$ mW

USING OHM'S LAW

There are four quantities that must be considered about any component: its resistance, the amount of voltage across it, the amount of current going through it, and the amount of power being dissipated by it. By using the formulas listed previously, you can find any two missing quantities provided that you know the other two. That is, if you know voltage and current, you can find the power by multiplying the voltage by the current, and you can find the resistance by dividing the voltage by the current. The same is true no matter which two quantities are missing. Usually, you will only need to use

$$E = I \times R \quad \text{and} \quad P = I \times E$$

in order to find the other two quantities. Sometimes it will be necessary to use

$$P = I^2 \times R$$

USING OHM'S LAW

The examples that follow show how the various missing components are found.

EXAMPLE 3-15

Given: $E = 20$ V
         $I = 6$ Ω
Find: R and P

Solution: To find R, use $E/I = R$:

$$\frac{20}{6} = 3.333 \text{ Ω}$$

To find P, use $E \times I = P$:

$$20 \times 6 = 120 \text{ W}$$

EXAMPLE 3-16

Given: $R = 7$ Ω
         $I = 4$ A
Find: E and P

Solution: To find E, use $I \times R = E$:

$$4 \times 7 = 28 \text{ V}$$

To find P, use $I \times E = P$:

$$4 \times 28 = 112 \text{ W}$$

EXAMPLE 3-17

Given: $P = 160$ W
         $R = 13$ Ω
Find: I and E

Solution: To find I, use $P/R = I^2$:

$$\frac{160}{13} = 12.31$$

$$\sqrt{12.31} = 3.508 \text{ A}$$

To find E, use $I \times R = E$

$$13 \times 3.508 = 45.61 \text{ V}$$

EXAMPLE 3-18

Given: $E = 100$ V
         $R = 20$ Ω
Find: I and P

Solution: To find I, use $E/R = I$:

$$\frac{100}{20} = 5 \text{ A}$$

To find P, use $I \times E = P$:

$$100 \times 5 = 500 \text{ W}$$

EXAMPLE 3-19

Given: $E = 75$ V
          $P = 12$ W
Find: I and R

Solution: To find I, use $P/E = I$:

$$\frac{125}{75} = 1.667 \text{ A}$$

To find R, use $E/I = R$:

$$\frac{75}{1.667} = 45 \text{ }\Omega$$

EXAMPLE 3-20

Given: $P = 45$ W
          $I = 3$ A
Find: E and R

Solution: To find E use $P/I = E$:

$$\frac{45}{3} = 15 \text{ V}$$

To find R use $E/I = R$:

$$\frac{15}{3} = 5 \text{ }\Omega$$

Remember, any time that two of the four parameters for a particular component are known, the other two can be calculated. Sometimes it will take a little thought to figure out the best way to proceed. The six examples provided show every possible type of calculation you will need to make when working Ohm's law–related problems.

CIRCUIT REQUIREMENTS

To operate, every circuit has three basic requirements:

1. A voltage source
2. A load or resistance
3. A continuous path for electricity

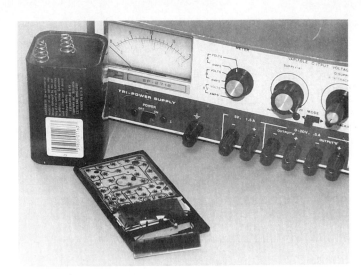

FIGURE 3-5 Typical Voltage Sources. Battery–left, solar cell (on calculator)–center, DC Power Supply–right.

The voltage source may be a battery or some sort of regulated power supply (Figure 3-5). In either case it has the responsibility of providing the voltage (force) for the circuit. Without this force there can be no current flow. Understand that it is possible for voltage to exist without providing any current flow. Think of it like this. When your water faucet is turned off, the water will not flow; however, there is still water pressure on the line. It is possible for a battery or other voltage source to provide the necessary emf for a circuit without producing any resultant current flow.

FIGURE 3-6 Typical Loads. Speaker–upper left, Light Bulb–center, Motor–upper right, Radio and Resistor lower left, Resistance Decade Box–lower right.

For our circuit to operate, we must have a load or resistance for our voltage to push against. This load might be anything: a motor, a light, or even a radio (Figure 3-6). Whatever it may be, the load is the component we want to operate. If we provided a continuous path for current without this load, we would short out the supply (Figure 3-7). This is an important fact to remember. Any time that we short out a voltage supply we draw an excessive amount of current from it. The supply can produce only a limited amount of current, and as a result the supply voltage will be pulled down, usually to zero volts.

For any circuit to operate, we must provide a continuous path for current (Figure 3-8). This path is made up of the wires that connect the load to the supply. In truth, these wires always have some resistance, but this resistance is so small that it can usually be neglected. Problems do sometimes arise when the circuit resistance (resistance of the wires and connectors) becomes too high. For the purpose of our discussion we consider these resistances as zero. If for some reason these wires do not form a continuous path for current, we have what is referred to as an *open circuit*. An open circuit has no current flow but does have a supply voltage. This is an important item to remember when troubleshooting an open circuit.

Again, to have an operating circuit it is necessary to provide (1) a voltage source, (2) a load, and (3) a continuous path for current.

FIGURE 3-7 Shorted Power Cord. Arrow points to short.

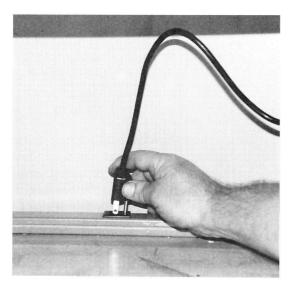

FIGURE 3-8 Hold plug firmly when plugging and unplugging appliances.

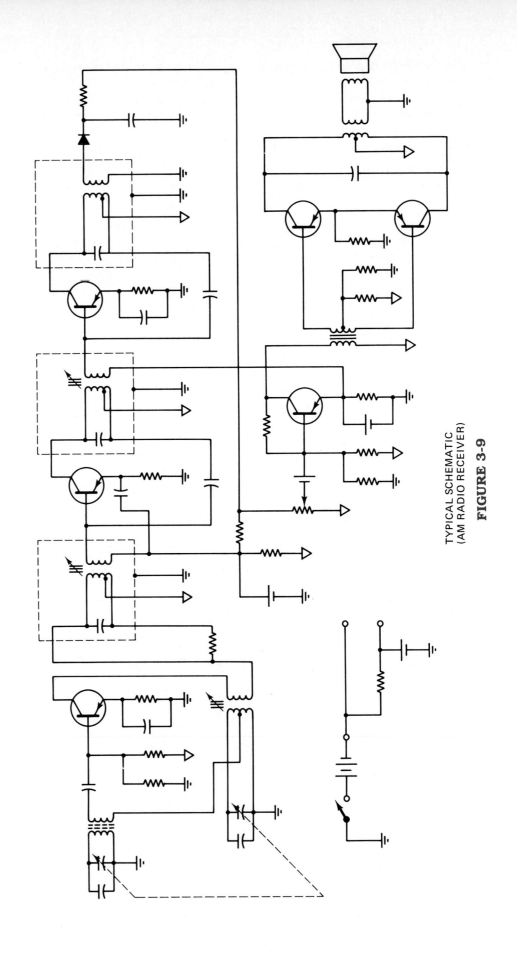

TYPICAL SCHEMATIC
(AM RADIO RECEIVER)
FIGURE 3-9

46

SCHEMATIC SYMBOLS

It is usually easier to analyze circuits by looking at schematics (Figure 3-9) than by looking at the actual components. To gain this ability, you must learn the basic circuit symbols. They are shown in Figure 3-10 and used extensively in the following chapters.

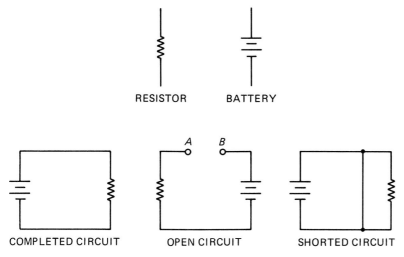

FIGURE 3-10 Typical circuit symbols and diagrams.

SUMMARY

Directly proportional refers to quantities that change linearly. When one quantity increases as another decreases, they are said to be inversely proportional.

Ohm's first law (E = IR) indicates that (1) E and I are directly proportional when R is held constant and (2) E and R are directly proportional when I is held constant. Ohm's second law (I = E/R) indicates that I and R are inversely proportional when E is held constant.

A short circuit implies that there is little or no resistance present in the circuit and that current is at a maximum level.

An open circuit implies that there is an infinite amount of resistance present in the circuit and that there is little or no current flowing.

The power rating of a component indicates its ability to dissipate heat. The formulas P = IE and P = E^2/R are used to calculate the amount of power a component will need to dissipate. The power rating for any component must equal or exceed the circuit requirement.

Any circuit must have (1) a voltage source, (2) a load, and (3) a continuous current path in order to operate.

QUESTIONS

3-1. All appliances have a tag or stamp. This tag is usually located on the back or lower front of the appliance. The tag contains the serial number, model number, and the electrical specifications for the appliance. Find this tag on four household appliances and copy down the electrical information given. Then apply Ohm's law to find the missing parameters for each appliance.

3-2. Describe how voltage relates to current.

3-3. Explain why current and heat are directly proportional in an electrical circuit.

3-4. Explain the difference between power and voltage.

3-5. Use Ohm's law to calculate current values for a circuit having a 20-V supply when the resistance is changed from the following values: 2 Ω, 4 Ω, 5 Ω, 10 Ω, 20 Ω. Construct a graph of the results. Is this a linear or an exponential relationship? Explain your choice.

3-6. Use Ohm's law to find the current values for a circuit having a 2-Ω resistor as the voltage is stepped up through the following sequence: 2 V, 4 V, 6 V, 8 V, 10 V, 20 V. Construct a graph of the results. Is this a linear or an exponential relationship? Explain your choice.

3-7. Use Ohm's law to calculate the voltage values for a 500-Ω resistor as the current is increased in the following steps: 10 mA, 20 mA, 30 mA, 50 mA, 100 mA. Construct a graph of the results. Is this a linear or an exponential relationship? Explain your choice.

3-8. A circuit has a 20-V supply. Find the current when a 100-Ω resistor is the load. Predict the current when the load is 50 Ω. Predict the current when the load is 200 Ω.

3-9. Explain what happens when the power rating of a component is too small for the circuit's operating conditions.

3-10. Explain what happens when the power rating of a component is too high for the circuit's operating conditions.

3-11. Draw a completed circuit having a 20-V battery and a 200-Ω resistor. Use schematic symbols.

3-12. Explain why the circuit must have a completed path.

3-13. Why is there voltage across an open circuit. What can this phenomenon be used for?

3-14. How many volts are across a shorted circuit? Explain your answer.

PRACTICE PROBLEMS

3-1. Given: $I = 100$ mA Find: $R = $ _____
$E = 20$ V $P = $ _____

3-2. Given: $R = 2.2$ kΩ Find: $P = $ _____
$I = 250$ mA $E = $ _____

3-3. Given: $P = 242$ mW Find: $E = $ _____
$I = 25$ mA $R = $ _____

3-4. Given: $R = 50$ kΩ Find: $I = $ _____
$P = 270$ mW $E = $ _____

3-5. Given: $R = 14$ Ω Find: $E = $ _____
$I = 37$ mA $P = $ _____

3-6. Given: $E = 35$ V Find: $I = $ _____
$R = 4.7$ kΩ $P = $ _____

3-7. Given: $P = 1.25$ W Find: $E = $ _____
$I = 0.56$ A $R = $ _____

3-8. Given: $E = 45$ V Find: $R = $ _____
$I = 0.025$ A $P = $ _____

3-9. Given: $P = 1500$ mW
$R = 22$ kΩ
Find: $I = $ _____
$E = $ _____

3-10. Given: $R = 120$ kΩ
$E = 15$ V
Find: $I = $ _____
$P = $ _____

3-11. Given: $E = 2.5$ V
$P = 67$ mW
Find: $I = $ _____
$R = $ _____

3-12. Given: $E = 4.5$ V
$I = 24$ μA
Find: $R = $ _____
$P = $ _____

3-13. Given: $P = 0.24$ W
$R = 2.7$ kΩ
Find: $E = $ _____
$I = $ _____

3-14. Given: $P = 0.45$ W
$E = 12$ V
Find: $R = $ _____
$I = $ _____

3-15. Given: $I = 3.2$ A
$R = 0.33$ Ω
Find: $E = $ _____
$P = $ _____

3-16. Given: $E = 22$ V
$R = 330$ kΩ
Find: $I = $ _____
$P = $ _____

3-17. Given: $P = 2500$ mW
$I = 45$ mA
Find: $E = $ _____
$R = $ _____

3-18. Given: $E = 25$ mV
$P = 35$ μW
Find: $R = $ _____
$I = $ _____

3-19. Given: $R = 56$ kΩ
$E = 6$ V
Find: $I = $ _____
$P = $ _____

3-20. Given: $E = 10$ V
$W = 10$ J
$T = 2.5$ s
Find: $P = $ _____
$I = $ _____
$R = $ _____

3-21. Given: $P = 350$ mW
$R = 2.7$ kΩ
$T = 6$ ms
Find: $I = $ _____
$E = $ _____
$W = $ _____

3-22. Given: $R = 2.2$ kΩ
$Q = 0.001$ C
$T = 20$ ms
Find: $P = $ _____
$I = $ _____
$E = $ _____

3-23. Given: $W = 10$ J
$T = 0.5$ s
$Q = 0.25$ C
Find: $P = $ _____
$I = $ _____
$R = $ _____
$E = $ _____
$e = $ _____

3-24. Given: $P = 1500$ mW
$I = 50$ mA
$e = 9.375 \times 10^{14}$ electrons
Find: $E = $ _____
$R = $ _____
$W = $ _____
$T = $ _____
$Q = $ _____

3-25. Given: $I = 0.04$ A
$R = 200$ Ω
$T = 2$s
Find: $P = $ _____
$E = $ _____
$W = $ _____
$Q = $ _____
$e = $ _____

4
Series Circuits

Chapter objectives

After reading this chapter and answering the questions and problems, you should be able to:

- Identify a series circuit.
- Draw the schematic symbol for a resistor and a battery. Label the positive and negative terminal of each when in a circuit.
- State the four series-circuit concepts.
- Given values of V, R, I, and P for a series circuit. Identify the key to the solution.
- Apply the circuit concepts and key to find the voltage, current, resistance, and power of series components.

The simplest circuit to analyze is the series circuit. It is one of the two basic circuit types. This chapter deals with the characteristics of a series circuit, how to identify series circuits, and ways to analyze individual volt, ohm, ampere, and watt values for each component. It is important that you understand the concepts behind series circuit analysis to proceed further in this book or, for that matter, for you to succeed in electronics at all. There are some things, in electronics, that you can get by without knowing well. This is not one.

DIRECTION OF CURRENT FLOW

For the purpose of this book, the direction of current flow will be from negative to positive. The true direction is a matter of convention rather than necessity. For consistency, we will use this direction throughout this book. Remember that in using this convention we are implying that the electrons are filling holes and are moving from the negative battery terminal toward the positive terminal. The negative terminal is shown as the short line of the battery symbol. The positive terminal is the long line of the battery symbol (see Figure 4-1).

In some circuits there are multiple paths for current to flow. In others there is only one path for current to flow. In the remainder of this chapter we will be concerned with single-path systems.

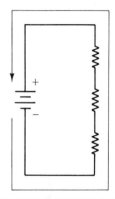

ELECTRON CURRENT FLOWS FROM THE − TERMINAL OF THE BATTERY TO THE + TERMINAL OF THE BATTERY

FIGURE 4-1

CIRCUIT IDENTIFICATION

A series circuit has only one path for current (Figure 4-2). If you start at the negative terminal of the battery and draw an arrow through the circuit, along the conductors, you will find only one continuous path (negative to positive) for current. The process is not unlike finding your way through the maze puzzles you solved as a child. This method seems a little silly at first, but it is the best way to begin to understand how current would flow through a circuit. When things are connected in series, the current will flow from one component to another in a serial fashion. It is like chaining the components together using only a single conductor. Below are several schematic representations of series circuits. All have one thing in common, their singular current path. Examine them closely. Learn to identify them.

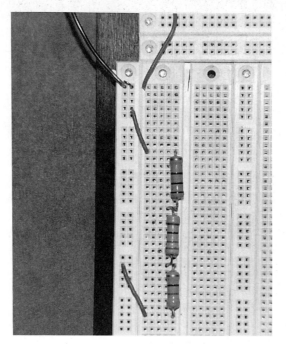

FIGURE 4-2 Series Circuit.

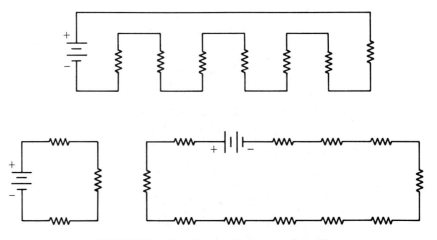

FIGURE 4-3 Typical series circuits.

In each of the circuits in Figure 4-3, indicate the direction of current flow, by sketching arrows from the negative terminal of the voltage source to the positive terminal.

SERIES-CIRCUIT CONCEPTS

The entire concept of series circuits can be summed up in four statements.

1. Since there is only one path for current, the same amount of current must pass through each component ($I_T = I_1 = I_2 = I_3 = \cdots$).
2. Voltage is additive ($V_T = V_1 + V_2 + V_3 + \cdots$).

3. Resistance is additive ($R_T = R_1 + R_2 + R_3 + \cdots$).
4. Power is additive ($P_T = P_1 + P_2 + P_3 + \cdots$).

These four concepts are used in analyzing individual component circuit values.

Same Current ($I_T = I_1 = I_2 = I_3 = \cdots$) Since all of the components are chained together, any current that flows through one must flow through all. The formula shown above indicates that the total current through the circuit is the same current that flows through the first component, and the second, and so on. In analyzing a series circuit this means that if we can calculate, or if we know, the value of the current through one component, this value will be the same for all the components in the circuit. It will also be the value of current supplied by the voltage source.

EXAMPLE 4-1

Suppose that the current through one of the resistors in a series circuit has been measured and is 45 mA. What is the current through the other resistors in that circuit?

Solution: Since this is a series circuit, the current is the same throughout. Therefore, the current through all components must be 45 mA. ∎

Voltage Is Additive ($V_T = V_1 + V_2 + V_3 + \cdots$) All of the voltage is supplied by the source. This source voltage is the total voltage of the circuit. Each individual component, on the other hand, has a resistance and current that provides an IR drop ($I \times R = E$). When we sum these individual IR drops, we get a value equal to the source voltage. This is why the source voltage is referred to as the total voltage of the circuit. A fine point—the voltage of the battery is correctly noted as E and the IR drops are correctly noted as V. For our purposes we will use the E's and V's interchangeably.

"Voltage is additive" is a most important concept in electronics—so much so that it has been named. It is called *Kirchhoff's voltage law* (KVL). The importance of your understanding this law cannot be overly stressed, as it will be used extensively throughout the remaining chapters. In short, Kirchhoff's voltage law states that *the sum of the voltage drops throughout any current path must equal the supply voltage*.

EXAMPLE 4-2

Find the source voltage in Figure 4-4.

Solution: First examine the circuit by tracing arrows around the circuit starting from the negative terminal of the battery and ending at the positive terminal of the battery. Notice that there is only one path for current flow. This is a series circuit.

Next remember that voltage is additive in a series circuit. Therefore, the source voltage is the sum of the individual voltage drops.

$$V_T = V_1 + V_2 + V_3$$
$$= 10 + 15 + 25$$
$$= 50 \text{ V}$$

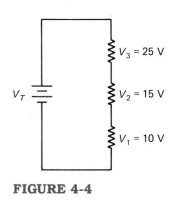

FIGURE 4-4

SERIES-CIRCUIT CONCEPTS

The supply voltage is 50 V.

Calculator Use: Enter the following sequence of keys:

$$[1][0][+][1][5][+][2][5][=]$$

The calculator should display

$$5.000 \quad 01$$

SELF-TEST

Find the missing voltages for each series circuit.

1. $V_1 = 20$ V
   $V_2 = 25$ V
   $V_T = $ _____

2. $V_T = 60$ V
   $V_1 = 15$ V
   $V_2 = 10$ V
   $V_3 = $ _____

3. $V_1 = 12$ V
   $V_T = 30$ V
   $V_2 = $ _____

4. $V_1 = 10$ V
   $V_2 = 25$ V
   $V_3 = 15$ V
   $V_T = $ _____

ANSWERS TO SELF-TEST

1. $V_T = 45$ V
2. $V_3 = 35$ V
3. $V_2 = 18$ V
4. $V_T = 50$ V

Resistance Is Additive ($R_T = R_1 + R_2 + R_3 + \cdots$) Chaining or linking the resistances of a series circuit together allows their resistances to add to each other. The sum of these resistances is the total circuit resistance. One thing to remember is that Ohm's law still applies to these circuits; however, we must apply it *only* to the individual levels. This will be demonstrated later.

EXAMPLE 4-3

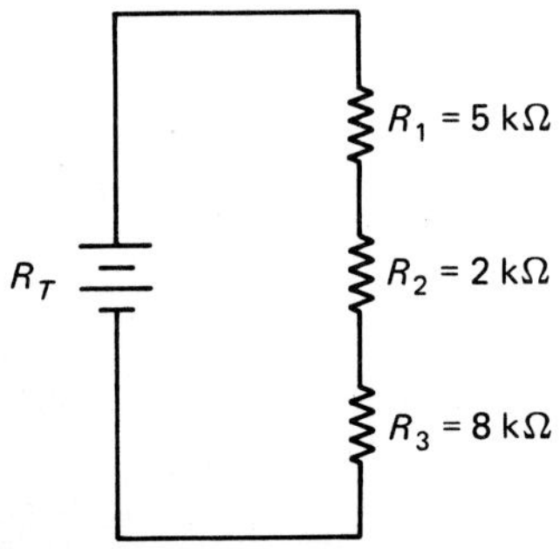

Find the total resistance of the circuit shown in Figure 4-5.

Solution: Again begin by tracing the current flow of this circuit. In this case the circuit is of a series construction. Remember that resistance is additive in a series circuit. So the total resistance would be the sum of the individual resistors.

$$R_T = R_1 + R_2 + R_3$$
$$= 5 \text{ k}\Omega + 2 \text{ k}\Omega + 8 \text{ k}\Omega$$
$$= 15 \text{ k}\Omega$$

FIGURE 4-5

The total circuit resistance is 15 kΩ.

Calculator Use: Enter the following sequence of keys:

$$[5][EE][3] + [2][EE][3] + [8][EE][3][=]$$

The calculator should display

$$1.500 \quad 04$$

SELF-TEST

Find the missing resistances for each of the series circuits.

1. $R_T = 1.2$ kΩ
   $R_1 = 400$ Ω
   $R_2 = 150$ Ω
   $R_3 = 350$ Ω
   $R_4 = $ _____
2. $R_1 = 150$ Ω
   $R_2 = 470$ Ω
   $R_3 = 560$ Ω
   $R_T = $ _____
3. $R_1 = 330$ Ω
   $R_T = 2000$ Ω
   $R_3 = 560$ Ω
   $R_4 = 270$ Ω
   $R_2 = $ _____
4. $R_1 = 45$ Ω
   $R_2 = 35$ Ω
   $R_T = $ _____

ANSWERS TO SELF-TEST

1. $R_4 = 300$ Ω
2. $R_T = 1180$ Ω
3. $R_2 = 840$ Ω
4. $R_T = 80$ Ω

Power Is Additive ($P_T = P_1 + P_2 + P_3 + \cdots$) Power is always additive. It makes no difference what type of circuit we are analyzing: The individual power dissipations will combine in an additive manner, and they equal the amount of power produced by the battery (Figure 4-6).

EXAMPLE 4-4

Find the total power dissipation of the circuit in Figure 4-6.

Solution: Power dissipation is always additive no matter what circuit construction is used. So the total power is the sum of the individual power dissipations.

$$P_T = P_1 + P_2 + P_3$$
$$= 10 \text{ mW} + 30 \text{ mW} + 40 \text{ mW}$$
$$= 80 \text{ mW}$$

The total power dissipation is 80 mW.
   Calculator Use: Enter the following sequence of keys:

[1] [0] [EE] [3] [+/−] [+]

[3] [0] [EE] [3] [+/−] [+]

[4] [0] [EE] [3] [+/−] [=]

The calculator should display

8.000   −02

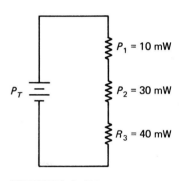

FIGURE 4-6

SELF-TEST

Find the missing power values.

1. $P_T = 450$ mW
   $P_1 = 100$ mW
   $P_2 = 330$ mW
   $P_3 = $ _____
2. $P_1 = 20$ mW
   $P_2 = 55$ mW
   $P_3 = 25$ mW
   $P_4 = 10$ mW
   $P_5 = 75$ mW
   $P_T = $ _____

SERIES-CIRCUIT CONCEPTS

3. $P_T = 1800$ mW
   $P_1 = 0.25$ W
   $P_3 = 0.35$ W
   $P_4 = 50$ mW
   $P_2 = $ _____

4. $P_T = 2$ W
   $P_1 = 500$ mW
   $P_2 = 0.1$ W
   $P_3 = $ _____

ANSWERS TO SELF-TEST

1. $P_3 = 20$ mW
2. $P_T = 185$ mW
3. $P_2 = 1.15$ W
4. 1.4 W

SERIES CIRCUIT ANALYSIS

The easiest way to analyze any circuit is by first building a matrix. A *matrix* is a method of laying out resistances, voltages, currents, and powers in an organized manner. One is shown below.

$V_T = $ $V_1 = $ $V_2 = $ $V_3 = $

$R_T = $ $R_1 = $ $R_2 = $ $R_3 = $

$I_T = $ $I_1 = $ $I_2 = $ $I_3 = $

$P_T = $ $P_1 = $ $P_2 = $ $P_3 = $

The matrix is used to show what values are known for each individual component, as well as what values are known for the voltage source. Ohm's law can be used in getting from one row to another. That is, if you know the value of V_2 and R_2, you can use Ohm's law to calculate values for P_2 and I_2. Notice that Ohm's law must be applied at individual levels only. In this case, Ohm's law was applied to the second level (V_2, R_2, I_2, and P_2). You could have also applied it to any of the other levels, including the total level.

You must use the circuit concepts described earlier to move from one column to another. That is, if you know values for V_T, V_1, and V_3, you can find the value for V_2 by subtracting V_1 from V_T, then by subtracting V_3 from your answer. Here you must work with all of the levels using only one quantity.

Some circuits are difficult to analyze. Using this matrix will allow you to view all the quantities for all the components simultaneously. When doing so, you need to look for individual levels where two of the four parameters are known. This is the key to solving for all the circuit values. The main thing to remember about series circuits is that the currents in the matrix are all the same current and thus will have the same value. Example 4-5 demonstrates this method of circuit analysis.

EXAMPLE 4-5

Find all the missing parameters for each of the components shown in Figure 4-7.

Solution: Begin by tracing arrows to indicate the path(s) for current flow. This will identify this circuit as having a series construction (one path). Next construct a matrix and fill in all known values.

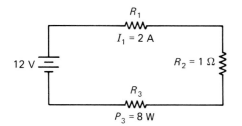

FIGURE 4-7

$V_T = 12$ V	$V_1 =$	$V_2 =$	$V_3 =$
$R_T =$	$R_1 =$	$R_2 = 1\ \Omega$	$R_3 =$
$I_T =$	$I_1 = 2$ A	$I_2 =$	$I_3 =$
$P_T =$	$P_1 =$	$P_2 =$	$P_3 = 8$ W

Then you should take note of what values are known. Notice that you know the amount of current flowing through component 1. Since this is a series circuit, we can assume that this is the current for all the components. Your matrix becomes

$V_T = 12$ V	$V_1 =$	$V_2 =$	$V_3 =$
$R_T =$	$R_1 =$	$R_2 = 1\ \Omega$	$R_3 =$
$I_T = 2$ A	$I_1 = 2$ A	$I_2 = 2$ A	$I_3 = 2$ A
$P_T =$	$P_1 =$	$P_2 =$	$P_3 = 8$ W

Calculator Use: Since current is the same throughout a series circuit, the current is a value that will be used continually when solving this problem. In general, it is easier to put this value in your calculator's memory rather than reentering it every time it is needed in a calculation.

Your calculator should have at least three buttons associated with its memory. They are labeled mem (sometimes M), sum (sometimes M+), and RCL (sometimes RM). The mem key stores the value shown on the display in your calculator's memory. The sum key adds the value shown on the display to the value stored in memory, then replaces the value stored in memory with this sum. The RCL key recalls the value stored in memory. This value is displayed and remains unchanged in memory.

At this time enter the following sequence of keys:

[2] [mem]

The calculator should display

2.000   00

Notice that you now have two values at both level 2 and level 3. Also, there are two values at the total level as well. You can use Ohm's law at these levels to find the other missing quantities.

Level 2:

$$V_2 = R_2 \times I_2$$
$$= 1 \times 2 = 2\ \text{V}$$

[1] [×] [mem] [=]   display   2.000   00

SERIES CIRCUIT ANALYSIS

$$P_2 = V_2 \times I_2$$
$$= 2 \times 2 = 4 \text{ W}$$

[2] [×] [mem] [=]   display   4.000   00

Level 3:
$$V_3 = \frac{P_3}{I_3}$$
$$= \frac{8}{2} = 4 \text{ V}$$

[8] [÷] [mem] [=]   display   4.000   00

$$R_3 = \frac{V_3}{I_3}$$
$$= \frac{4}{2} = 2 \text{ }\Omega$$

[4] [÷] [mem] [=]   display   2.000   00

Level T:
$$R_T = \frac{V_T}{I_T}$$
$$= \frac{12}{2} = 6 \text{ }\Omega$$

[1] [2] [÷] [mem] [=]   display   6.000   00

$$P_T = V_T \times I_T$$
$$= 12 \times 2 = 24 \text{ W}$$

[1] [2] [×] [mem] [=]   display   2.400   01

The use of the memory function does not save much work on this particular problem, but on most, the use of the memory function can be a big time-saver.

Placing these values in the matrix gives you

$V_T = 12$ V	$V_1 =$	$V_2 = 2$ V	$V_3 = 4$ V
$R_T = 6$ Ω	$R_1 =$	$R_2 = 1$ Ω	$R_3 = 2$ Ω
$I_T = 2$ A	$I_1 = 2$ A	$I_2 = 2$ A	$I_3 = 2$ A
$P_T = 24$ W	$P_1 =$	$P_2 = 4$ W	$P_3 = 8$ W

Notice that the only level which is not complete is the first level. Using the other series circuit concepts will provide solutions for these missing values.

Voltage:
$$V_1 = V_T - V_2 - V_3$$
$$= 12 - 2 - 4 = 6 \text{ V}$$

Resistance:
$$R_1 = R_T - R_2 - R_3$$
$$= 6 - 1 - 2 = 3 \text{ }\Omega$$

Power:
$$P_1 = P_T - P_2 - P_3$$
$$= 24 - 4 - 8 = 12 \text{ W}$$

These values complete the matrix and also end our need to calculate further. ∎

EXAMPLE 4-6

Three components are connected in series to a 100-V source. The first component has a voltage drop of 24 V, the second has a voltage drop of 18 V, and the third has a resistance of 28 Ω. Find the circuit values for each component.

Solution: Begin by drawing the circuit (Figure 4-8). Next build a matrix and fill in the known values.

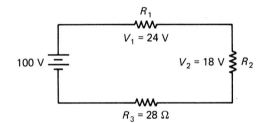

FIGURE 4-8

$V_T = 100$ V	$V_1 = 24$ V	$V_2 = 18$ V	$V_3 =$
$R_T =$	$R_1 =$	$R_2 =$	$R_3 = 28$ Ω
$I_T =$	$I_1 =$	$I_2 =$	$I_3 =$
$P_T =$	$P_1 =$	$P_2 =$	$P_3 =$

Inspection should dictate that the voltage quantity is the key to solving this circuit.

Voltage:
$$V_3 = V_T - V_1 - V_2$$
$$= 100 - 24 - 18 = 58 \text{ V}$$

Placing this value in the matrix gives you

$V_T = 100$ V	$V_1 = 24$ V	$V_2 = 18$ V	$V_3 = 58$ V
$R_T =$	$R_1 =$	$R_2 =$	$R_3 = 28$ Ω
$I_T =$	$I_1 =$	$I_2 =$	$I_3 =$
$P_T =$	$P_1 =$	$P_2 =$	$P_3 =$

Now you have two quantities at level 3. Use Ohm's law to find I_3. Since this is a series circuit, all current values are equal. The completed matrix would look as follows:

$V_T = 100$ V	$V_1 = 24$ V	$V_2 = 18$ V	$V_3 = 58$ V
$R_T = 48.28$ Ω	$R_1 = 11.59$ Ω	$R_2 = 8.69$ Ω	$R_3 = 28$ Ω

$I_T = 2.071$ A $\quad I_1 = 2.071$ A $\quad I_2 = 2.071$ A $\quad I_3 = 2.071$ A

$P_T = 207.1$ W $\quad P_1 = 49.71$ W $\quad P_2 = 37.29$ W $\quad P_3 = 120.1$ W ∎

USING THE CIRCUIT FOR THE MATRIX

It is a good practice for you to draw any circuit that you are solving. Always draw the circuit large enough for each of the four quantities to be written by their respective component. This will help you to visualize how the circuit is operating. By doing so, the matrix can be built around the circuit. It is easiest for most students to find the key to solving a circuit when the circuit values are placed in matrix form; however, it is equally important for you to become proficient at finding the key when the circuit values are placed around the circuit. An example of this type of matrix and circuit construction is shown in Figure 4-9.

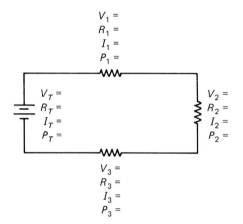

FIGURE 4-9

SUMMARY

In summary, the prime consideration when analyzing a series circuit is the same current. The first thing you should associate with a series circuit is that the current is the same value for all the components in the circuit. Second, the concept of adding the voltages together to find the total voltage is true for a series circuit. Also, the resistance has the same additive qualities as voltage. In any circuit construction the power is additive.

Although the matrix was stressed in this discussion, it is still a crutch. You must work toward seeing the key to solving the circuit without its use. The main idea is to find two of the four quantities for one component. Once this is done, the other quantities can be found by applying the four concepts discussed in the previous paragraphs.

Most people have difficulty remembering formulas. Rather than memorizing formulas, try to remember phrases. Learn the mathematical meaning of these phrases. These phrases will be used throughout this book and can ease the difficulty of learning electronics.

QUESTIONS

4-1. Explain what is meant by *additive resistance*. How does this relate to a series circuit?

4-2. Explain what is meant by *additive voltage*. How does this relate to a series circuit?

4-3. Explain what is meant by *additive power*. How does this relate to a series circuit? To other circuit types?

4-4. Why is current the same through all components in a series circuit?

4-5. Describe what happens to circuit conductance when more resistance is added to a series circuit.

4-6. Draw a pictorial sketch of two light bulbs in series with a battery. Suppose that one of the light bulbs burns out. Would this be a shorted- or an open-circuit condition?

4-7. Discuss the difference between a shorted and an open circuit in terms of voltage, resistance, and current flow.

4-8. What is more damaging to a circuit—a short or an open? Why?

4-9. A circuit requires a certain component to dissipate 350 mW of power. Which of the following power ratings would be sufficient? $\frac{1}{8}$ W, $\frac{1}{4}$ W, $\frac{1}{2}$ W, 1 W. Why?

4-10. If two equal resistors are placed in series with each other, how much of the supply voltage would appear across each of them?

4-11. Why is voltage sometimes referred to as potential difference?

4-12. A television set typically requires 4 A to operate. How much is its load resistance?

4-13. When we say that current flow moves from negative to positive, are we referring to electron or hole movement?

4-14. Suppose that four resistors—1 MΩ, 50 kΩ, 500 Ω, and 1 Ω—are placed in series. What is the approximate total resistance of the circuit? Why?

PRACTICE PROBLEMS

4-1. The supply voltage to a series circuit is 25 V. The total resistance is 200 Ω. The first resistor in a chain of four has a resistance of 47 Ω. What is the current through the 47-Ω resistor?

4-2. The voltage drops around a series circuit containing three resistors are 10 V, 25 V, and 5 V. The resistance corresponding to the 10-V *IR* drop is 300 Ω. How much current is the voltage providing?

4-3. It is found that the total resistance of a five-resistor series circuit is 1.5 kΩ. The first three resistors are 220 Ω, 180 Ω, and 150 Ω, respectively. The two remaining resistors are equal. What is the resistance of each of the remaining resistors?

4-4. It has been determined that the total current produced by the battery of a series circuit is 500 mA. The supply voltage is 25 V. The first three resistors dissipate a total of 10 W of power. How much power does the remaining resistor dissipate?

4-5. The first two resistors in a series circuit containing three resistors have voltage drops of 2.5 V and 4.5 V, respectively. The supply voltage is 10 V. What is the voltage drop of the third resistor?

For the following problems, refer to Figure 4-9 and solve for the missing quantities.

4-6. Given: $V_T = 20$ V
$R_1 = 150\ \Omega$
$R_2 = 330\ \Omega$
$R_3 = 470\ \Omega$

4-7. Given: $I_2 = 250$ mA
$R_1 = 1.2$ kΩ
$R_2 = 2.7$ kΩ
$R_3 = 5.6$ kΩ

4-8. Given: $V_T = 10$ V
$V_1 = 2.1$ V
$V_2 = 4.3$ V
$I_3 = 140$ mA

4-9. Given: $I_1 = 25$ mA
$R_1 = 56$ kΩ
$R_2 = 27$ kΩ
$P_3 = 56$ mW

4-10. Given: $R_1 = 220\ \Omega$
$R_2 = 1.2$ kΩ
$I_2 = 2.5$ mA
$R_T = 2.7$ kΩ

4-11. Given: $R_T = 4.7$ kΩ
$V_T = 30$ V
$R_2 = 1.8$ kΩ
$R_3 = 860\ \Omega$

4-12. Given: $P_T = 450$ mW
$I_T = 2$ mA
$V_1 = 100$ V
$R_2 = 27$ kΩ

4-13. Given: $R_1 = 120\ \Omega$
$R_2 = 390\ \Omega$
$R_3 = 500\ \Omega$
$V_3 = 3$ V

4-14. Given: $V_T = 35$ mV
$R_1 = 2$ kΩ
$R_2 = 4.7$ kΩ
$R_3 = 3.3$ kΩ

4-15. Given: $I_T = 20$ mA
$R_1 = 8.6$ kΩ
$R_2 = 4.7$ kΩ
$V_3 = 20$ V

Parallel Circuits

Chapter objectives

After reading this chapter and answering the questions and problems, you should be able to:

- Identify series and parallel circuits and components.
- State the four parallel-circuit concepts.
- Use both the reciprocal method and the product-over-sum method to find total parallel resistance.
- Use supposition to find the total resistance of a parallel circuit when voltage and current are not known.
- Use the parallel-circuit concepts to calculate V, I, R, and P values for any parallel component or circuit.
- Use two or more parallel resistors to fabricate nonstandard resistor values.
- Define *branch*, *junction*, and *mainline* as they refer to parallel circuits and currents.
- Given any number of equal resistors, find their parallel resistance without the use of a calculator.

While most systems use only one voltage source, complex systems may have several circuits operating from the same voltage source. These circuits may perform independently of each other. An example of this can be found in your automobile's electrical system. It has only one battery that supplies power to the ignition system, the radio, the headlights, and all the other accessories. The type of circuit where each load is fed by the battery separately is called a *parallel circuit*. Parallel circuits are the topic of this chapter. Information about their identification, circuit concepts, and circuit analysis will be provided. Systems are classified as series, parallel, or a combination of the two. It is important that you become proficient at identifying and solving parallel circuits.

CIRCUIT IDENTIFICATION

Parallel circuits have more than one path for current (Figure 5-1). They can be identified if you start tracing the current flow through the circuit along the conductors. Do this by drawing arrows around the circuit. Remember that these arrows must point from the negative terminal toward the positive terminal of the battery. On a parallel circuit, you will find junctions or branches where there are different ways for current to flow. This process is like solving a maze puzzle that has more than one solution.

FIGURE 5-1 Parallel Circuit.

This method may seem a little silly at first but it is the best way to begin to understand how current would flow through a circuit. When things are connected in parallel, the current will flow through all the components in an independent manner.

Figure 5-2 contains several schematic representations of parallel circuits. All have one thing in common. That is their multiple current paths. Examine them closely. Learn to identify where the junctions are located. Take the time to sketch arrows for current flow in each of the figures and highlight each of the junctions as they are encountered.

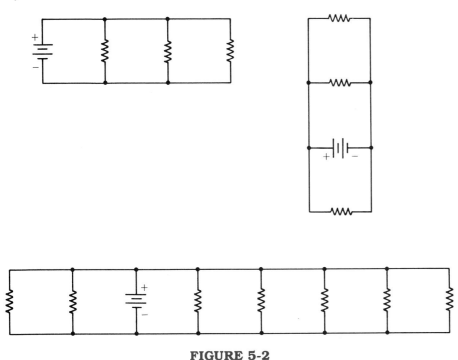

FIGURE 5-2

PARALLEL-CIRCUIT CONCEPTS

Again there are four basic concepts associated with parallel circuits. You will need to become familiar with them if you are to succeed in electronics. They are very similar to the series concepts discussed in Chapter 4. The parallel-circuit concepts are:

1. Only one voltage source, so voltage is the same everywhere ($V_T = V_1 = V_2 = V_3 = \cdots$).
2. Current is additive ($I_T = I_1 + I_2 + I_3 + \cdots$).
3. Conductance is additive ($1/R_T = 1/R_1 + 1/R_2 + 1/R_3 + \cdots$).
4. Power is additive ($P_T = P_1 + P_2 + P_3 + \cdots$).

Notice that with parallel circuits you use the additive conductance instead of resistance (resistance was added on a series circuit). Also, a parallel circuit and a series circuit have the roles of voltage and current exchanged. There is one voltage on a parallel circuit and one current on a series circuit. A discussion of each of these four concepts follows.

One Voltage Source ($V_T = V_1 = V_2 = V_3 = \cdots$) There is only one voltage source, so the same voltage value is fed to all the parallel components. In analyzing this type of circuit, if you know the voltage of the supply or the voltage across any of the components, you can assume that the voltage supplied to all of the others is the same value. It should be. If it is not, there is a fault in the circuit.

The portion of the circuit where all the current flows is called the *mainline*. An open here can affect the circuit as a whole. In general, opens have an effect only on the portion of the circuit in which they are contained. That is, when there is an open in one branch, only that branch is affected. If any portion of a parallel circuit is shorted, the whole circuit is shorted. These are important troubleshooting facts. They will help in checking parallel circuit operation.

SELF-TEST

Answer the following questions.

1. Suppose that one branch of a parallel circuit had a resistance of 25 Ω and that its current was 50 mA. How large is the source voltage?
2. The source voltage is 25 V. How many volts are there across the second branch?
3. The voltage source provides 5 mA of current to the circuit as a whole. The total resistance of the circuit is 500 Ω. What is the voltage of the supply?
4. Suppose that there are three 120-V light bulbs in parallel with each other. What will the voltage be across the other two when the first light bulb opens?

ANSWERS TO SELF-TEST

1. 1.25 V
2. 25 V
3. 2.5 V
4. 120 V

Current Is Additive ($I_T = I_1 = I_2 = I_3 + \cdots$) The battery must supply all of the current to the circuit. This current is called the *total current*, because it is the sum of each of the currents received by the separate loads (Figure 5-3). In other words, adding the load currents provides the amount of current being provided by the voltage source. Ohm's law will work on an individual level and can be used to calculate individual load currents. It is important to note that any fluctuation of individual load current will affect the total current being supplied. If one load shorts out, it will short out the supply and all the other components. This is something to remember when troubleshooting a parallel circuit.

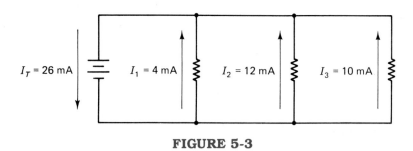

FIGURE 5-3

SELF-TEST

Answer the following questions.

1. There are four resistors in parallel. The current through the first is 20 mA. The current through the second is 30 mA. The current through the third is 15 mA. The total current for the supply is 120 mA. How much current flows through the remaining resistor?
2. It is found that the mainline current in a parallel circuit is 500 mA. It is also known that the first of two resistors has a current of 238 mA. What is the current through the second parallel resistor?
3. Suppose that there are three resistors in a parallel circuit. The current through these resistors is 12.5 mA, 14.9 mA, and 5.8 mA. What is the total current for the circuit?
4. The total current in a two-resistor parallel circuit is 0.45 mA. The current through one of these resistors is 235 μA. What is the current through the other?

ANSWERS TO SELF-TEST

1. 55 mA
3. 33.2 mA
2. 262 mA
4. 215 μA

Conductance Is Additive ($1/R_T = 1/R_1 + 1/R_2 + 1/R_3 + \cdots$) Probably the most confusing part of learning parallel circuit analysis is learning about conductance. Remember, conductance is the lack of opposition to current flow. It is the reciprocal of the resistance value and is measured in mhos (Figure 5-4). Combining resistances in parallel lessens the total amount of circuit resistance. Think of it as like adding more lanes to a highway. More lanes mean more traffic flow and fewer traffic jams. In electronics, parallel paths allow more current flow. More current flow means less circuit resistance. For this reason it is necessary for us to look at the lack of resistance rather than the resistance of parallel circuits.

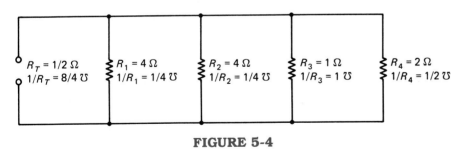

FIGURE 5-4

Let's examine the total resistance formula:

$$\frac{1}{R_T} = \frac{1}{R_1} + \frac{1}{R_2} + \frac{1}{R_3} + \cdots$$

Understand that $1/R_T$ is a conductance value and not a resistance value. This formula, at first, seems awesome. It is not. Your calculator should have a reciprocal key. It is the one labeled $1/x$. When calculating the total resistance of a parallel cir-

PARALLEL CIRCUIT CONCEPTS

cuit, you will need to push this key after each resistance has been entered and again after the equal sign has been pushed.

EXAMPLE 5-1

Suppose that we are trying to find the total resistance of two 10-Ω resistors connected in parallel. (Use the reciprocal method.)

Solution: Use the parallel resistance formula to get

$$\frac{1}{R_T} = \frac{1}{10} + \frac{1}{10}$$

$$= \frac{2}{10}$$

$$= \frac{1}{5}$$

So $1/R_T = 1/5$ and $R_T = 5$.

Calculator Use: Enter the following key strokes:

[1] [0] [1/x] [+] [1] [0] [1/x] [=] [1/x]

The calculator should display

5.000   00

As you can see, this calculation is not as complicated as the formula indicates. Notice that the reciprocal key is pushed after each resistance has been entered. This causes your calculator to convert the resistances into their equivalent conductances. Then when the equal key on your calculator is pressed, the calculator displays the total conductance. It is necessary to press the reciprocal key one more time to convert the total conductance into total resistance. ■

It is strongly suggested that you use the reciprocal method for all of your parallel calculations. It will work on parallel circuits regardless of the number of resistors in the circuit. Also, it will provide an answer with the fewest number of keystrokes. Two other methods are discussed in the following paragraphs. Understand that these methods are included only to prepare you for more advanced work.

A more traditional way of finding total resistance is by using the times-plus formula (also called the product-over-sum method). It is called times-plus for obvious reasons. This formula is shown below.

$$R_T = \frac{R_1 \times R_2}{R_1 + R_2}$$

Although this formula can only be used on two-resistor circuits, it is important that you become familiar with it, because more advanced texts will use this formula to explain and verify many of their calculations.

EXAMPLE 5-2

Suppose that we are trying to find the total resistance of two 10-Ω resistors connected in parallel. (Use the times-plus formula.)

Solution: Use the parallel resistance formula to get

$$R_T = \frac{10 \times 10}{10 + 10}$$

$$= \frac{100}{20} = 5$$

Calculator Use: You will notice that this is an easier method to understand mathematically, but most students have difficulty using their calculators to find the answers in this manner. The problem arises because of the hierarchy used by the calculator in solving equations. Basically, your calculator does multiplication and division before it does addition and subtraction. When you key this problem into your calculator, you should key it like this:

[1] [0] [×] [1] [0] [÷] [(] [1] [0] [+] [1] [0] [)] [=]

Most people forget to use the parentheses. Without the parentheses the calculator reads or understands the instructions as "multiply 10 by 10 and then divide the product by 10." This would give an answer of 10. The calculator would then add 10 to this answer to give a final result of 20. To avoid this, either include the parentheses or make the addition calculation first and save it in memory. Then do the multiplication and recall the sum for the division. The keystrokes for this would look as follows:

[1] [0] [+] [1] [0] [=] [mem]
[1] [0] [×] [1] [0] [=] [÷] [rcl] [=]

When two equal resistors are placed in parallel, the total is half of the value of the individual resistances. Also, this concept can be expanded to apply to larger numbers. If you combine three equal resistances in parallel, you will wind up with one-third of the individual amount. Four would be one-fourth, and so on. In other words, the individual resistance can be divided by the number of equal resistors to find the total parallel resistance.

What about unequal resistances? Their combined parallel resistance is always less than the smaller of the set.

EXAMPLE 5-3

If we were to combine three resistances of 2, 6, and 10 Ω in parallel, the total would be less than 2 Ω. Try this calculation yourself.

$$\frac{1}{R_T} = \frac{1}{2} + \frac{1}{6} + \frac{1}{10}$$

$$= 0.766666667$$

$$R_T = \frac{1}{0.766666667} = 1.304347826 \, \Omega$$

Calculator Use: Enter the following keystrokes:

[2] [1/x] [+] [6] [1/x] [+] [1] [0] [1/x] [=] [1/x]

The calculator should display

1.3043   00

PARALLEL CIRCUIT CONCEPTS

As you noticed, the answer was less than 2 Ω, the smallest of the three resistors. This will always be the case. You can use this concept to check your work.

Any time calculations are involved, don't just punch keys. Think about what you are doing. Use approximations to predict your answers. For example, if your answer would have been 2.5 in the preceding problem, you should have recognized 2.5 as an invalid answer and reworked the problem to find the error.

SELF-TEST

Answer the following questions.

1. Three resistors are placed in parallel. They have resistances of 2.5, 33, and 4.7 kΩ, respectively. What is their total resistance?
2. Find the conductance of the following resistors: 500 Ω, 2500 Ω, 20 kΩ, and 3.3 kΩ.
3. What is the approximate resistance of a circuit, when 300, 800, and 20 Ω are placed in parallel?
4. When solving a parallel resistance problem, a student gets an answer of 450 Ω. The parallel resistances were 500 Ω, 2500 Ω, 20 kΩ and 3.3 kΩ. Does his answer seem correct? Is it?

ANSWERS TO SELF-TEST

1. 1.555 kΩ
2. 2 m℧ 400 μ℧ 50 μ℧ 303 μ℧
3. Approximately 20 Ω
4. The approximate resistance would be less than 500, so 450 would seem correct. The exact answer, however, is 363 Ω.

Power Is Additive ($P_T = P_1 + P_2 + P_3 + \cdots$) Power on any electrical circuit is additive (Figure 5-5). Since power is a statement of the amount of work being performed in a given length of time, the circuit configuration does not affect its total value. The only quantities that do affect power are the individual currents and resistances. This is predicted in the $P = I^2 \times R$ formula.

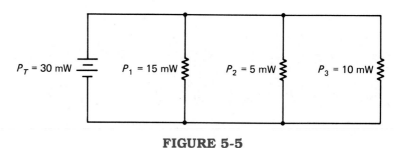

FIGURE 5-5

The easiest manner to calculate the power consumption of any component is first to measure the voltage ($I \times R$). Then read the resistance value and calculate the power by using E^2/R formula.

Other than the resistance value, the power rating is the most crucial specification. A value smaller than the required amount will lead to a shortened life for the resistor. Always replace a resistor with one of equal or higher power rating. That is, a $\frac{1}{2}$-W resistor could be replaced with a 1-W resistor but not with a $\frac{1}{4}$-W resistor.

Again, heat dissipation capabilities are the reason for this action. A 1-W resistor can dissipate more heat than a $\frac{1}{2}$-W resistor. A $\frac{1}{4}$-W resistor cannot dissipate as much as can a $\frac{1}{2}$-W resistor. If a $\frac{1}{4}$-W resistor was used to replace a $\frac{1}{2}$-W resistor, it would overheat and soon burn up. If you try using a resistor of too low a power rating in the lab, you will notice an accompanying bad smell.

PARALLEL CIRCUIT ANALYSIS

The matrix method also works when analyzing a parallel circuit. Make a table for laying out values of individual resistances, voltages, currents, and powers in an organized manner. One is shown below.

$V_T =$	$V_1 =$	$V_2 =$	$V_3 =$
$R_T =$	$R_1 =$	$R_2 =$	$R_3 =$
$I_T =$	$I_1 =$	$I_2 =$	$I_3 =$
$P_T =$	$P_1 =$	$P_2 =$	$P_3 =$

This matrix is used to show what quantities are known for each individual component, as well as what quantities are known for the voltage source. Ohm's law can be used in solving for the quantities in each row. That is, if you know the value for V_2 and R_2, you can use Ohm's law to calculate values for P_2 and I_2.

Notice that Ohm's law must be applied at individual levels only. In this case you applied Ohm's law to the second level. You could have also applied it to any of the other levels, including the total level.

You must use the circuit concepts described earlier to find the quantities in each column. For instance, if you know the value for V_T, then V_1, V_2, and V_3 must also be of this value, because all the voltages are the same in a parallel circuit. Here we must work with all the levels using only one quantity.

Some circuits are difficult to analyze. Using this matrix will allow you to view all of the quantities for all the components simultaneously. When doing so, you need to look for individual levels where two of the four parameters are known. That is the key to solving for all the other circuit values.

EXAMPLE 5-4

Calculate the voltage, current, resistance, and power parameters for each of the components in Figure 5-6.

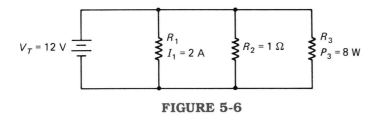

FIGURE 5-6

Solution: First place the schematic values in the matrix and take note of what values are known.

$V_T = 12$ V	$V_1 =$	$V_2 =$	$V_3 =$
$R_T =$	$R_1 =$	$R_2 = 1\ \Omega$	$R_3 =$

$$I_T = \quad I_1 = 2\text{ A} \quad I_2 = \quad I_3 =$$
$$P_T = \quad P_1 = \quad P_2 = \quad P_3 = 8\text{ W}$$

Notice that you know the amount of supply voltage. Since this is a parallel circuit, you can assume that this is the voltage for all the components. Your matrix becomes

$$V_T = 12\text{ V} \quad V_1 = 12\text{ V} \quad V_2 = 12\text{ V} \quad V_3 = 12\text{ V}$$
$$R_T = \quad R_1 = \quad R_2 = 1\ \Omega \quad R_3 =$$
$$I_T = \quad I_1 = 2\text{ A} \quad I_2 = \quad I_3 =$$
$$P_T = \quad P_1 = \quad P_2 = \quad P_3 = 8\text{ W}$$

Now you have two values at all the component levels. You can use Ohm's law at each level to find the other missing quantities.

Level 1:

$$R_1 = \frac{V_1}{I_1}$$
$$= \frac{12}{2}$$
$$= 6\ \Omega$$

[1] [2] [÷] [2] [=]   display   6.0000   00

$$P_1 = V_1 \times I_1$$
$$= 12 \times 2$$
$$= 24\text{ W}$$

[1] [2] [×] [2] [=]   display   2.4000   01

Level 2:

$$I_2 = \frac{V_2}{R_2}$$
$$= \frac{12}{1}$$
$$= 12\text{ A}$$
$$P_2 = V_2 \times I_2$$
$$= 12 \times 12$$
$$= 144\text{ W}$$

Level 3:

$$R_3 = \frac{(E)^2}{P_3}$$
$$= \frac{(12)^2}{8}$$
$$= 18\ \Omega$$

[1] [2] [x²] [÷] [8] [=]   display   1.8000   01

$$I_3 = \frac{V_3}{R_3}$$

$$= \frac{12}{18}$$

$$= 0.6667 \text{ A}$$

Placing these values in the matrix gives you

$V_T = 12$ V	$V_1 = 12$ V	$V_2 = 12$ V	$V_3 = 12$ V
$R_T =$	$R_1 = 6\ \Omega$	$R_2 = 1\ \Omega$	$R_3 = 18\ \Omega$
$I_T =$	$I_1 = 2$ A	$I_2 = 12$ A	$I_3 = 0.6667$ A
$P_T =$	$P_1 = 24$ W	$P_2 = 144$ W	$P_3 = 8$ W

The total level values can be calculated by any of the parallel circuit concepts. It is usually easier to total the current and use it to calculate the other values.

$$I_T = I_1 + I_2 + I_3$$

$$= 2 + 12 + 0.6667$$

$$= 14.67 \text{ A}$$

[2] [+] [1] [2] [+] [.] [6] [6] [6] [6] [6] [6] [7] [=]
     [mem]   display   1.4667   01

This calculation provides the key to solving for the total level values.

Level T:

$$R_T = \frac{V_T}{I_T}$$

$$= \frac{12}{14.67}$$

$$= 0.8182\ \Omega$$

[1] [2] [÷] [rcl] [=]   display   8.1818   −01

$$P_T = V_T \times I_T$$

$$= 12 \times 14.67$$

$$= 176 \text{ W}$$

[1] [2] [×] [rcl] [=]   display   1.7600   01

These values complete the matrix and also end our need to calculate further. ∎

Let's try another problem, but this time let's start with a circuit description.

EXAMPLE 5-5

Three components are connected in parallel to a 100-V source. The first component has a current of 2 A, the second has a resistance of 20 Ω, and the third has a resistance of 25 Ω. Find the circuit values for each component.

Solution: Begin by drawing a schematic of the circuit (Figure 5-7). Next build a matrix and fill in the known values for components.

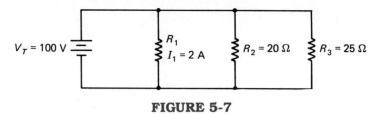

FIGURE 5-7

$V_T = 100$ V $V_1 =$ $V_2 =$ $V_3 =$

$R_T =$ $R_1 =$ $R_2 = 20$ Ω $R_3 = 25$ Ω

$I_T =$ $I_1 = 2$ A $I_2 =$ $I_3 =$

$P_T =$ $P_1 =$ $P_2 =$ $P_3 =$

Inspection should indicate that the voltage quantity is the key to solving this circuit, as voltage is the same for all the components. Placing this value in the matrix gives

$V_T = 100$ V $V_1 = 100$ V $V_2 = 100$ V $V_3 = 100$ V

$R_T =$ $R_1 =$ $R_2 = 20$ Ω $R_3 = 25$ Ω

$I_T =$ $I_1 = 2$ A $I_2 =$ $I_3 =$

$P_T =$ $P_1 =$ $P_2 =$ $P_3 =$

There are now three levels that have two quantities. Ohm's law will provide values to fill each of these individual levels.

Level 1:
$$R_1 = \frac{V_1}{I_1}$$
$$= \frac{100}{2}$$
$$= 50 \text{ Ω}$$
$$P_1 = V_1 \times I_1$$
$$= 100 \times 2$$
$$= 200 \text{ W}$$

Level 2:
$$I_2 = \frac{V_2}{R_2}$$
$$= \frac{100}{20}$$
$$= 5 \text{ A}$$
$$P_2 = \frac{V_2}{I_2}$$
$$= 100 \times 5$$
$$= 500 \text{ W}$$

Level 3:

$$I_3 = \frac{V_3}{R_3}$$

$$= \frac{100}{25}$$

$$= 4 \text{ A}$$

$$P_3 = V_3 \times I_3$$

$$= 100 \times 4$$

$$= 400 \text{ W}$$

Inserting these values in the matrix, we have

$V_T = 100$ V	$V_1 = 100$ V	$V_2 = 100$ V	$V_3 = 100$ V
$R_T =$	$R_1 = 50\ \Omega$	$R_2 = 20\ \Omega$	$R_3 = 25\ \Omega$
$I_T =$	$I_1 = 2$ A	$I_2 = 5$ A	$I_3 = 4$ A
$P_T =$	$P_1 = 200$ W	$P_2 = 500$ W	$P_3 = 400$ W

Again any of the methods for finding a total value can be used as a key to completing the matrix. The author prefers using the additive current. However, traditionally the parallel calculation of total resistance has been the standard method for completing the table.

$$\frac{1}{R_T} = \frac{1}{R_1} + \frac{1}{R_2} + \frac{1}{R_3}$$

$$= \frac{1}{50} + \frac{1}{20} + \frac{1}{25}$$

$$= 0.11\ \mho$$

$$R_T = \frac{1}{0.11}$$

$$= 9.091\ \Omega$$

[5] [0] [1/x] [+] [2] [0] [1/x] [+] [2] [5] [=] [1/x]

display   9.0909   00

You will notice that this is a more complicated way of calculating the total resistance. It is usually easier to find total resistance by first finding total current.

$$I_T = I_1 + I_2 + I_3$$

$$= 2 + 5 + 4$$

$$= 11 \text{ A}$$

$$R_T = \frac{V_T}{I_T}$$

$$= \frac{100}{11}$$

$$= 9.091\ \Omega$$

PARALLEL CIRCUIT ANALYSIS

Generally, this method will lead to fewer button pushes. The number of button pushes is directly proportional to the probability of obtaining a correct answer. The more buttons pushed, the more likely a mistake will be made.

∎

There will be times when a voltage will not be known. No voltage leads one to believe that the additive property of current could not be used. This is not the case. Notice the conductance in the circuit above, 0.11 ℧. Look at the current 11 A. Do you see a similarity? (0.11 × 100 = 11.) When we calculate conductance what we are really doing is using an advanced network theorem called *supposition*. We are supposing the voltage to be 1 and calculating the resultant current. We could suppose any voltage we wish.

In some circuits 1 V would not be the best voltage. When using this method, always try to pick a value easy to work with.

EXAMPLE 5-6

Use supposition to find the total resistance of the following parallel resistances:

$$R_1 = 50 \ \Omega$$

$$R_2 = 20 \ \Omega$$

$$R_3 = 25 \ \Omega$$

Solution: Pick a supposed voltage that all three of the resistances can easily be divided into; such a number might be 200. Use Ohm's law to find the resultant current flow through each branch.

$$I_1 = \frac{200}{50} = 4 \text{ A}$$

$$I_2 = \frac{200}{20} = 10 \text{ A}$$

$$I_3 = \frac{200}{25} = 8 \text{ A}$$

Then find the total current.

$$I_T = 4 + 10 + 8 = 22 \text{ A}$$

The total resistance would be the supposed voltage, 200 V, divided by the total current 22 A.

$$R_T = \frac{200}{22} = 9.091 \ \Omega$$

∎

FABRICATING RESISTOR VALUES

There are times when a particular resistor value either will not be available or will cost too much. At these times, it may be possible to use two or more resistors in a parallel combination to provide the necessary resistance. The reciprocal and times-minus methods are used to calculate these combinations. An explanation of both follows.

The reciprocal method is a method of calculation that you should try to master. It is just the reverse of the reciprocal equation already explained.

EXAMPLE 5-7

Use the reciprocal method to fabricate a 7.5-Ω resistance.

Solution: Begin by picking a resistance value greater than 7.5 Ω. You will use this resistance as a pole piece. Let's use 20 Ω. Set up the reciprocal equation as follows:

$$\frac{1}{R_T} = \frac{1}{R_1} + \frac{1}{R_2}$$

$$\frac{1}{7.5} = \frac{1}{20} + \frac{1}{R_{unknown}}$$

If you rearrange this, it becomes

$$\frac{1}{R_{unknown}} = \frac{1}{7.5} - \frac{1}{20}$$

$$= 0.1333 - 0.05$$

$$= 0.08333 \; \mho$$

and

$$R_{unknown} = \frac{1}{0.08333}$$

$$= 12 \; \Omega$$

This is a more direct way of finding the answer.

Calculator Use: Using your calculator, the keystrokes would be

[7] [.] [5] [1/x] [−] [2] [0] [1/x] [=] [1/x]

The calculator would display

1.2000   01   ∎

Sometimes the times-minus formula is used to make this calculation. It would look like this:

$$\frac{20 \times 7.5}{20 - 7.5} = 12 \; \Omega$$

The particular manner that is used to find the unknown resistor is a matter of preference rather than convention. You need to be familiar with both of the various methods. Then choose one, the one that is best and easiest for you, to work with.

SUMMARY

In summary, a parallel circuit is a circuit that has several paths for current. It is one of the two basic circuit types. There are four circuit concepts associated with parallel circuits: (1) one voltage, (2) additive current, (3) additive conductance, and (4) additive power. With parallel circuits, as with series circuits, Ohm's law will only apply for an individual component.

In troubleshooting parallel circuits it is important to remember the following points: (1) The voltage across all components must be equal. When voltage is not equal, a fault is present. (2) Any component that opens will affect only one path. All of the other components will function normally. (3) Also, the presence of a short in any branch will render the entire circuit inoperable.

Hard-to-find or expensive resistance values can be fabricated by combining two or more standard values.

QUESTIONS

5-1. Describe what is meant by a parallel circuit having only one voltage.

5-2. Explain the concept behind the currents being additive in a parallel circuit.

5-3. List two methods for finding the total parallel resistance of a circuit.

5-4. What resistance must be placed in parallel with 500 Ω in order for the total resistance to be 100 Ω?

5-5. Describe the difference between the reciprocal method for finding the missing resistance in Question 5-4 and the times-minus method.

5-6. Why is conductance used to find the total resistance in a parallel circuit?

5-7. Discuss the manner in which a total resistance can be approximated. What can this be used for?

5-8. Describe the effect of replacing a faulty resistor with one of a greater power rating and equal resistance.

5-9. Describe the effect of replacing a faulty resistor with one of less power rating and equal resistance.

5-10. Discuss several reasons why Ohm's law can be used only at an individual component level.

5-11. List some of the advantages and disadvantages of using the matrix to solve parallel problems.

5-12. Describe what happens to the total resistance of a parallel circuit when another parallel resistance is added to the circuit. Explain your answer.

5-13. Describe the effect of shorts and opens on parallel circuits.

PRACTICE PROBLEMS

5-1. Five resistors all having a resistance of 5 kΩ are placed in a parallel circuit. What is their total resistance?

5-2. Four resistors, of 12 kΩ, 22 kΩ, 6.8 kΩ, and 15 kΩ, make up a parallel circuit. The power dissipation of the 6.8-kΩ resistor is 1470 mW. Find the current through each resistor.

5-3. The total current in a parallel circuit is 250 mA. The current through the first of three resistors is 85 mA. The current through the second is 45 mA. Find the resistance of each of these resistors if the supply voltage is 100 V.

5-4. The supply voltage of a parallel circuit is 25 V. The current through the first resistor is 12.5 mA. The resistance of each of the other two is 1.2 and 2.7 kΩ. Find the power dissipated by each resistor and the amount of power produced by the battery.

5-5. The resistors in a parallel circuit are 8.2, 5.6, 4.7, and 2.7 kΩ. The supply voltage is 18 V. Find the total current supplied by the battery.

5-6. The total resistance of a parallel circuit is 1.6569 kΩ. The resistance of the first two resistors is 3.3 and 8.2 kΩ, respectively. What is the resistance of the remaining resistor?

5-7. The total current supplied by a 12-V battery is 0.45 A. Two of the three resistors that make up this circuit have resistances of 120 and 330 Ω. Find the resistance of the remaining resistor.

5-8. An automobile battery has a terminal voltage of 13.2 V. The resistance of the ignition system is 3 Ω. The load resistance of the windshield wipers is 0.7 Ω. The load resistance of the headlights is 0.88 Ω. Find the total resistance of this circuit and the total current of this circuit when all three of these systems are activated.

5-9. The television set requires 3 A of current to operate. The refrigerator requires 5 A. A lamp that has three 100-W light bulbs requires 2.5 A. Find the total current that would be required of a circuit having these three appliances operating at the same time. Household voltage is 120 V.

5-10. The power supply for a home computer system must provide current for a disk drive (2.5 A), a monitor (3.5 A), and a microprocessor (5 A). Find the total current and power this 5-V supply must produce.

Refer to Figure 5-8 to solve for the missing quantities for each of the following problems.

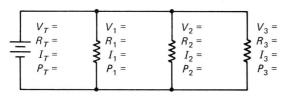

FIGURE 5-8

5-11. Given: $V_T = 50$ V
$R_1 = 1.5$ kΩ
$R_2 = 3.5$ kΩ
$R_3 = 1.2$ kΩ

5-12. Given: $I_1 = 4.166$ mA
$I_2 = 2.273$ mA
$I_3 = 0.625$ mA
$R_T = 10.62$ kΩ

5-13. Given: $R_1 = 680$ Ω
$R_2 = 470$ Ω
$R_3 = 1.2$ kΩ
$P_T = 2.7697$ W

5-14. Given: $V_T = 10$ V
$R_1 = 3.3$ Ω
$R_2 = 12$ kΩ
$R_3 = 4.7$ kΩ

5-15. Given: $I_T = 3.7157$ mA
$I_2 = 3.214$ mA
$R_1 = 56$ kΩ
$P_T = 66.883$ mW

5-16. Given: $V_T = 120$ V
$I_3 = 21.428$ mA
$R_2 = 4.7$ kΩ
$P_1 = 3.6923$ W

5-17. Given: $V_1 = 37$ V
$R_2 = 47$ kΩ
$P_T = 88.958$ mW
$P_3 = 50.704$ mW

5-18. Given: $V_T = 5$ V
$R_T = 177$ Ω
$R_2 = 1.2$ kΩ
$R_3 = 330$ Ω

5-19. Given: $I_T = 531.79$ mA
$I_2 = 0.25532$ A
$I_3 = 100$ mA
$R_1 = 68$ Ω

5-20. Given: $V_T = 20$ V
$R_1 = 6.8$ kΩ
$R_2 = 4.7$ kΩ
$R_3 = 3.3$ kΩ

5-21. Given: $V_1 = 30$ V
$I_1 = 36.585\ \mu\text{A}$
$I_3 = 53.57\ \mu\text{A}$
$R_2 = 470\ \text{k}\Omega$

5-22. Given: $I_1 = 12.5$ mA
$R_1 = 1.2\ \text{k}\Omega$
$R_2 = 3.9\ \text{k}\Omega$
$R_3 = 3.3\ \text{k}\Omega$

5-23. Given: $R_2 = 15\ \text{k}\Omega$
$P_1 = .1875$ W
$P_2 = 57.69$ mW
$P_3 = 68.18$ mW

5-24. Given: $V_T = 12$ V
$I_1 = 1.2$ A
$R_2 = 15\ \Omega$
$P_3 = 16.74$ W

5-25. Given: $V_3 = 120$ V
$I_2 = 20$ mA
$R_1 = 47\ \text{k}\Omega$
$P_T = 4.1463$ W

Series and Parallel Combination Circuits

Chapter objectives

After reading this chapter and answering the questions and problems, you should be able to:

- State Kirchhoff's voltage law. Include an example.
- State Kirchhoff's current law. Include an example.
- Given any series–parallel combination circuit, redraw the circuit with all (or most) resistors placed horizontally, providing easy identification of series and parallel sections.
- Given any series–parallel combination circuit, calculate the total circuit resistance.
- Given any series–parallel combination circuit, find the total circuit resistance and use Ohm's law and Kirchhoff's laws to find the missing parameters.
- Describe the construction and use of a Wheatstone bridge.
- Use the balanced bridge theory to calculate the unknown resistance value being measured by a Wheatstone bridge.

There are many circuits that are combinations of series circuits and parallel circuits. In this chapter we discuss these circuits and common misconceptions about such circuits. A demonstration of the correct way to analyze each of six basic forms of series–parallel circuits is also included.

KIRCHHOFF'S LAWS

For you to understand series–parallel circuits, it is necessary to be familiar with Kirchhoff's laws. They are discussed fully in a later chapter; however, their use is helpful in analyzing series–parallel combination circuits.

Kirchhoff's voltage law (KVL) says simply that *the total voltage dropped around any single current path must equal the source voltage.* Consider the circuit shown in Figure 6-1. There are two paths for current flow from the negative battery terminal through the circuit back to the positive battery terminal. One path is through R_1 and R_2; the other is through R_1 and R_3. Either of these two paths will drop a total of 20 V. KVL can be used for any type of circuit. Remember that any single current path must drop all of the applied voltage.

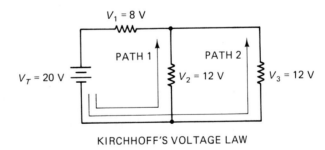

KIRCHHOFF'S VOLTAGE LAW

FIGURE 6-1

EXAMPLE 6-1

Use Kirchhoff's voltage law to find the voltage across R_2 and R_3 in Figure 6-2.

Solution: Current path 1 consists of R_1, R_2, and R_5. So

$$V_T = V_1 + V_2 + V_5$$
$$90 = 20 + V_2 + 40$$
$$= 60 + V_2$$
$$V_2 = 90 - 60 = 30 \text{ V}$$

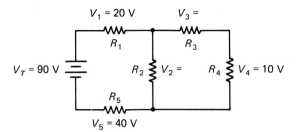

FIGURE 6-2

Current path 2 consists of R_1, R_3, R_4, and R_5. So

$$V_T = V_1 + V_3 + V_4 + V_5$$
$$90 = 20 + V_3 + 10 + 40$$
$$= 70 + V_3$$
$$V_3 = 90 - 70 = 20 \text{ V}$$

Or since R_2 is in parallel with the series string R_3 and R_4,

$$V_2 = V_3 + V_4$$
$$30 = V_3 + 10$$
$$V_3 = 30 - 10 = 20 \text{ V} \quad \blacksquare$$

Kirchhoff's current law (KCL) works in a similar manner. It states that *the current entering a point must be equal to the current leaving the same point*. Consider the circuit shown in Figure 6-3. Here you find 50 mA entering a point (the connection between R_1, R_2, and R_3). There must also be a total of 50 mA leaving this point. Notice that 20 mA leaves the point through R_2 and 30 mA leaves the point through R_3, for a total of 50 mA leaving the point. Remember that the current entering and leaving any point must be equal.

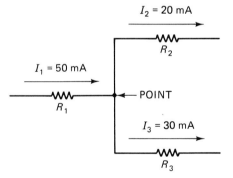

KIRCHHOFF'S CURRENT LAW   **FIGURE 6-3**

EXAMPLE 6-2

Use Kirchhoff's current law to find the current through R_2 and R_5 in Figure 6-4.

Solution: Point 1 consists of the junction between R_1, R_2, R_3, and R_4. So

$$I_1 = I_2 + I_3 + I_4$$
$$0.100 = I_2 + 0.040 + 0.030$$

$$= I_2 + 0.070$$
$$I_2 = 0.100 - 0.070 = 30 \text{ mA}$$

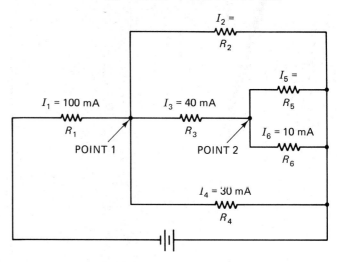

FIGURE 6-4

Point 2 consists of the junction between R_3, R_5, and R_6. So

$$I_3 = I_5 + I_6$$
$$0.040 = I_5 + 0.010$$
$$I_5 = 0.040 - 0.010 = 30 \text{ mA} \quad \blacksquare$$

SELF-TEST

Use KVL and KCL to answer each of the following questions.

1. The voltage across R_2 in Figure 6-5 is 12 V and the supply voltage is 20 V. What would be the value of the voltage across R_3?

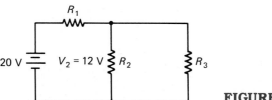

FIGURE 6-5

2. Referring to Figure 6-6, suppose that the voltage across R_2 was 6 V. Suppose also that the voltage across R_4 was 4 V. What would be the voltage across R_1 and R_3?

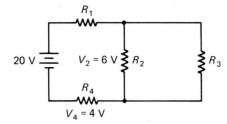

FIGURE 6-6

3. Referring to a circuit similar to Figure 6-7, suppose that the voltage across R_1 is 10 V. What is the voltage across R_2, R_3, and R_4? (The supply voltage is 40 V.)

4. Suppose that through calculations you had found the voltage across R_7 to be 5.5 V, the voltage across R_8 to be 6.5 V, the voltage across R_5 to be 12 V, the voltage across R_2 to be 3.5 V, and the voltage across R_1 to be 7.5 V in Figure 6-8. What would be the supply voltage?

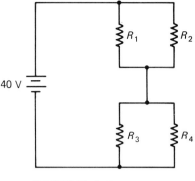

FIGURE 6-7

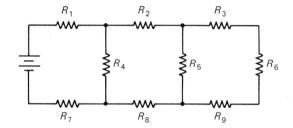

FIGURE 6-8

5. It is found that the current though R_1 in a circuit similar to the one shown in Figure 6-9 is 25 mA. What is the current through R_3 in that circuit?

6. It is found that the current through R_6 in a circuit of similar construction to that shown in Figure 6-10 is 125 mA. Also, the current through R_5 is 45 mA. What is the current through R_3 and R_4?

7. In a circuit like that shown in Figure 6-8, it is known that the supply is producing 250 mA and that the current through R_2 is 75 mA. What is the current through R_1, R_7, R_8, and R_4?

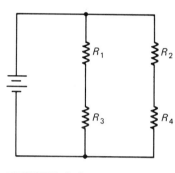

FIGURE 6-9

ANSWERS TO SELF-TEST

1. 12 V
2. $V_{R1} = 10$ V, $V_{R3} = 6$ V
3. $V_{R2} = 10$ V, $V_{R3} = V_{R4} = 30$ V
4. 35 V
5. 25 mA
6. $I_{R3} = 80$ mA, $I_{R4} = 45$ mA
7. $I_{R1} = I_{R7} = 250$ mA, $I_{R4} = 175$ mA, $I_{R8} = 75$ mA

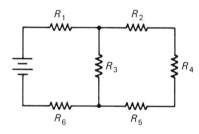

FIGURE 6-10

FINDING TOTAL RESISTANCE

Analyzing series–parallel circuits requires a large amount of thought. It is important that you learn to think in a logical manner. It takes a great deal of organized planning to analyze one of these types of circuits. It is not an exceptionally difficult process but it takes organizational skills and thought to arrive at an answer correctly. The first step, of many, is to find the total resistance of the circuit in question. Students usually learn this easily and in short order.

Begin by identifying what type of circuit is used for each of the combined sections. Again using arrows for current flow will help you decide what circuit type is being used. The basic rule for finding the total resistance is to start at the parallel branch farthest away from the battery. Work backward toward the voltage source, by calculating the resistance of each series and parallel combination.

EXAMPLE 6-3

Find the total resistance of the circuit shown in Figure 6-11.

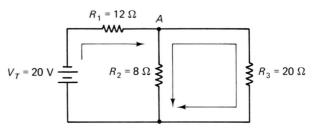

FIGURE 6-11

Solution: Notice the arrows leaving the battery. The current leaves the battery through the negative terminal and travels through R_1. At point A a portion of the current flows through R_2 while the rest of it flows through R_3. Any arrow that indicates a current flow through a single resistor, or series of resistances, indicates a series circuit. R_1 is a series component. Any current that is broken up into two or more separate currents will indicate a parallel section. R_2 and R_3 are parallel sections.

To find total resistance you must combine R_2 and R_3 in parallel ($R_2 \| R_3$). This resistance value must be added to the series resistance of R_1. The calculations would look like this:

$$R_2 \| R_3 = 8 \| 20$$
$$= \frac{8 \times 20}{8 + 20}$$
$$= 5.714 \; \Omega$$

This is the total resistance of the parallel combination of R_2 and R_3. You must add the series resistance of R_1 to this value to find the total resistance of the circuit.

$$R_T = R_1 + (R_2 \| R_3)$$
$$= 12 + 5.714$$
$$= 17.71 \; \Omega$$

The total resistance of the circuit of Figure 6-11 is 17.71 Ω.

Calculator Use: There are several ways to use the calculator to find parallel resistance. Again, as a rule, the times-plus formula is the easiest to understand and perform by hand; but when a calculator is used, the reciprocal method is much faster. Enter the following sequence of keystrokes:

$$[8] \; [1/x] \; + \; [2] \; [0] \; [1/x] \; = \; [1/x]$$

This finds the parallel resistance of R_2 and R_3.

Then continue by pressing the following sequence:

$$[+] \; [1] \; [2] \; [=]$$

The calculator should display

1.7714   01   ∎

Let's examine another circuit combination, Figure 6-12. This one is similar to the one in Figure 6-11, although it is a bit more complicated.

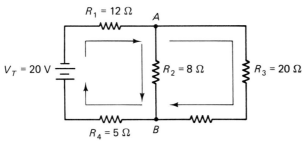

FIGURE 6-12

EXAMPLE 6-4

Find the total resistance of the circuit shown in Figure 6-12.

Solution: Begin by sketching in arrows to indicate current flow. Current flows through R_1 and breaks up at point A, where a portion of it (the current) flows through R_2; the rest flows through R_3, to point B, where the currents unite to flow through R_4. This means that R_2 and R_3 are in parallel with each other and their combined parallel resistance is in series with R_1 and R_4.

Again start the circuit analysis at the parallel branch, which is farthest from the voltage source. In this example you have R_2 in parallel with R_3. This parallel combination is $8 \| 20 = 5.714 \, \Omega$, as before. The combined resistance is then added to the resistance values of R_1 and R_4 to give

$$R_T = (R_2 \| R_3) + R_1 + R_4$$
$$= 5.714 + 12 + 5$$
$$= 22.71 \, \Omega$$

■

Trunk resistances such as R_1 and R_4 of Figure 6-12 are often a confusing factor, especially to students as they begin learning circuit analysis. One method that sometimes makes understanding easier is to redraw the circuit as a long string of components. The circuit of Figure 6-12 is redrawn in this manner in Figure 6-13. Drawn in this manner, it is readily visible that R_2 and R_3 are in parallel, while R_1 and R_4 are in series with the parallel combination.

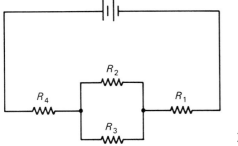

FIGURE 6-13

Another circuit, which causes some confusion, is the one shown in Figure 6-14. It is not really confusing; however, it requires an organized thought process.

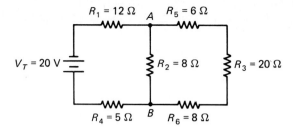

FIGURE 6-14

EXAMPLE 6-5

Find the total resistance of the circuit shown in Figure 6-14.

Solution: Begin by sketching in arrows to indicate current flow. Current flows through R_1 and breaks up at point A, where a portion flows through R_2, the rest through R_3, R_5, and R_6, to point B, where the currents unite to flow through R_4. This means that R_3, R_5, and R_6 share the same current and are thus in series. This series combination is in parallel with R_2 and the combined parallel resistance is in series with R_1 and R_4 (see Figure 6-15).

Again start analyzing the circuit at the parallel branch farthest away from the voltage source. In this case the parallel branch itself forms a series circuit. Combining the resistances of R_3, R_5, and R_6, you have

$$R_3 + R_5 + R_6 = 20 + 6 + 8$$
$$= 34 \ \Omega$$

This resistance value is in parallel with R_2 and can be calculated using the parallel resistance formula:

$$(R_3 + R_5 + R_6) \| R_2 = 34 \| 8$$
$$= 6.476 \ \Omega$$

This resistance is then combined in a serial fashion with the resistance values of R_1 and R_4 to give

$$R_T = 6.476 + 12 + 5$$
$$= 23.48 \ \Omega$$

23.48 Ω is the total resistance for this parallel–series combination. ∎

The first step to analyzing any circuit is to figure out which components are in series and which components are in parallel with each other. After this is accomplished, you must find the total resistance of the circuit. This is almost always the second step in analyzing a series–parallel circuit.

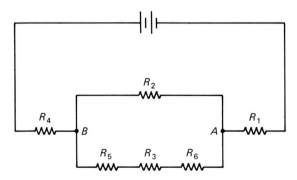

FIGURE 6-15

SERIES–PARALLEL CIRCUIT ANALYSIS

This book has broken down series–parallel circuits into six basic forms. Any series–parallel circuit can be broken down into one or more of these basic forms. Also, each of these forms has specific problem areas for students. These problem areas will be discussed, and methods of avoiding *wrong thinking* will be discussed for each. It is important to understand that these six forms are only examples of series–parallel circuits. It is not intended for you to memorize these forms. It is intended for you to gain an insight into the thought process of analyzing series–parallel circuits through the use of these six forms and their examples.

Form 1

EXAMPLE 6-6

Find the missing parameters for each component in Figure 6-16.

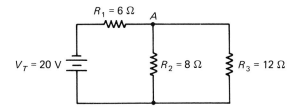

FIGURE 6-16

Solution:

1. *Analyze.* This is a simple series–parallel circuit. With this form, as with all others, the first step is to figure out which components are in series and which are in parallel. That is done easiest by sketching in arrows to indicate the direction of current flow. The current leaves the battery from the negative terminal of the battery and flows through R_1 to point A, where the current separates. Current always separates at a junction. Here the junction is called point A. Look for the junctions in a circuit. They indicate that there are parallel components present. At point A a portion of the current flows through R_2, while the rest flows through R_3. This means that R_2 is in parallel with R_3 and that the parallel combination of these two resistors is in series with R_1.

2. *Calculate R_T.* The calculations for total resistance would appear as this:

$$R_2 \| R_3 = 8 \| 12 = 4.8 \ \Omega$$

$$R_T = 4.8 + 6 = 10.8 \ \Omega$$

3. *Construct a matrix.* A matrix will help in further analysis of this circuit. You will make columns for voltage, resistance, and current only. At this time, for simplicity, you will not worry about the power parameters.

$$V_T = 20 \text{ V} \qquad R_T = 10.8 \ \Omega \qquad I_T =$$
$$V_1 = \qquad R_1 = \ \ 6 \ \Omega \qquad I_1 =$$
$$V_2 = \qquad R_2 = \ \ 8 \ \Omega \qquad I_2 =$$
$$V_3 = \qquad R_3 = 12 \ \Omega \qquad I_3 =$$

Filling in the known values for the components would provide a matrix like the one shown above.

4. *Calculate I_T.* (Use OHM'S law.) Notice that the total level has values which will allow us to find the total current supplied by the voltage source. With series–parallel circuits, as with other circuits, Ohm's law still does apply; however, you must be careful to use it at an individual level only. A common mistake made by students is mixing and matching individual levels. That is, students will invariably try to use the 20-V battery as the voltage for all the resistors: 20/6 for I_1, 20/8 for I_2, and 20/12 for I_3. *This is not the correct way to solve for these values.* The correct way is to use Ohm's law on the total level, which gives

$$\frac{V_T}{R_T} = \frac{20}{10.8} = 1.852 \text{ A}$$

This is the total amount of current flowing out of the battery. Since all of the current must flow through R_1, this is also the value for the current flowing through R_1 and the matrix becomes

$$V_T = 20 \text{ V} \qquad R_T = 10.8 \text{ } \Omega \qquad I_T = 1.852 \text{ A}$$
$$V_1 = \qquad R_1 = 6 \text{ } \Omega \qquad I_1 = 1.852 \text{ A}$$
$$V_2 = \qquad R_2 = 8 \text{ } \Omega \qquad I_2 =$$
$$V_3 = \qquad R_3 = 12 \text{ } \Omega \qquad I_3 =$$

5. *Calculate voltages.* Notice now, if you will, that the first level has enough parameters for completion.

$$R_1 \times I_1 = 6 \times 1.852 = 11.11 \text{ V}$$

Inserting this value in the matrix will give

$$V_T = 20 \text{ V} \qquad R_T = 10.8 \text{ } \Omega \qquad I_T = 1.852 \text{ A}$$
$$V_1 = 11.11 \text{ V} \qquad R_1 = 6 \text{ } \Omega \qquad I_1 = 1.852 \text{ A}$$
$$V_2 = \qquad R_2 = 8 \text{ } \Omega \qquad I_2 =$$
$$V_3 = \qquad R_3 = 12 \text{ } \Omega \qquad I_3 =$$

6. *Use KVL or KCL.* To get the remaining values you must use KVL, which states that the total voltage around any one current path must equal the supply voltage or, in this case,

$$V_T - V_1 = V_2$$
$$20 - 11.11 = 8.89 \text{ V}$$

Thus the voltage across R_2 is 8.89 V. Remember from our calculation of total resistance that R_2 and R_3 are in parallel. Parallel resistances share the same voltage, so the voltage across R_3 is also 8.89 V. Then the matrix becomes

$$V_T = 20 \text{ V} \qquad R_T = 10.8 \text{ } \Omega \qquad I_T = 1.852 \text{ A}$$
$$V_1 = 11.11 \text{ V} \qquad R_1 = 6 \text{ } \Omega \qquad I_1 = 1.852 \text{ A}$$
$$V_2 = 8.89 \text{ V} \qquad R_2 = 8 \text{ } \Omega \qquad I_2 =$$
$$V_3 = 8.89 \text{ V} \qquad R_3 = 12 \text{ } \Omega \qquad I_3 =$$

7. *Use OHM'S law.* Ohm's law can be used to solve for the remainder of the circuit parameters. That is,

$$\frac{V_2}{R_2} = \frac{8.89}{8} = 1.111 \text{ A}$$

$$\frac{V_3}{R_3} = \frac{8.89}{12} = 0.741 \text{ A}$$

Notice that $1.111 + 0.741 = 1.852$, which, except for calculation errors, is the same value as the total current supplied by the battery. ∎

It is important to realize that the rules which apply to series circuits still apply to the series portion of this circuit, and the rules which apply to parallel circuits still apply to the parallel portion of this circuit. That is, as far as the series portion of this circuit is concerned, this circuit has only one current, the current flowing from the battery. All of the series components, in this case R_1, will share this current. In a parallel circuit, the source voltage and the parallel component voltage are the same voltage. Here is a point that causes students great difficulty. The parallel portion of this circuit has its voltage sourced from R_1 and the positive terminal of the battery. For that reason, R_2 and R_3 are provided with the same voltage. They share the same source, R_1, and the positive terminal of the battery. This is an important concept and will be used throughout this book.

Form 2

EXAMPLE 6-7

Find the missing parameters for each component in Figure 6-17.

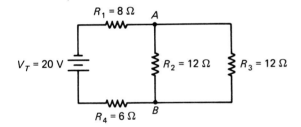

FIGURE 6-17

Solution:

1. *Analyze.* Sketching arrows for current flow will show that current leaves the battery from the negative terminal and travels through R_1. At point A this current separates one portion and goes through R_2, the balance goes through R_3. At point B these separate currents recombine to flow through R_4. Thus R_2 and R_3 are in parallel and their combined resistance is in series with R_1 and R_4.

2. *Calculate R_T.* Solving for total resistance, you have

$$R_2 \| R_3 = 12 \| 12 = 6 \text{ Ω} \quad \text{(equal resistances in parallel)}$$

$$R_T = 8 + 6 + 6 = 20 \text{ Ω}$$

3. *Construct a matrix.* Placing these known values in a matrix gives

$V_T = 20$ V	$R_T = 20$ Ω	$I_T =$
$V_1 =$	$R_1 = 8$ Ω	$I_1 =$
$V_2 =$	$R_2 = 12$ Ω	$I_2 =$

$$V_3 = \qquad R_3 = 12\ \Omega \qquad I_3 =$$
$$V_4 = \qquad R_4 = 6\ \Omega \qquad I_4 =$$

4. *Calculate I_T.* Again, examining the matrix indicates which step would be next. In this case, the total level has two values; you can use Ohm's law to find the total current.

$$\frac{V_T}{R_T} = \frac{20}{20} = 1\ \text{A}$$

This is the total amount of current both leaving the battery and returning to the battery. Most students overlook this fact. They see that this total current is the current flowing through R_1, but overlook that it also flows through R_4. It is the total current supplied by the battery and as such must flow through all of the series components. At any rate, $I_T = 1$ A, $I_1 = 1$ A, and $I_4 = 1$ A Placing these values in the matrix gives you

$$V_T = 20\ \text{V} \qquad R_T = 20\ \Omega \qquad I_T = 1\text{A}$$
$$V_1 = \qquad R_1 = 8\ \Omega \qquad I_1 = 1\ \text{A}$$
$$V_2 = \qquad R_2 = 12\ \Omega \qquad I_2 =$$
$$V_3 = \qquad R_3 = 12\ \Omega \qquad I_3 =$$
$$V_4 = \qquad R_4 = 6\ \Omega \qquad I_4 = 1\ \text{A}$$

5. *Use Ohm's law.* Now you have values that will allow completion of levels 1 and 4. For level 1,

$$V_1 = R_1 \times I_1$$
$$= 8 \times 1 = 8\ \text{V}$$

and for level 4,

$$V_4 = R_4 \times I_4$$
$$= 6 \times 1 = 6\ \text{V}$$

and the matrix becomes

$$V_T = 20\ \text{V} \qquad R_T = 20\ \Omega \qquad I_T = 1\text{A}$$
$$V_1 = 8\ \text{V} \qquad R_1 = 8\ \Omega \qquad I_1 = 1\text{A}$$
$$V_2 = \qquad R_2 = 12\ \Omega \qquad I_2 =$$
$$V_3 = \qquad R_3 = 12\ \Omega \qquad I_3 =$$
$$V_4 = 6\ \text{V} \qquad R_4 = 6\ \Omega \qquad I_4 = 1\text{A}$$

6. *Use KVL or KCL.* There are two ways to proceed from this point. One is intuitive. That is, by looking at the circuit you should realize that I_2 is equal to I_3, since each has an equal resistance and voltage (parallel resistances). Therefore, both must equal one-half of the total current, 0.5 A.

A method that you will find works more often is as follows. The voltage column allows us to perform KVL to find V_2 and V_3. That is, $20 - 8 - 6 = 6$ V. Remember, the total voltage dropped around any single current path must equal the total voltage of the supply, so

$$V_T = V_1 + V_2 + V_4 = V_1 + V_3 + V_4$$

In any case, the matrix becomes

$$V_T = 20 \text{ V} \qquad R_T = 20 \text{ Ω} \qquad I_T = 1 \text{ A}$$
$$V_1 = 8 \text{ V} \qquad R_1 = 8 \text{ Ω} \qquad I_1 = 1 \text{ A}$$
$$V_2 = 6 \text{ V} \qquad R_2 = 12 \text{ Ω} \qquad I_2 =$$
$$V_3 = 6 \text{ V} \qquad R_3 = 12 \text{ Ω} \qquad I_3 =$$
$$V_4 = 6 \text{ V} \qquad R_4 = 6 \text{ Ω} \qquad I_4 = 1 \text{ A}$$

7. *Use Ohm's law.* Ohm's law will quickly solve the remaining section of the matrix.

$$\frac{6}{12} = 0.5 \text{ A}$$

for the current through both R_2 and R_3. ∎

The most missed concept concerning this type of circuit is that *the current leaving the battery must equal the current returning to the battery*. Remember, that is how the value for I_4 was found. It is an extremely important concept.

Form 3

EXAMPLE 6-8

Find the missing parameters for each component in Figure 6-18.

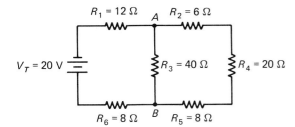

FIGURE 6-18

Solution:

1. *Analyze.* Begin by sketching in arrows for current flow. This will show that the current leaves the battery from the negative terminal of the battery, where it then flows through R_1 to point A. At point A the current separates to follow two distinct paths one through the single resistance R_3. The other through the multiple resistances R_2, R_4, and R_5. At point B these currents reunite to form one current through R_6.

An alternative method for analyzing series–parallel circuits is to redraw them. Start by drawing the first resistor in series with the battery. Then at each junction use a "T" and place the resistors side by side. This will allow you a better view of the circuit construction. See Figure 6-19 for an example of this process. The redraw method works well on circuits that are more complex.

2. *Calculate R_T.* To find the total resistance you must start at the parallel section farthest away from the battery. That would be the parallel section, made up of the series string R_2, R_4, and R_5 and R_3. R_2, R_4, and R_5 are in series with each other and their total resistance would be

$$6 + 20 + 8 = 34 \text{ Ω}$$

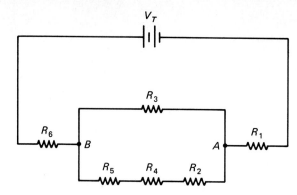

FIGURE 6-19

This total resistance is in parallel with R_3. When combined you can find the total resistance of the parallel section by using

$$\frac{34 \times 40}{34 + 40} = 18.38 \, \Omega$$

This resistance is then combined in serial form with the series resistances, R_1 and R_6, to find the total resistance of the circuit.

$$R_T = 18.38 + 8 + 12$$
$$= 38.38 \, \Omega$$

3. *Construct a matrix.* The matrix for this circuit would be constructed like this:

$V_T = 20 \text{ V}$ $R_T = 38.38 \, \Omega$ $I_T =$

$V_1 =$ $R_1 = 12 \, \Omega$ $I_1 =$

$V_2 =$ $R_2 = 6 \, \Omega$ $I_2 =$

$V_3 =$ $R_3 = 40 \, \Omega$ $I_3 =$

$V_4 =$ $R_4 = 20 \, \Omega$ $I_4 =$

$V_5 =$ $R_5 = 8 \, \Omega$ $I_5 =$

$V_6 =$ $R_6 = 8 \, \Omega$ $I_6 =$

4. *Calculate I_T.* Completing the total level will provide the total current supplied by the battery.

$$I_T = \frac{20}{38.38}$$
$$= 0.5211 \text{ A}$$

This is the current that is supplied by the battery. Any current leaving the battery must flow through R_1, and in order for it to return to the battery, it must flow through R_6. Therefore, $I_1 = I_6 = 0.5211$ A, and the matrix becomes

$V_T = 20 \text{ V}$ $R_T = 38.38 \, \Omega$ $I_T = 0.5211$ A

$V_1 =$ $R_1 = 12 \, \Omega$ $I_1 = 0.5211$ A

$V_2 =$ $R_2 = 6 \, \Omega$ $I_2 =$

$V_3 =$ $R_3 = 40 \, \Omega$ $I_3 =$

$V_4 =$ $R_4 = 20 \ \Omega$ $I_4 =$
$V_5 =$ $R_5 = 8 \ \Omega$ $I_5 =$
$V_6 =$ $R_6 = 8 \ \Omega$ $I_6 = 0.5211$ A

5. *Use Ohm's Law*. Use Ohm's law to find the voltage across R_1 and R_6.

$$V_1 = R_1 \times I_1$$
$$= 12 \times 0.5211$$
$$= 6.254 \text{ V}$$

and

$$V_6 = R_6 \times I_6$$
$$= 8 \times 0.5211$$
$$= 4.169 \text{ V}$$

When these values are placed in the matrix, it becomes

$V_T = 20$ V $R_T = 38.38 \ \Omega$ $I_T = 0.5211$ A
$V_1 = 6.254$ V $R_1 = 12 \ \Omega$ $I_1 = 0.5211$ A
$V_2 =$ $R_2 = 6 \ \Omega$ $I_2 =$
$V_3 =$ $R_3 = 40 \ \Omega$ $I_3 =$
$V_4 =$ $R_4 = 20 \ \Omega$ $I_4 =$
$V_5 =$ $R_5 = 8 \ \Omega$ $I_5 =$
$V_6 = 4.169$ V $R_6 = 8 \ \Omega$ $I_6 = 0.5211$ A

6. *Use KVL*. The matrix becomes a bit more difficult to analyze at this point of the calculations. Remember, the arrows showed two current paths: one through R_3, the other through R_2, R_4, and R_5. You can use KVL to find the voltage across R_3.

$$20 - 6.254 - 4.169 = 9.577 \text{ V}$$

7. *Use Ohm's law*. If you then use Ohm's law on this voltage, you can find the current through R_3.

$$\frac{9.577}{40} = 0.2394 \text{ A}$$

Placing this value in the matrix gives you

$V_T = 20$ V $R_T = 38.38 \ \Omega$ $I_T = 0.5211$ A
$V_1 = 6.254$ V $R_1 = 12 \ \Omega$ $I_1 = 0.5211$ A
$V_2 =$ $R_2 = 6 \ \Omega$ $I_2 =$
$V_3 = 9.577$ V $R_3 = 40 \ \Omega$ $I_3 = 0.2394$ A
$V_4 =$ $R_4 = 20 \ \Omega$ $I_4 =$
$V_5 =$ $R_5 = 8 \ \Omega$ $I_5 =$
$V_6 = 4.169$ V $R_6 = 8 \ \Omega$ $I_6 = 0.5211$ A

8. *Use KCL.* Using KCL at point A, you find 0.5211 A entering A and 0.2394 A leaving point A through R_3. The balance must follow the other current path. Mathematically, then,

$$0.5211 - 0.2394 = 0.2817 \text{ A}$$

is the current through the series string formed by R_2, R_4, and R_5. The matrix becomes

$V_T = 20$ V	$R_T = 38.38\ \Omega$	$I_T = 0.5211$ A
$V_1 = 6.254$ V	$R_1 = 12\ \Omega$	$I_1 = 0.5211$ A
$V_2 =$	$R_2 = 6\ \Omega$	$I_2 = 0.2817$ A
$V_3 = 9.577$ V	$R_3 = 40\ \Omega$	$I_3 = 0.2394$ A
$V_4 =$	$R_4 = 20\ \Omega$	$I_4 = 0.2817$ A
$V_5 =$	$R_5 = 8\ \Omega$	$I_5 = 0.2817$ A
$V_6 = 4.169$ V	$R_6 = 8\ \Omega$	$I_6 = 0.5211$ A

9. *Use Ohm's law.* Ohm's law can be used for the remainder of the values. That is,

$$6 \times 0.2817 = 1.690 \text{ V for } R_2$$
$$20 \times 0.2817 = 5.634 \text{ V for } R_4$$

and

$$8 \times 0.2817 = 2.254 \text{ V for } R_5$$

Notice that the sum of these three voltages,

$$1.690 + 5.634 + 2.254 = 9.578 \text{ V}$$

except for round-off error, is the same voltage as that across R_3. ∎

People have the most difficulty with the idea of three resistances R_2, R_4, and R_5 being in series and in parallel with another resistance. They generally arrive at the correct total resistance; however, when calculating the voltage values for these three, they invariably want to use the voltage value for R_3 as a parallel voltage for all three. That is, they think *incorrectly* that V_2, V_3, V_4, and V_5 all have the same voltage. This is not correct. It is the series combination of R_2, R_4, and R_5 which has a total voltage drop that is equal to the voltage across R_3.

Form 4 The circuit shown in Figure 6-2 is an important type of circuit. It is called a *bridge*. The bridge can be in a balanced state or an unbalanced state. The state of the bridge is determined by the voltage or potential difference between point C and point D. In a balanced bridge the potential between these two points would be zero. Stated another way, if the voltage across R_3 is equal to the voltage across R_4, the bridge is in a *balanced state*.

EXAMPLE 6-9

Find the missing parameters for each component in Figure 6-20.

Solution:

1. *Analyze.* The analysis of this circuit is started in the same manner as all of the other circuits examined so far. Sketch arrows in the circuit to show

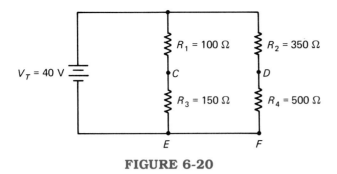

FIGURE 6-20

the paths for current. The current leaves the negative battery and flows through the circuit to point E, where it separates. A portion of it flows upward through R_3, and the rest flows to point F. In an actual circuit, point E and point F may very well be at the same location, since there are no components (resistors) between them. There are, then, two currents, one that flows through R_3 and R_1, and another that flows through R_4 and R_2. R_3 and R_1 are series components, as are R_4 and R_2. These two separate series sections are in parallel with each other.

2. *Calculate RT.* Solving for total resistance should be done in a manner similar to this:

$$R_1 + R_3 = 100 + 150 = 250 \ \Omega$$
$$R_2 + R_4 = 350 + 500 = 850 \ \Omega$$
$$R_T = 250 \parallel 850$$
$$= \frac{250 \times 850}{250 + 850}$$
$$= 193.2 \ \Omega$$

3. *Construct a matrix.* Again, it is easier to see what the next step is to be by using a matrix. Constructing the matrix and filling in the known values would leave

$V_T = 40 \text{ V}$	$R_T = 193.2 \ \Omega$	$I_T =$
$V_1 =$	$R_1 = 100 \ \Omega$	$I_1 =$
$V_2 =$	$R_2 = 350 \ \Omega$	$I_2 =$
$V_3 =$	$R_3 = 150 \ \Omega$	$I_3 =$
$V_4 =$	$R_4 = 500 \ \Omega$	$I_4 =$

4. *Calculate I_T.* Total current would be

$$\frac{40}{193.2} = 0.2071 \text{ A}$$

This is not the amount of current that flows through either of the two series sections. It is the total amount of current produced by the battery.

5. *Use Ohm's law.* In order to find the current through each of the separate series sections, you must use Ohm's law on the total of each of the branch resistances. That is,

$$\frac{40}{250} = 0.16 \text{ A}$$

SERIES-PARALLEL CIRCUIT ANALYSIS

for the current through R_1 and R_3, and

$$\frac{40}{850} = 0.04706 \text{ A}$$

for the current through R_2 and R_4. Inserting these values in the matrix gives

$V_T = 40$ V	$R_T = 193.2 \, \Omega$	$I_T = 0.2071$ A
$V_1 =$	$R_1 = 100 \, \Omega$	$I_1 = 0.16$ A
$V_2 =$	$R_2 = 350 \, \Omega$	$I_2 = 0.04706$ A
$V_3 =$	$R_3 = 150 \, \Omega$	$I_3 = 0.16$ A
$V_4 =$	$R_4 = 500 \, \Omega$	$I_4 = 0.04706$ A

6. *Use KCL*. At this time, KCL can be used to verify the results. Notice that the combined current through the series sections is equal to the total supplied by the battery. That is, 0.04706 + 0.16 is approximately equal to 0.2071 A. It is a good practice to check your work periodically. KVL and KCL are simple ways to do this.

7. *Use Ohm's law*. Ohm's law will provide the voltage parameters for the matrix.

Level 1:

$$R_1 \times I_1 = 100 \times 0.16$$
$$= 16 \text{ V}$$

Level 2:

$$R_2 \times I_2 = 350 \times 0.04706$$
$$= 16.47 \text{ V}$$

Level 3:

$$R_3 \times I_3 = 150 \times 0.16$$
$$= 24 \text{ V}$$

Level 4:

$$R_4 \times I_4 = 500 \times 0.04706$$
$$= 23.53 \text{ V}$$

8. *Use KVL*. At this time KVL can be used to verify the results of your work. Notice when the voltage across R_1 is added to the voltage across R_3, the answer is the supply voltage, 40 V. The same is true when the voltage drops across the other two resistors are combined. Notice also that the voltage across R_3 does not equal the voltage across R_4, so this is an unbalanced bridge circuit. ■

Keep in mind that this circuit is effectively two separate series circuits that work in conjunction with each other. Each of these circuits is being supplied by the same source. It is possible to analyze one half, say the R_1 and R_3 side, and then do the same for the other half. This is sometimes less confusing.

Form 5

EXAMPLE 6-10

Find the missing parameters for each component in Figure 6-21.

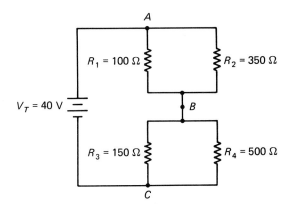

FIGURE 6-21

Solution:

1. *Analyze*. Sketch arrows to show the current flow. Current flows from the negative terminal of the battery to point C, where current separates and a portion flows upward through R_3 while the rest flows through R_4. At point B the current reunites only to separate again, a portion of which continues to flow upward through R_1 while the remainder flows upward through R_2. The indications are that R_3 and R_4 are in parallel with each other, that R_1 and R_2 are in parallel with each other, and that each of these parallel combinations is in series with each other.

2. *Calculate R_T*. The total resistance is calculated in the following manner:

$$R_3 \| R_4 = \frac{150 \times 500}{150 + 500}$$

$$= 115.4 \ \Omega$$

$$R_1 \| R_2 = \frac{100 \times 350}{100 + 350}$$

$$= 77.78 \ \Omega$$

$$R_T = 115.4 + 77.78$$

$$= 193.2 \ \Omega$$

3. *Construct a matrix*. When these values are inserted in the matrix they produce

$V_T = 40$ V	$R_T = 193.2 \ \Omega$	$I_T =$
$V_1 =$	$R_1 = 100 \ \Omega$	$I_1 =$
$V_2 =$	$R_2 = 350 \ \Omega$	$I_2 =$
$V_3 =$	$R_3 = 150 \ \Omega$	$I_3 =$
$V_4 =$	$R_4 = 500 \ \Omega$	$I_4 =$

SERIES-PARALLEL CIRCUIT ANALYSIS

4. *Calculate I_T*. Using Ohm's law on the total level provides

$$\frac{40}{193.2} = 0.2071 \text{ A}$$

This type of circuit requires an especially organized thought pattern. The total current is 0.2071 A. This total current is separated before it flows through any of the components.

5. *Use Ohm's law*. (Find parallel voltages.) To solve for the other circuit parameters, it is necessary to use the total parallel section resistances along with this total current. That is, 115.4 Ω, the total resistance of the parallel combination of R_3 and R_4, must be multiplied by the total current provided by the battery.

$$115.4 \times 0.2071 = 23.89 \text{ V}$$

The same must be done with the parallel combination of R_1 and R_2.

$$77.78 \times 0.2071 = 16.11 \text{ V}$$

This concept is very useful on a lot of series–parallel circuits. It should be remembered. The basic idea is that you are using resistance values of combined resistances which in the previous examples have been ignored (see Figure 6-22). Sometimes the only way to solve a particular circuit is by recalling these values and applying Ohm's law to them. At any rate, the resulting matrix would be

$V_T = 40$ V	$R_T = 193.2$ Ω	$I_T = 0.2071$ A
$V_1 = 16.11$ V	$R_1 = 100$ Ω	$I_1 =$
$V_2 = 16.11$ V	$R_2 = 350$ Ω	$I_2 =$
$V_3 = 23.89$ V	$R_3 = 150$ Ω	$I_3 =$
$V_4 = 23.89$ V	$R_4 = 500$ Ω	$I_4 =$

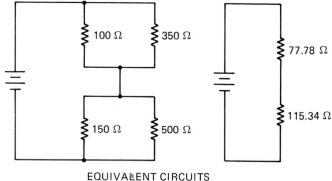

EQUIVALENT CIRCUITS

FIGURE 6-22

6. *Use Ohm's law*. (Find parallel currents.) Since R_1 and R_2 are in parallel, they have the same voltage drop. The same is true of R_3 and R_4. Ohm's law can be used to solve for the remaining currents:

Level 1:

$$\frac{V_1}{R_1} = \frac{16.11}{100}$$

$$= 0.1611 \text{ A}$$

Level 2:

$$\frac{V_2}{R_2} = \frac{16.11}{350}$$

$$= 0.04602 \text{ A}$$

Level 3:

$$\frac{V_3}{R_3} = \frac{23.89}{150}$$

$$= 0.1593 \text{ A}$$

Level 4:

$$\frac{V_4}{R_4} = \frac{23.89}{500}$$

$$= 0.04779 \text{ A}$$

Form 6 This circuit requires that you reuse values of resistances calculated earlier in the analysis of the circuit, which is a concept that must be mastered. Figure 6-23 is very similar to the circuit in Form 3, shown previously. In fact, it differs only in that it has an extra section. This particular circuit causes students an unusual amount of difficulty, because it requires reusing values calculated previously.

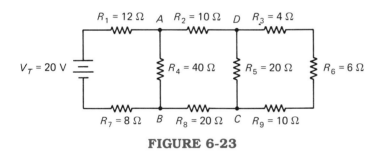

FIGURE 6-23

Students start falling into the habit of punching numbers into their calculators and not writing them down. This particular circuit will require you to write values down or constantly be recalculating them. Trying to be a *super calc* will be disastrous when trying to solve this circuit.

EXAMPLE 6-11

Find the missing parameters for each component in Figure 6-23.

Solution:

1. *Analyze*. Begin by drawing arrows in the direction of current flow or by redrawing the circuit (see Figure 6-24). Current leaves the negative terminal and flows through R_7 to point B, where a portion flows upward through R_4 and the rest flows on through R_8 to point C. At point C this portion of the current breaks up once again. Some of it flows upward through R_5, while the remainder flows through R_9, R_6, and R_3 to point D. At point D a portion of the current reunites to flow through R_2 to point A. At point A the current combines to form a single current that flows through R_1 to the positive battery terminal.

SERIES-PARALLEL CIRCUIT ANALYSIS

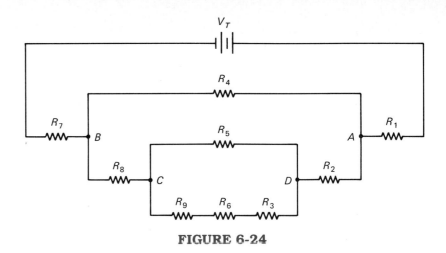

FIGURE 6-24

2. *Calculate R_T*. To simplify this circuit, start with the parallel section farthest from the battery. That would be the series string of R_3, R_6, and R_9. Their total is

$$4 + 6 + 10 = 20 \ \Omega$$

This resistance is in parallel with R_5. And the combined amount of resistance is 10 Ω since these two values are equal ($20 \| 20 = 10$). Thus the total resistance for the portion of the circuit between points C and D is 10 Ω. You can think of the circuit as Figure 6-25.

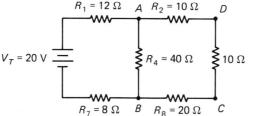

FIGURE 6-25

This step is called *simplification*. When analyzing complex circuits it is sometimes easier to redraw the circuit in a more familiar or simpler fashion. This simplified circuit is exactly like the one in Form 3. Combining the value just calculated with R_2 and R_8, in a serial manner will give

$$10 + 10 + 20 = 40 \ \Omega$$

If this value is placed in parallel with R_4, you will get

$$40 \| 40 = 20 \ \Omega$$
$$R_T = R_1 + 20 + R_7$$
$$= 12 + 20 + 8$$
$$= 40 \ \Omega$$

The total resistance is then 40 Ω, or you can simplify one more time to get Figure 6-26.

In any case, the total resistance of the circuit would be 40 Ω. The difference lies in the way you proceed in circuit analysis. You can construct the entire matrix. That would be time consuming and confusing, since there are 10

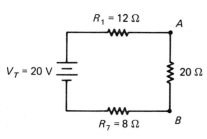

FIGURE 6-26

CHAP. 6 / SERIES AND PARALLEL COMBINATION CIRCUITS

levels, including the total level, to work with. Or you can work these simplified circuits, one at a time until you know all the circuit values. This is the recommended method for solution.

3. *Solve simplified circuits.*

$$\frac{20}{40} = 0.5 \text{ A}$$

The total current would be 0.5 A.

$$0.5 \times 12 = 6 \text{ V}$$

The voltage across R_1 would be 6 V.

$$0.5 \times 8 = 4 \text{ V}$$

The voltage across R_7 would be 4 V.

$$0.5 \times 20 = 10 \text{ V}$$

The voltage from point A to point B would be 10 V. Then the circuit above becomes Figure 6-27.

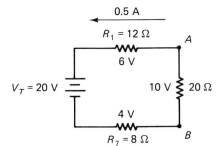

FIGURE 6-27

Applying these values to the previous circuit, you get Figure 6-28. Notice R_4. This resistance has a voltage of 10 V across it. You can use this value to find the current through R_4.

$$\frac{10}{40} = 0.25 \text{ A}$$

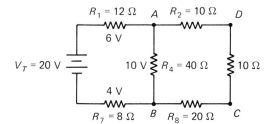

FIGURE 6-28

There are 0.5 A flowing out of point A. There is only 0.25 A flowing up through R_4 into point A, which means that there must be 0.25 A flowing through R_2 into point A (0.5 − 0.25 = 0.25 A). Since R_2 and R_8 are in series with the resistance from point C to point D, you can calculate the following values. The voltage across R_2 is 2.5 V (10 × 0.25 = 2.5 V). The voltage across R_8 is 5 V (20 × 0.25 = 5 V). The voltage from point C to point D is 2.5 V (10 × 0.25 = 2.5 V). Inserting these values into the circuit, you get Figure 6-29.

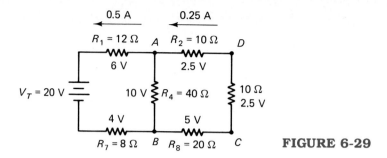

FIGURE 6-29

Applying these values to the original circuit, you have Figure 6-30.

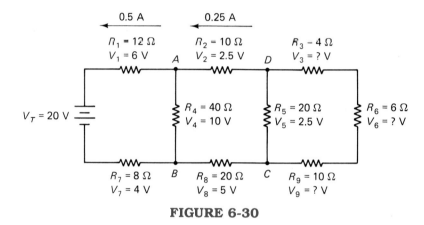

FIGURE 6-30

Now, the currents through R_5 and R_3 can be calculated. The current through R_5 is 0.125 A (2.5/20 = 0.125 A). Point D has 0.25 A flowing out of it, while there are only 0.125 A flowing up through R_5 into point D. Therefore, there must be 0.125 A flowing through R_3 into point D (0.25 − 0.125 = .125 A). Using Ohm's law, you can find the remaining voltages.

$$V_3 = 4 \times 0.125 = 0.5 \text{ V}$$

$$V_6 = 6 \times 0.125 = 0.75 \text{ V}$$

$$V_9 = 10 \times 0.125 = 1.25 \text{ V}$$

∎

In Example 6-11, notice how KCL was used to find the missing current. This is an *often* overlooked calculation. Also the matrix method of analysis was abandoned. The matrix method works well for simple circuits (circuits which have a maximum of five to six resistors). For more complicated circuits it is easier to use the simplification method shown in Example 6.11.

A PRACTICAL CIRCUIT

When first learning electronics, it is hard to see any practical good to come from the material being learned. In the following section we describe a method of resistance measurement using the bridge circuit explained earlier. This bridge circuit is especially useful in the measurement of very small resistances and very large resistances. It has no particular advantage for common resistance values, however. It is called a Wheatstone bridge. It was invented by Samuel Christie around 1835. It was Wheatstone's idea; however, he lacked the mathematical background to design the circuit and called upon his friend Christie to help in its design. The circuit is shown in Figure 6-31.

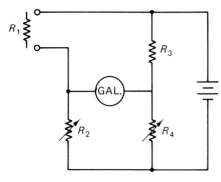

WHEATSTONE BRIDGE   **FIGURE 6-31**

You will notice that this circuit uses a galvanometer. A galvanometer is an extremely sensitive current meter. It not only measures small currents but also has a needle that points to the middle of the scale when no current is flowing. It measures the amount of current flow and indicates the direction of current flow. R_1 is an unknown resistance value. It is the resistance that is being measured. R_2 is an adjustable resistance called a potentiometer. R_3 is generally a fixed resistance, while the R_4 resistance is generally a decade box. A decade box is a device that contains several resistors. Each of these resistors would be a multiple of the others. That is it could contain resistors of the following values: 1 Ω, 10 Ω, 100 Ω, 1000 Ω, and so on. These resistors are connected to a switch that can choose which value resistor is inserted in the circuit. The ratio between R_3 and R_4 determines the multiplier.

When the bridge is balanced, no current will flow through the galvanometer. By adjusting R_4 and R_2 you can reach a balanced state, which will be indicated by a zero reading on the galvanometer. This zero reading on the galvanometer means that the ratio between R_1 and R_2 is the same as the ratio between R_3 and R_4. In equation form this would be

$$\frac{R_1}{R_2} = \frac{R_3}{R_4}$$

or

$$R_1 = R_2 \times \frac{R_3}{R_4}$$

Remember, the values for R_3, R_4, and R_2 are known values. So the value for R_1 is to be calculated.

EXAMPLE 6-12

Suppose that the values, for R_2, R_3, and R_4 are 250 Ω, 750 kΩ, 35 kΩ, respectively. Find the value of R_1.

Solution: You can find the value of R_1 by finding the multiplier,

$$\frac{R_3}{R_4} = \frac{750{,}000}{35{,}000}$$

$$= 21.43$$

A PRACTICAL CIRCUIT

then using the multiplier and R_2 to find R_1.

$$\frac{R_1}{R_2} = 21.43$$

$$\frac{R_1}{750} = 21.43$$

$$R_1 = 21.43 \times 750$$
$$= 16070 \ \Omega$$

Calculator Use: Enter the following sequence of keys:

[7] [5] [0] [EE] [3] [÷] [3] [5] [EE] [3] [×] [7] [5] [0] [=]

The calculator should display

1.6071   04   ∎

Some other uses for the bridge circuit are as an *LCR* bridge, a circuit used to measure the alternating-current resistance (impedance) of a component, and as a Weinbridge, a circuit used in communications to produce circuit oscillations. The bridge circuit is an important concept in electronics.

SELF-TEST

Use Figure 6-31 to solve for R_1.

1. Given: $R_2 = 4.5 \ k\Omega$
   $R_3 = 72 \ k\Omega$
   $R_4 = 1.2 \ M\Omega$

2. Given: $R_2 = 90 \ k\Omega$
   $R_3 = 200 \ \Omega$
   $R_4 = 2.2 \ k\Omega$

3. Given: $R_2 = 200 \ k\Omega$
   $R_3 = 1.5 \ k\Omega$
   $R_4 = 27 \ \Omega$

4. Given: $R_2 = 10 \ \Omega$
   $R_3 = 150 \ \Omega$
   $R_4 = 150 \ k\Omega$

ANSWERS TO SELF-TEST

1. 270 Ω 2. 8.182 MΩ 3. 11.11 kΩ 4. 0.01 Ω

SUMMARY

1. Sketch in arrows to show current flow or redraw the circuit. This will indicate which components are in series and which components are in parallel.
2. Analyze total resistance starting at the parallel branch farthest from the battery and work toward the battery. Keep in mind which resistances are in parallel and which are in series.
3. Use Ohm's law only at an individual level. Do not mistakenly mix and match particular level values.
4. All concepts that are true for series circuits are still true for the components located in a series section.
5. All concepts that are true for parallel circuits are still true for the components located in a parallel section.

6. When there is no way to use Ohm's law, remember Kirchhoff's voltage and current laws. They will provide the other values.
7. Start slow and think about what is being done. Most circuits can be broken down into a simple form. Thought and comparison should provide the correct values.

QUESTIONS

6-1. Briefly explain Kirchhoff's voltage law as it pertains to series–parallel circuits.

6-2. Briefly explain Kirchhoff's current law as it pertains to series–parallel circuits.

6-3. Describe how Kirchhoff's laws can be used to solve series–parallel circuits.

6-4. Refer to Figure 6-32; describe the operation of the circuit. Include series and parallel connections and methods for calculating current flow and voltage levels.

6-5. Briefly describe the process used to find the total resistance of a series–parallel circuit.

6-6. Describe the operation of the circuit shown in Figure 6-33. Include series and parallel connections and methods for calculating current flow and voltage levels.

6-7. Briefly describe the operation of a Wheatstone bridge; be sure to include applications.

6-8. Describe the method you would use to find the total resistance of the circuit shown in Figure 6-34.

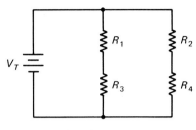

FIGURE 6-32

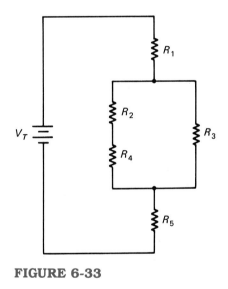

FIGURE 6-33

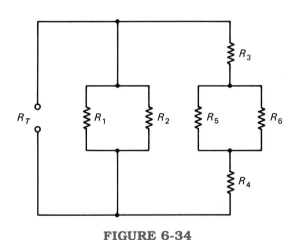

FIGURE 6-34

6-9. Suppose that there are two paths for current into point A and three paths for current out of point A. The two paths into the point are carrying 150 and 250 mA, respectively. It is known that two of the three paths leaving point A are carrying 80 and 120 mA. What is the total amount of current leaving point A? How much current is the third path exiting point A carrying?

6-10. Why is it important to write all your work down as you are doing the calculations, not just the answers?

PRACTICE PROBLEMS

For Problems 6-1 and 6-2, refer to Figure 6-35.

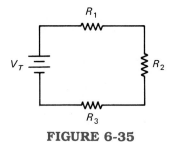

FIGURE 6-35

6-1. Given $R_1 = 6\ \Omega$, $R_2 = 5\ \Omega$, $E_2 = 8$ V, and $E_3 = 12$ V, find R_T, P_T, E_T, R_3, E_1, I_T, P_1, P_2, and P_3.

6-2. Given $R_1 = 8.5$ kΩ, $R_2 = 9.6$ kΩ, $E_2 = 10.2$ V, and $E_3 = 5.6$ V, find R_T, P_T, E_T, R_3, E_1, I_T, P_1, P_2, and P_3.

For Problem 6-3, refer to Figure 6-36.

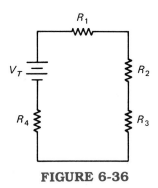

FIGURE 6-36

6-3. Given $I_T = 56$ mA, $P_1 = 110$ mW, $P_2 = 215$ mW, $P_3 = 460$ mW, and $P_T = 1.5$ W, find P_4, R_1, R_2, R_3, R_4, E_1, E_2, E_3, E_4, and V_A.

6-4. Indicate whether Figure 6-37 is series or parallel.

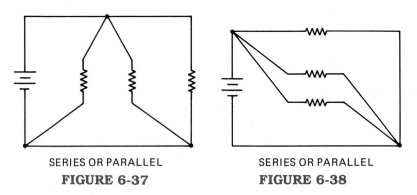

SERIES OR PARALLEL
FIGURE 6-37

SERIES OR PARALLEL
FIGURE 6-38

6-5. Indicate whether Figure 6-38 is series or parallel.
6-6. Indicate whether Figure 6-39 is series or parallel.

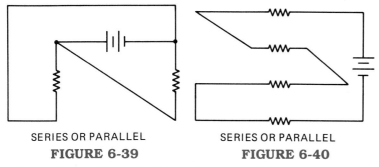

SERIES OR PARALLEL
FIGURE 6-39

SERIES OR PARALLEL
FIGURE 6-40

6-7. Indicate whether Figure 6-40 is series or parallel.

6-8. Indicate whether Figure 6-41 is series or parallel.

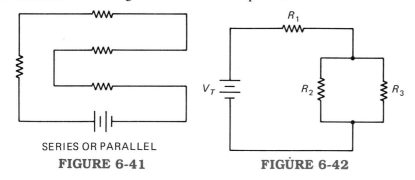

SERIES OR PARALLEL
FIGURE 6-41

FIGURE 6-42

For Problems 6-9 and 6-10, refer to Figure 6-42.

6-9. Given $V_T = 9$ V, $I_1 = 1.5$ mA, $I_3 = 1$ mA, and $P_3 = 4$ mW, find R_T, R_1, R_2, R_3, V_1, V_2, V_3, I_2, I_T, P_1, P_2, and P_T.

6-10. Given $V_T = 9$ V, $R_2 = 10$ kΩ, $V_2 = 4$ V, and $I_3 = 0.6$ mA, find R_T, R_1, R_3, V_1, V_3, I_1, I_2, I_T, P_1, P_2, P_3, and P_T.

For Problems 6-11, 6-12, and 6-13, refer to Figure 6-43.

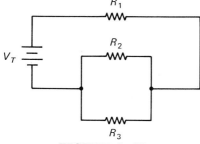

FIGURE 6-43

6-11. Given $R_1 = 10$ kΩ, $V_T = 12$ V, $V_1 = 10$ V, and $I_2 = 0.55$ mA, find R_2, R_3, V_2, V_3, R_T, I_1, I_3, I_T, P_1, P_2, P_3, and P_T.

6-12. Given $R_3 = 25$ kΩ, $I_2 = 1.2$ mA, $P_2 = 1$ mW, and $V_T = 12$ V, find R_1, R_2, R_T, V_1, V_2, V_3, I_1, I_3, I_T, P_1, P_3, and P_T.

6-13. Given $V_T = 12$ V, $R_1 = 3$ kΩ, $I_T = 3$ mA, and $P_2 = 3$ mW, find R_2, R_3, R_T, V_1, V_2, V_3, I_1, I_2, and I_3.

6-14. Given $V_T = 100$ V, $R_1 = 6$ kΩ, $R_2 = 2$ kΩ, $R_3 = 30$ kΩ, $R_4 = 10$ kΩ, $R_5 = 4$ kΩ, $R_6 = 6$ kΩ, refer to Figure 6-44 to find the missing voltages and currents.

PRACTICE PROBLEMS

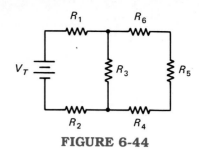

FIGURE 6-44

6-15. Given $V_T = 60$ V, $I_1 = 2.2$ mA, $R_4 = 4.8$ kΩ, $R_3 = 6.7$ Ω, $V_3 = 10$ V, and $V_2 = 20$ V, refer to Figure 6-45 to find the missing voltage, resistances, and currents.

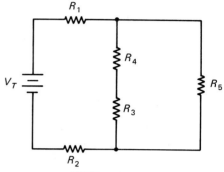

FIGURE 6-45

6-16. Given $R_1 = 4$ kΩ, $R_2 = 16$ kΩ, $R_3 = 6$ kΩ, $R_4 = 6$ kΩ, and $I_3 = 10$ mA, refer to Figure 6-46 to find V_A, I_4, V_4, I_2, V_2, and V_3.

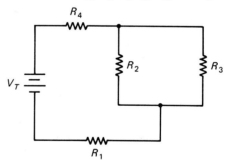

FIGURE 6-46

6-17. Given $V_T = 404.15$ V, $P_1 = 200$ mW, $R_2 = 4$ kΩ, $R_1 = 5$ kΩ, $R_4 = 12$ kΩ, $R_5 = 6$ kΩ, $R_6 = 13$ kΩ, and $R_7 = 12$ kΩ, refer to Figure 6-47 to find V_2, V_4, V_5, V_6, and V_7.

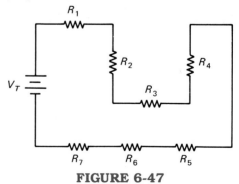

FIGURE 6-47

6-18. Given $V_T = 100$ V, $V_1 = 50$ V, $V_4 = 5$ V, $V_5 = 10$ V, and $V_2 = 20$ V, refer to Figure 6-48 to find V_6.

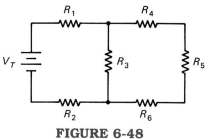

FIGURE 6-48

6-19. Given $I_1 = 20$ mA, $I_3 = 6.7$ mA, and $I_6 = 5.3$ mA, refer to Figure 6-49 to find I_T, I_4, I_5, I_9, I_8, and I_7.

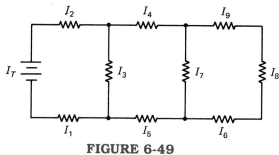

FIGURE 6-49

6-20. Given $R_1 = 8$ kΩ, $R_2 = 6$ kΩ, $R_3 = 5$ kΩ, $R_4 = 7$ kΩ, $R_5 = 2$ kΩ, $R_6 = 1$ kΩ, $R_7 = 6$ kΩ, $R_8 = 7$ kΩ, $R_9 = 10$ kΩ, and $V_T = 90$ V, refer to Figure 6-50 to find all the voltages.

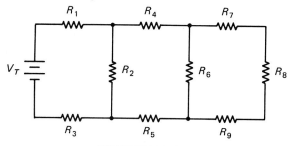

FIGURE 6-50

6-21. Given $R_1 = 150$ Ω, $R_2 = 450$ Ω, $R_3 = 500$ Ω, $R_4 = 1$ kΩ, $R_5 = 500$ Ω, $R_6 = 1000$ Ω, $R_7 = 2$ kΩ, $R_8 = 250$ Ω, $R_9 = 1$ kΩ, and $V_T = 100$ V, refer to Figure 6-51 to find all the voltages.

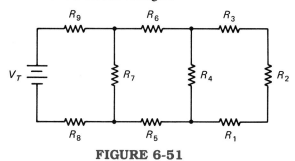

FIGURE 6-51

7

Troubleshooting DC Circuits

Chapter objectives

After reading this chapter and answering the questions and problems, you should be able to:

- Describe the use of a voltmeter, an ohmmeter, and a current meter.
- Given a circuit with a ground connection, correctly state the expected voltage to ground for various points around the circuit.
- Describe the operation and function of each of the oscilloscope controls.
- Indicate the proper method of oscilloscope hookup.
- State, in order, the six steps to troubleshooting any system.
- Describe each of the special tests, including advantages and disadvantages and when each is most applicable.
- Describe how to test resistors and relays using an ohmmeter.

The most important thing for anyone to learn is how to determine where the problem (malfunction) is. In this chapter we give an explanation of basic dc troubleshooting procedures. This will be valuable when projects that have been constructed with no apparent errors still do not work. This is the time when you, as a technician, will learn the tools of the trade and gain a greater understanding of circuit operation. In this chapter we discuss meter use, ground and measurement from ground, a six-step method for troubleshooting any circuit, and some specific troubleshooting tests and their applications. These will help in troubleshooting problem circuits.

METER USE

There are three basic meters that are used in troubleshooting electronic circuits: the voltmeter, the ampmeter, and the ohmmeter. Usually, all three of these meters are combined into one unit, called a multimeter or VOM (volt-ohm-milliampere) meter (Figure 7-1). This section of the book will cover how to use each of these meters. The most used of these meters is the voltmeter, so let's discuss it first.

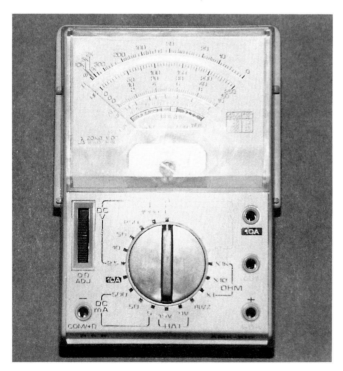

FIGURE 7-1 Typical VOM meter.

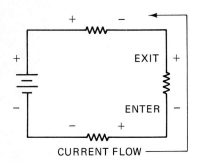

FIGURE 7-2

Voltmeter Usage Voltmeters are attached (connected) to the circuit in a parallel fashion. That is, they are connected across the component. Since a voltmeter has two leads, one lead is attached to one side of the component and the other lead is attached to the opposite side of the component. It is important to connect the leads with the proper polarity. The positive lead must connect to the positive side of the component being measured, and vice versa. The determination of the polarity of a circuit component is done in a manner similar to the way you would determine whether a component is in series or parallel. Instead of drawing arrows from the negative terminal of the schematic around to the positive terminal, you trace (follow) the circuit from the negative terminal to the positive terminal of the supply. The current will then enter the negative side of the component and exit the positive side of the component. The schematic shown in Figure 7-2 should clarify this method for you.

When connecting a voltmeter to a circuit, always connect the red lead to the positive side of the component and the black lead to the negative side of the component. It is a matter of convention that the red is positive. Most meters utilize red as the color for the positive meter lead. The few that do not use red will use either green or white for positive, but black will be used for negative. Obviously, if you work with electronics you will find exceptions to all rules. It is best to read the operator's manual for an unfamiliar meter before you attempt to use it. The standard way of connecting a voltmeter to a circuit is shown in Figure 7-3.

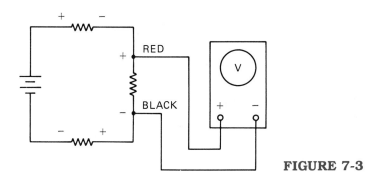

FIGURE 7-3

Notice that the meter is connected in parallel. The meter itself has a resistance; ideally, that resistance is a very large number, so when the meter is connected in this manner it changes the resistance of the circuit very little. Remember, in a parallel connection the total resistance must be less than the resistance of the smallest resistor. It is important that the meter resistance be substantially larger than the resistance of the component whose voltage is being measured. This will be discussed in detail in a later chapter. For now, let's say that a voltmeter should have a sensitivity of at least 20,000 Ω per volt. Also, a meter with a sensitivity of 50,000 Ω per volt will measure the voltage more accurately than one with a sensitivity of 20,000 Ω per volt. The higher the sensitivity, the more accurate the meter (Figure 7-4).

Ohmmeter Usage The next most used meter is the ohmmeter. The first and foremost thing to remember about an ohmmeter is that it has its own voltage source and must be used on a dead circuit (one that has had the power disconnected). The ohmmeter is more difficult to use than a voltmeter because it must be calibrated—not just on initial use but also any time that the resistance range is changed. The calibration is generally simple, but it is the thing that is most often forgotten.

FIGURE 7-4 VOM sensitivity.

To calibrate an ohmmeter, turn the meter to the desired range, connect the leads to each other, and turn the adjustment knob until the meter reads zero (Figure 7-5). Remember that this must be done every time the resistance range is changed.

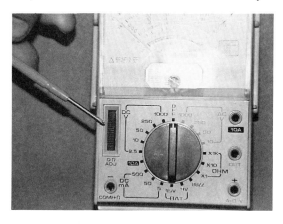

FIGURE 7-5 VOM Zero Ohms Adjust.

The ohmmeter has two leads. On a VOM meter the same two leads are used for measuring volts and ohms. The leads have a polarity; however, capacitors and semiconductors are the only components that require a specific polarity hookup when testing. In general, the ohmmeter is connected in the same manner as the voltmeter: in a parallel fashion. Some things to remember are:

1. The in-circuit resistance will always be lower than the out-of-circuit resistance. This is caused by parallel resistance paths in the circuit. For an accurate resistance measurement, one end of the component must be unsoldered and removed from the circuit before the resistance is measured (Figure 7-6).
2. Your body has resistance. If you touch both leads of the ohmmeter while you are taking a resistance reading, you will read your body resistance in parallel with the resistance being measured. For small resistances (less than 1000 Ω) the difference between readings will be unnoticeable; however, for large resistance values (in the neighborhood of 100,000 Ω) there will be an appreciable difference. So remember not to hold the component leads while you are testing its resistance.

METER USE

FIGURE 7-6 Measuring Resistance. Remember to unsolder resistor before measuring resistance.

Current Meter Usage The least used of the meter types is the current meter. It is the least used, but it is the meter most often misused. That is, the current meter is connected improperly more often than the other meter types. The current meter is designed to have an exceptionally small amount of meter resistance, and as such it can easily be damaged by an incorrect connection. It has two leads (a VOM uses the same two leads for all its meters). They have polarity and must be connected with that polarity in mind. A current meter must be inserted in the circuit in a serial fashion (Figure 7-7). This is why it is the least often used of the meter types.

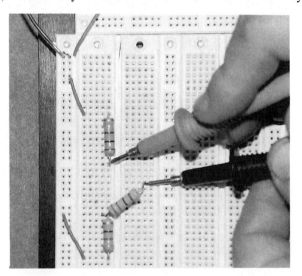

FIGURE 7-7 Measuring Current. Remember to break the circuit and insert the current meter.

To take a current reading of a component, the circuit must have operating voltage and one end of the component must be disconnected from the circuit (unsoldered). The current meter is used to complete the circuit, by attaching one end of the meter to the circuit (where the component was disconnected) and attaching the other to the disconnected end of the component itself. Remember that the polarity must be followed. This type of serial connection is demonstrated in Figure 7-8.

Notice that these connections are made in a manner that allows the current to enter the negative lead of the current meter and exit the positive lead of the current meter. This is true for all components that are connected in series. Remember that the current meter (ammeter) must always be connected in series. The most common mistake in using the ammeter is connecting the meter in parallel. This allows a large

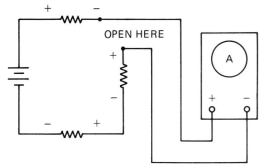

FIGURE 7-8

current to flow through the meter (the meter has zero resistance, which shorts out the component). The large current will either damage the meter movement, blow the fuse protecting the meter movement, or both. There is also the possibility that this improper connection will damage the circuit further.

SELF-TEST

Please check your answers with the instructor before continuing.

1. How can the polarity of a component be determined?
2. In a series circuit, is the positive terminal of one component connected to the (positive or negative) terminal of the next component?
3. Referring to question 2, is this true for the battery?
4. Is a voltmeter connected in (series or parallel) to measure voltage?
5. Is the red lead of a VOM meter the positive or negative lead?
6. What colors, other than red, could be used to indicate the positive lead of a VOM meter?
7. Sketch a four-resistor series circuit. Include a battery. Show how a voltmeter would be connected to measure the voltage of each resistor.
8. List the steps for zeroing an ohmmeter.
9. Why shouldn't you hold a component when you are measuring its resistance?
10. Is a current meter connected in series or parallel to measure amperage?
11. Sketch a four-resistor parallel circuit. Include a battery. Show how a current meter would be connected to measure the current of each resistor. Also show how it would be connected to measure the total circuit current.
12. Why is the current meter the least used of the meter types?
13. Why is it important for you to know how to use a current meter if it is seldom used?
14. Does the ohmmeter have polarity? Is the respective polarity important when measuring the resistance of a resistor?
15. Does the voltmeter have a large or a small internal resistance?
16. Does the current meter have a large or a small internal resistance?

Oscilloscope Usage Another test instrument used in troubleshooting electronic circuits is the oscilloscope. The oscilloscope is nothing more than a fancy voltmeter which displays voltage levels vertically on the screen and time horizontally across the screen. In the following paragraphs we describe each of the primary controls and the basic hookup procedure for the oscilloscope.

Oscilloscope Controls

The oscilloscope controls are located on the front of the oscilloscope (Figure 7-9). This provides easy access and visibility. The following outline lists the major control knobs (and switches) and gives a brief description of each knob's function.

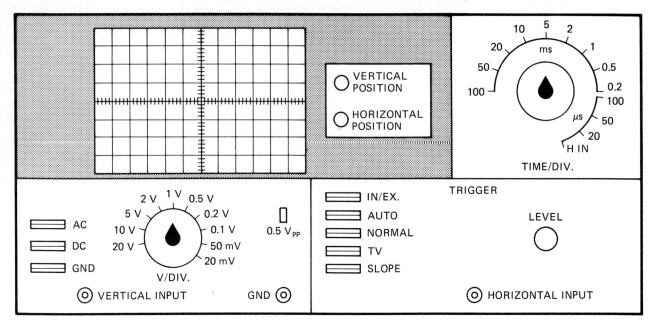

FIGURE 7-9

1. *Vertical position.* This knob will move the pattern up or down the screen, which allows you to position (place) the pattern anywhere on the screen. Repositioning the pattern can make reading voltage levels faster and easier (see Figure 7-9).

2. *Horizontal position.* This positions the pattern (both patterns on a dual-trace scope) to the desired horizontal position. The horizontal position control moves the pattern back and forth across the screen. This movement can make time and frequency (discussed in a later chapter) readings simpler (see Figure 7-9).

3. *Volts/division.* The volts/division (read "volts per division") control is very much like the range switch on a VOM. It allows you to choose how many volts are represented by each vertical square on the oscilloscope screen. Turning this knob to the 5-V position will cause each square to be equal to 5 volts (see Figure 7-9).

4. *Time/division.* The time/division (read "time per division") control is the horizontal counterpart to the volts/division knob. It allows you to choose the time frame represented by each horizontal square of the oscilloscope screen. Turning this control to 50 μs will cause each square to be equal to 50 microseconds (see Figure 7-9).

Basic Oscilloscope Hookup

Every oscilloscope has three basic attachments to the circuit: a ground, a vertical input, and a horizontal input. Depending on the number of traces (individual patterns an oscilloscope can display at the same time), there may be more than one vertical input.

1. *Ground.* To read voltages correctly, an oscilloscope must be connected to ground. Most oscilloscopes use a fixed ground system. A fixed ground requires that the ground of the oscilloscope be connected to the ground of the circuit that is being

checked (especially if they are connected to the same ac source). All measurements are with respect to ground. The very least that can happen if for some reason the ground lead is not connected to the ground of the circuit is that the readings will be incorrect. In extreme cases, misconnecting the ground lead can result in damaged leads and components. Systems that utilize a positive ground (hot chassis) need to be connected to an isolation transformer before the oscilloscope is connected to the circuit. Consider the circuit shown in Figure 7-10. It demonstrates that the oscilloscope ground and the signal ground are at the same potential because they are both connected to the ac circuit ground (common).

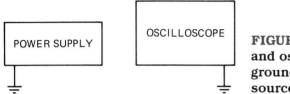

FIGURE 7-10 The power supply and oscilloscope share the same ground through the ac voltage source.

Figure 7-11 shows the effect of connecting the oscilloscope to an incorrect grounding point. This shorts out several of the components (the ones between the scope ground and the signal ground are removed from the circuit). It is important to understand that voltage reading with a fixed-ground oscilloscope cannot be done in the same manner as when done with a VOM meter. That is, you cannot connect across a component to read its voltage level using a fixed-ground oscilloscope.

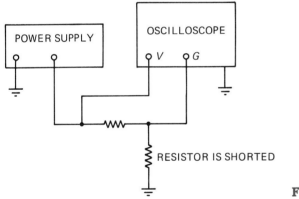

FIGURE 7-11

2. *Vertical input*. The vertical input is used primarily for reading the voltage levels at particular test points. It is connected to the test point to show the voltage level from that point to ground (see Figure 7-12). The vertical input will usually have a voltage probe connected to it. Voltage probes allow the technician to check the voltage at several points by moving the probe quickly from one point to another.

The voltage probe used most often is the ×10 probe, which contains a voltage divider that allows only one-tenth of the signal voltage to enter the oscilloscope (Figure 7-13). This in itself is not important, but along with the reduced voltage, it provides increased internal resistance and less circuit loading. A normal oscilloscope will have an input impedance of about 1 MΩ. The same oscilloscope will have an input impedance of 10 MΩ when a ×10 probe is used to read the voltage.

3. *Horizontal input*. The horizontal input is used to measure phase relations between waveforms (discussed in a later chapter). Single-trace oscilloscopes have their horizontal inputs located in the triggering section of the scope (see Figure 7-14). Dual-trace oscilloscopes will generally provide a switch that will allow the operator to use the B input as either a vertical or a horizontal input.

FIGURE 7-12 Measuring voltage. Remember the probe measures the voltage to ground.

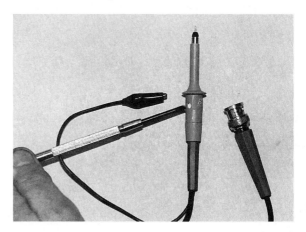

FIGURE 7-13 ×10 probe. The ×10 probe is adjusted to remove excess capacitance.

GROUND

What is *ground?* In an electrical circuit, ground is a common connection between all components. Usually, this common connection has a zero potential (no voltage is applied to ground), and for safety purposes this common connection is connected to the earth through a water pipe or other easily found conductor. Ground sometimes is the metal chassis of the electrical apparatus (TV set, radio, computer, etc.). This allows the manufacturer to use only half of the wire necessary to complete the circuit.

The automobile is a perfect example of this type of wiring. The negative terminal is connected to the body of the car. Wires and switches are connected to the positive terminal of the battery and to one side of the load. The other side of the load is also connected to ground. From the negative terminal the electricity flows through ground (the body of the car) to the load (radio, fan, headlights, etc.). The electricity then flows through the wires and switches, where it returns to the positive terminal of the battery. In this manner the manufacturer need only run a wire to the load. The chassis provides the source path for current. The two schematic drawings in Figure 7-15 demonstrate this concept.

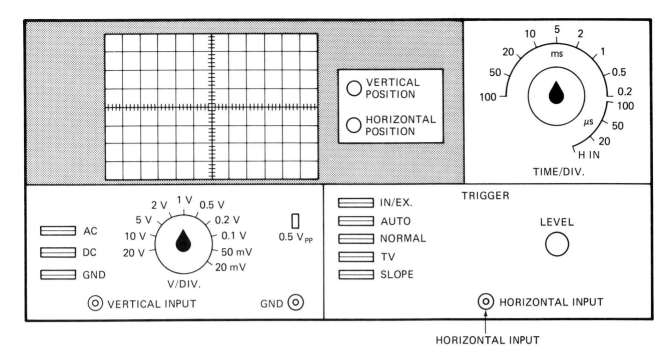

FIGURE 7-14

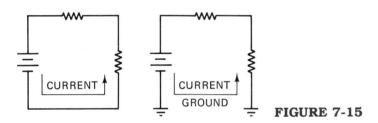

FIGURE 7-15

Schematics used in troubleshooting electronic circuits will have voltage test points (Figure 7-16). These test points are shown on the schematic along with the voltage specified for that point. These voltages are voltage levels with respect to ground. For that reason it is necessary to understand how a differing ground point will change the voltage readings. Also, by using voltages with respect to ground, a technician can connect one lead of the voltmeter to ground and take all of the readings by touching the other lead to the points specified. (For safety, remember the one-hand-in-the-pocket rule.) The basic rule for taking voltage readings with respect to ground is that positive voltages have negative grounds and negative voltages have positive grounds.

Consider the schematic shown in Figure 7-17. The current flows from the negative terminal to ground. It then flows out of ground to point D, where it enters the negative side of the resistor. This indicates that R_3 has a negative ground and as such will have a positive voltage across it. If you would connect a voltmeter to ground and to point D, you would read 0 volts, because there is no resistance to drop any voltage ($I \times R = I \times 0 = 0$ V). If you connected a voltmeter to point C, you would read 10 V, the voltage dropped across R_3. Connecting a voltmeter to point B would show 25 V, the voltage dropped across R_2 and R_3. Point A would show 45 V. These are all positive voltages because the negative side of each resistor is closest to ground.

This is the recommended way to measure voltages. One lead is connected to a common point, while the other is moved to different test periods.

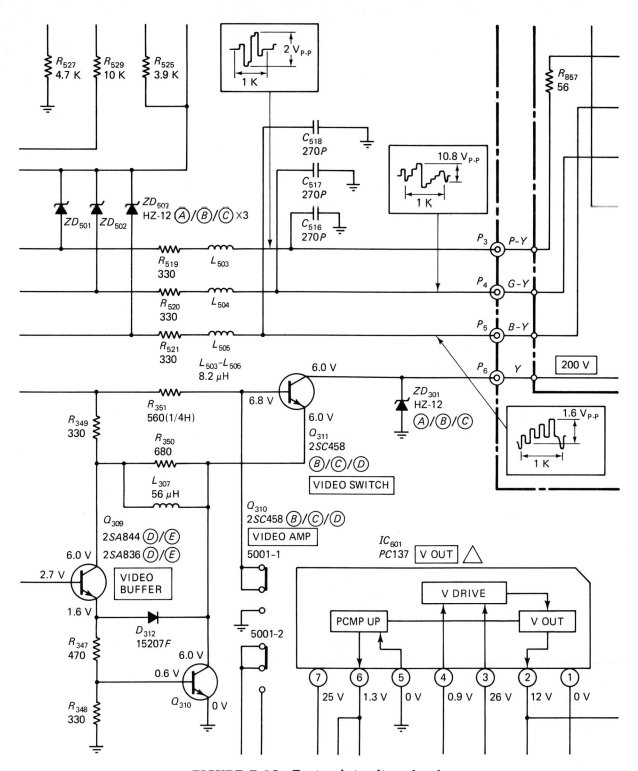

FIGURE 7-16 Test point voltage levels.

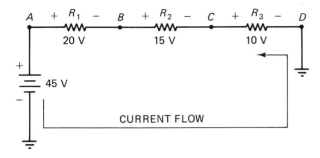

FIGURE 7-17

EXAMPLE 7-1

Refer to Figure 7-17. Given that the voltage reading at point C is 10 V and that the voltage reading at point B is 25 V, find the voltage dropped across R_2.

Solution: The voltage dropped across R_2 is equal to the difference between the voltage levels at point C and point B.

$$25 - 10 = 15 \text{ V}$$

The voltage across R_2 is 15 V. ∎

Let's examine the effects of moving the ground point to point C (See Table 7-1). The schematic circuit would look as shown in Figure 7-18. Notice that with point C as ground, R_3 now has a positive ground. So the voltage dropped across R_3 would be -10 V, and if you measured from point D to ground it would be necessary for you to reverse the leads. That is, in Example 7.1 you would connect the negative lead of the voltmeter to the ground (this is the correct connection for positive voltages). In this case, however, you would be reading a negative voltage and would have to reverse the leads, in other words, connect the positive voltmeter lead to ground. Connecting the voltmeter to ground and point C would provide a reading of 0 V (no resistance between the leads). Point B and ground would give a reading of 15 V., while point A and ground would show 35 V. Both of these voltages are positive voltages. So in taking the three readings it would be necessary to reverse the leads one time.

Basically speaking, any time the voltmeter is connected with incorrect polarity, you will notice a slight reverse needle movement as the voltmeter is connected, reversing the leads will provide a forward reading. If the negative lead is connected to ground, the reading is positive. Let's examine, now, what effects a ground at point B would have on the circuit. This is shown in Figure 7-19.

Now when you measure from point D to ground, you get -25 V, the combined voltage drop of R_2 and R_3. This voltage is negative because the ground for these resistors is positive. Point C and ground would give a reading of -15 V. Point B would give 0 volts and point A would give $+20$ V. The point A reading would be a positive voltage since its ground is negative. Finally, let's move the ground to point A and examine its effect on circuit readings. The circuit for this ground placement is shown in Figure 7-20.

All of the readings are now negative voltages, because the positive side of each resistor is closest to ground. The voltmeter would read -45 V when connected between ground and point D, -35 V when connected between ground and point C, -20 V when connected between ground and point B, and 0 V when connected between ground and point A. It is helpful to examine these effects all at the same time. Table 7-1 shows these changes in this manner.

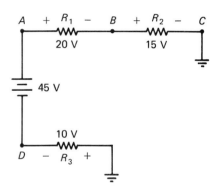

FIGURE 7-18

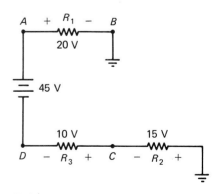

FIGURE 7-19

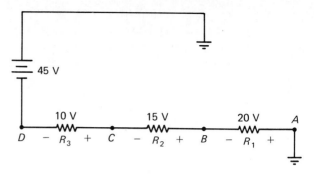

FIGURE 7-20

TABLE 7-1 Ground/Voltage Relationships

Voltage at:	Ground at:			
	Point A	Point B	Point C	Point D
Point A	0	+20	+35	+45
Point B	−20	0	+15	+25
Point C	−35	−15	0	+10
Point D	−45	−25	−10	0

One important item to notice is the potential difference between points. The circuit operation is the same no matter which point is considered ground. That is, the potential difference between point A and point B is 20 V no matter where ground is connected.

EXAMPLE 7-2

Demonstrate that the potential difference between point A and point B is 20 V for all circuit ground placings.

Solution: Potential difference, as the name implies, refers to the difference (subtraction) of voltage levels.

Ground at point A:

$$V_A - V_B = 0 - (-20) = +20 \, V$$

Ground at point B:

$$V_A - V_B = 20 - 0 = +20 \, V$$

Ground at point C:

$$V_A - V_B = 35 - 15 = +20 \, V$$

Ground at point D:

$$V_A - V_B = 45 - 25 = +20 \, V$$

The same is true for any of the other points in the circuit. The only thing that changes is the amount of voltage read on the voltmeter. The voltage drops remain the same. Important items to remember are:

1. Voltage readings are more easily taken with respect to ground.
2. Negative voltages have positive grounds.
3. Positive voltages have negative grounds.
4. Voltage levels are not affected by ground points, but their readings are.

SELF-TEST

Please check your answers with your instructor before you continue.

1. Would a positive voltage have a positive or negative ground?
2. How can the ground of a component be determined by using a voltmeter?
3. Does *potential difference* indicate adding or subtracting voltage levels?
4. Why do manufacturers use grounds in systems?
5. What is normally the correct voltage level for ground?
6. Why is ground sometimes referred to as earth?
7. Describe the recommended manner for using a voltmeter to measure voltage.
8. Describe how schematics show voltage levels.
9. List some materials that can be used for ground. List some materials that cannot be used for ground.
10. Sketch a circuit having four resistors, a 10-kΩ resistor, a 15-kΩ resistor, a 7.2-kΩ resistor, and a 27-kΩ resistor. All of these resistors are powered by a 100-V supply and they are connected in series. Find the voltages to ground as the ground is moved around the circuit.

STEPS IN TROUBLESHOOTING

The most important thing to have when troubleshooting is an orgainzed manner of checking first systems, and finally, components. In electronics today, any device can be broken down into sections or systems called *blocks*. The overall device is checked for operation, which will generally indicate what block is causing a particular problem. An essential item for determining and locating each block is the schematic diagram. For this reason, you should not start without the proper schematic—to do so would be nothing more than a guessing game. This section is devoted to the troubleshooting of dc circuits and testing resistors and relays. Other component (capacitor, inductors, etc.) testing is explained later.

Complaint Verification The first step in any organized troubleshoot is to verify the complaint. This is done for two reasons; first, it is difficult to understand exactly what a complaint is until you have seen it. Also, a technician has a better idea of what else to check for. For example, on a tube-type TV set a technician would check to see if any tubes lighted up when the set was malfunctioning. What else to check for depends greatly on what device is being checked and on experience. By the way, there is only one way to get experience, and that is by doing. So don't be afraid to try; however, also expect failures. Failures are time consuming but do have a good side—without them there really can't be any valuable experience. Remember that Babe Ruth, who for a number of years held the record for home runs, also struck out more times than any other baseball player.

Visual Inspection The second step is a visual inspection. Since the advent of solid-state electronic components, the odds of an electronic malfunction have been reduced. More often than not the malfunction in an electronic device is mechanical in nature: a broken belt, lever, or other component which controls a switch, for example. A quick visual inspection will show these problems. If there is an electronic problem, the visual inspection will show overheating conditions. Look for resistors and other components that have been too hot, are physically broken, or are leaking in the case of capacitors. The circuit should be checked for possible high-voltage arcing and for shorted-circuit current flow. These will appear as discolorations on the circuit board.

Power Supply Testing The third step is to check the power supply. It has happened that a technician has been asked to repair an item that needed only to be plugged in or turned on. Don't spend days troubleshooting a device that is not plugged in. Power supplies also include the fuse. If a particular device is plugged in, remember to check the fuse or circuit breaker. An ohmmeter can be used to test the fuse. A good fuse will have little or no resistance.

If the fuse is blown, examine it. A dark color indicates that a large current drain was made on the circuit. In this case, don't replace the fuse until after a current measurement has been taken. In the event that a short circuit is present, use a current-limiting load in place of the fuse. A light bulb can be used as such a load (Figure 7-21). This load will usually limit the current enough for further troubleshooting to be done. The light bulb will burn brightly if there is a short present; when the short is removed, the light will go out. Remove the loads one at a time until the shorted one is located. Refer to the diagram in Figure 7-22.

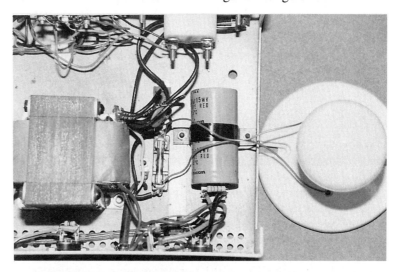

FIGURE 7-21 Finding a short. The light bulb is used as a load in place of the blown fuse. The bulb will glow brightly when the circuit is shorted.

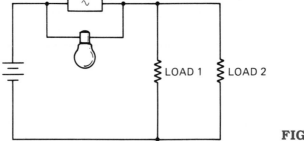

FIGURE 7-22

Block Testing The fourth step, block testing, can be done by one of two methods, but is best done in the form of a combination of the two methods. The first is called *signal (voltage level) tracing*. Here the voltage or signal is traced systematically through the circuit. In doing this you would use what is called the *half-step check*. The half-step check is a method of checking the signal at a point approximately at the middle of the circuit. If the signal is good, you would check the signal at a point halfway between this half-point and the end point, or a point three-fourths of the way through the circuit. By stepping forward or backward, as the checks indicate, you can narrow down the troubled section.

EXAMPLE 7-3

Describe how the half-step method would be used to troubleshoot the circuit shown in Figure 7-23.

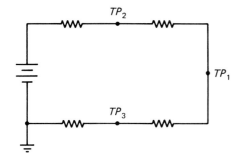

FIGURE 7-23

Solution: Notice the negative ground on this system. You would test this system using a voltmeter. The negative lead would be connected to ground and you would use the positive lead as the test probe. The first point to check for a signal is test point 1. If the signal (voltage) is present here, you would move to test point 3. If not, you would move to test point 2. The presence of a signal at test point 1 indicates that there is a continuous path for current from the battery to test point 1. Testing the voltage at test point 2 is redundant, since this point lies between the battery and the point where the circuit has a signal. In this case the next point to check is test point 3. If there is no signal there, the fault lies between test point 1 and test point 3. ∎

Understand that this is the correct procedure; however, Example 7-3 is overly simplified. It is intended that you gain a concept of how this troubleshooting step is accomplished. The only way to master it is to practice. Like any other process, there will be mistakes made. Don't let them discourage you. Remember that Thomas Edison had 1000 failures before he succeeded with the light bulb.

The second method of block checking is called *signal injection*. The idea behind signal injection is similar to that of signal tracing, except that instead of measuring the voltage level, you inject the proper signal and check for proper operation. Signal injection requires that you know exactly what voltage levels are required at a particular point and provide exactly that voltage level for a signal. If the worng voltage levels are injected, the device can be severely damaged. For this reason, technicians, especially inexperienced ones, tend not to use signal injection. The simple truth of the matter is that with signal injecting or signal tracing, the correct voltage levels must be known. There is no reason to shy away from signal injection just because you must know the correct voltage levels. There are times when signal injection is faster and easier than signal tracing.

EXAMPLE 7-4

The circuit of Figure 7-24 shows a battery, a switch, a relay, and a load. The load is a fan. A relay must be used because the fan requires more current than

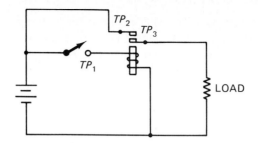

FIGURE 7-24

the switch can handle. The relay is an easy way to allow a low-amperage switch to control a large current load. In this example the fan does not operate. The problem is not the battery. Describe how you can use signal injection to find what component does not operate.

Solution: Connecting a jumper wire from the positive battery terminal to test point 1 will test the operation of the relay, and the load. If the fan functions, the problem must be in the switch circuit. Connecting the jumper wire to test point 3 will test the operation of the fan. If signal injection at test point 1 does not cause the fan to operate, and signal injection at test point 3 does, there is a problem in the relay circuit. Note that this does not rule out a switching circuit problem also. With this method of testing it is important to remember the possibility of multiple problems. ■

Component Testing The fifth step is component testing. For now, this phase of troubleshooting will be discussed only as it relates to relays and resistors. In Example 7.4, signal injection was used to test the relay. An ohmmeter could also be used.

There are two tests made to determine the condition of a relay. Internally, a coil of wire (the winding) is used as an electromagnet to open or close a set of contacts. The first test checks the condition of the coil winding. It is performed by taking the resistance reading of the coil. There is a specified amount of resistance (typically, 50 to 100 Ω) this coil should have. An excessive amount indicates that the coil is open. Too little would indicate that the coil windings are shorted. Either fault may make the relay inoperable. This check is made with the relay disconnected from the circuit.

The second test checks the condition of the contacts. Here you read the amount of resistance across the contacts when they are closed. The relay may need to be energized to perform this test. The contact resistance should be very small (typically, less than 1 Ω) with the contacts closed. An excessive amount would indicate that the point contacts have been damaged by overheating. Resistors are checked for proper resistance with an ohmmeter. They should be within the specified tolerance of the resistor.

Repair The sixth and final step is the repair of the faulty component. Remember that replacing faulty components is not the end of troubleshooting. It is important to consider the original cause of the malfunction. Replacing any component without proper consideration of the cause of failure will undoubtedly lead to another failure in the future. Possibly a fix could be something as simple as using a component with a larger power rating. Whatever the case may be, spend some time thinking of the possible cause of any malfunction.

The six steps in troubleshooting are as follows, listed in order:

1. Complaint verification
2. Visual inspection
3. Power supply testing
4. Block or system testing
5. Component testing
6. Repair

These are valid troubleshooting steps no matter what is being tested. Anything from a problem on an automobile to a problem with a swimming pool can be tested in this manner. It is a standard method of troubleshooting taught in all forms of repair, and thus should be memorized.

SELF-TEST

Check your answers with your instructor before continuing.

1. Define the term *block* as it refers to a schematic diagram.
2. Why is a schematic necessary?
3. List and briefly describe each of the six steps in correct troubleshooting.
4. Which is more common in electronic systems today, electronic malfunctions or mechanical ones?
5. List some of the things a technician should look for when performing a visual inspection.
6. Describe how to test a fuse using an ohmmeter.
7. What are some things that could cause a fuse to open?
8. What is meant by a *current-limiting load*? How can it be used to test and locate a shorted circuit?
9. Describe the difference between signal tracing and signal injecting. Include advantages and disadvantages for each method.
10. What is meant by the *half-step method* of troubleshooting?
11. When signal injection and signal tracing, why is it important that you know what the correct voltage levels are supposed to be?
12. Why is it necessary to find the cause of a malfunction rather than just repairing an appliance?
13. Describe some visual conditions which would indicate that the circuit was subjected to too much current flow.
14. When troubleshooting a component that seems to have a power supply problem, what things should be checked first?
15. Why is it necessary to memorize the six troubleshooting steps?

SPECIAL TESTS

There are several special tests that can be made. These tests are in addition to the standard troubleshooting voltage tests. They usually check the wires for excessive resistance, or merely for correct connection. The simplest of these is the test for continuity.

Continuity Test The continuity test is made with an ohmmeter and checks only for proper connection. The ohmmeter is adjusted to a low scale. It can be calibrated, but it is not necessary since you are interested only in whether there is a reading at all, and not with what the particular reading happens to be.

First you examine the schematic to find what components are connected together. You would then use the ohmmeter to check to see if there is a connection between the points or components. When there is a continuous path for electricity the ohmmeter will show a low reading—hence the name *continuity*. This testing procedure is shown in Figure 7-25.

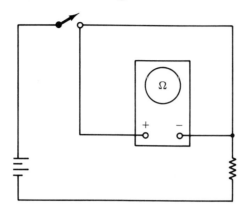

FIGURE 7-25

Remember that to use the ohmmeter, the power must be disconnected. The battery would either need to be disconnected or the switch would need to be in the open position. Preferably the battery would be disconnected. As the schematic indicates, there should be little, if any, resistance between these points. If a large amount of resistance were measured, the resultant voltage drop might affect the circuit performance. Cold solder joints may cause this type of problem. Dirty or corroded connections may also be a source of this unwanted circuit resistance. In any case, the continuity test can be used to find this circuit resistance.

Circuit Resistance Tests Another test that can sometimes locate excessive voltage drop is the circuit resistance test. A voltmeter is used to make this resistance check. This test is made primarily on circuits that pull a large amount of current. If you are dealing with a current of 100 A or so, a small amount of resistance (0.1 Ω) could cause a voltage drop of 10 V. An ohmmeter will not generally show a resistance this small. Also, a resistance of this type is usually of a dynamic nature (it has a value that changes with the circuit current). That means that the resistance measured with an ohmmeter would not be the same as the resistance of the circuit when it is in operation. One thing to remember about this test is that the circuit must be in operation when the test is made. Ideally, the voltage drop per connection should be 1% of the supply voltage or less. The voltmeter connection for this test is shown in Figure 7-26.

One percent of 10 V is 0.1 V. There are five connections between the battery and the resistor (the switch counts as one connection, as do relays). If the voltmeter were to show a reading of 0.5 V or less, there would not be an excessive amount of circuit resistance.

Ground Circuit Resistance Test

An often-overlooked source of circuit resistance is the ground circuit. It is important to check the ground circuit resistance. This is done in a similar manner. The

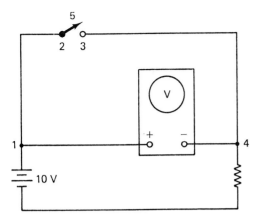

FIGURE 7-26

only basic difference is that the voltmeter is connected to the negative side of the circuit. The rules for voltage drop per connection are still 1% of the supply voltage. However, there should be no more than two ground connections, which means that the total acceptable voltage drop is 2% of the supply voltage.

Short-Circuit Voltage Test

The last of the special tests is a short-circuit voltage test. This test uses a voltmeter to find a shorted switch or relay. The voltmeter is installed in the circuit in series with the battery. All of the circuits are turned off and the voltage is read. The voltage reading should be low, in the neighborhood of 10% of the supply voltage. A reading of the supply voltage would indicate a shorted circuit. This test is usually made on a circuit that has a small current draw that is draining the battery even when the circuit is turned off. The current draw would be so small that a current reading could not easily be taken. You use the voltmeter as a load. It will react to a small amount of current. The meter is used in a manner similar to the way the light bulb was used in a previous test. The voltmeter connection for this test is shown in Figure 7-27.

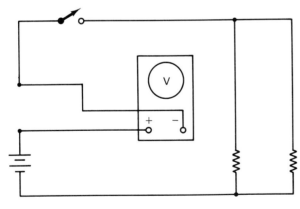

FIGURE 7-27

Understand that a high meter reading (supply voltage) indicates an amperage draw. The load causing this problem can be found by disconnecting the loads, one at a time, until the shorted circuit is found. This will be indicated by a low meter reading (10% of the supply voltage).

These are not tests that need to be made on every single circuit; however, in the event that the normal troubleshooting procedures do not turn up any problem,

SPECIAL TESTS

one of these tests usually will. Also these are tests that are not generally listed in repair manuals, but they do test and locate circuit resistance, which can be a headache to find otherwise.

SELF-TEST

Please check your answers with your instructor before continuing.

1. Why isn't it necessary to zero-adjust the ohmmeter before performing a continuity test?
2. What precautions should be made before an ohmmeter is used in circuit testing?
3. List some possible causes of high circuit resistance.
4. High circuit resistance would have what effect on circuit voltage?
5. In what instance can a voltmeter better find resistance that an ohmmeter cannot find?
6. What is meant by *dynamic resistance*?
7. Describe how to test a circuit for circuit resistance. Include the circuit requirements.
8. Why is it also necessary to check ground circuit resistance?
9. Describe how a short-circuit voltage test is performed.
10. Why is it necessary to have all of the circuits off when performing a short-circuit voltage test?
11. Describe the manner in which one can locate a fault when performing a short-circuit voltage test.
12. What is the rule-of-thumb maximum circuit resistance in volts?
13. How does the voltmeter connection differ from the previous test and the short-circuit voltage test?

SUMMARY

A VOM meter is a combination of an ohmmeter, a voltmeter, and an ammeter all built into one unit. It is important that you know how to use each of these meters. Three important things to remember about the use of these meters are:

1. A voltmeter is usually connected in parallel with the component whose voltage is being checked.
2. A current meter is connected in series with the component whose current is being checked.
3. An ohmmeter must be used on a dead circuit.

Troubleshooting any system is done in an orderly fashion following a six-step procedure. The steps are (1) complaint verification, (2) visual inspection, (3) power supply testing, (4) block or system testing, (5) component testing, and (6) repair.

Unwanted circuit resistance is a major cause of circuit malfunctions. Remember that a voltmeter can be used to measure the resistance under high current conditions where the resistance may be dynamic in nature.

QUESTIONS

7-1. Describe the use of the voltmeter, the ohmmeter, and the current meter.

7-2. List each of the basic oscilloscope controls and briefly describe the function of the control.

7-3. What is meant by *horizontal position*?

7-4. What is meant by *vertical position*?

7-5. To what does *volts/division* refer?

7-6. To what does *time/division* refer?

7-7. List the inputs that are typically used when connecting an oscilloscope to a circuit.

7-8. What is the vertical input used for?

7-9. What is the horizontal input used for?

7-10. Why would a relay be used on some circuits instead of a switch?

7-11. Refer to Figure 7-23. Suppose that the resistors were 2.2, 4.7, 6.8, and 7.2 kΩ going in a counterclockwise direction away from the 100-V battery. Calculate the correct voltage to ground at each of the test points. What would be the test point voltages if the 6.8-kΩ resistor were to open?

7-12. The circuit in Figure 7-24 has a 12-V supply. When signal tracing, a technician finds that test point 1 has 12 V when the switch is energized, as does point 2. Test point 3 shows a voltage reading of 6 V. What could be causing this problem?

7-13. When testing the circuit shown in Figure 7-25, a technician finds that the ohmmeter gives a 100-Ω reading. What should be done to locate the cause of this resistance correctly? How could it be reduced?

7-14. The supply voltage is 20 V in Figure 7-26. A technician finds the total circuit resistance to be 0.5 V. Is there a problem with this circuit? Explain.

7-15. The circuit in Figure 7-27 has a 6-V supply. The voltmeter reads 6 V whether the switch is open or closed. What could be at fault?

PRACTICE PROBLEMS

7-1. Referring to Figure 6-11, a technician measures 1 mA as the current flowing through R_1. What could be the malfunction?

7-2. Referring to Figure 6-11, a technician measures 1.7 mA as the current flowing through R_1. What could be the malfunction?

7-3. Referring to Figure 6-12, a technician measures 13.5 V across R_1. What could be the malfunction?

7-4. Referring to Figure 6-20, a technician measures 40 V across R_3. What could be the malfunction?

7-5. Referring to Figure 6-25, a technician measures 1 mA flowing through R_7. What could be the malfunction?

7-6. A technician working on the circuit shown in Figure 6-26 finds the potential difference between point A and point B to be 0 V. What could be at fault?

7-7. Referring to Figure 6-34, the following are listed as resistor values: $R_1 = 10$ kΩ, $R_2 = 20$ kΩ, $R_3 = 6.8$ kΩ, $R_4 = 2.7$ kΩ, $R_5 = 8.2$ kΩ, and $R_6 = 3.9$ kΩ. A technician measuring total circuit resistance finds it to equal 7.6 kΩ. What could be the cause of this reading?

7-8. Referring to Figure 6-34, the following are listed as resistor values. $R_1 = 10$ kΩ, $R_2 = 20$ kΩ, $R_3 = 6.8$ kΩ, $R_4 = 2.7$ kΩ, $R_5 = 8.2$ kΩ, and $R_6 = 3.9$ kΩ. A technician measuring total circuit resistance finds it to equal 3.9 kΩ. What could be the cause of this reading?

7-9. Referring to Figure 6-47, a technician connects the negative lead of a voltmeter to the negative terminal of the battery. When she touches the positive side of R_4, she reads battery voltage. When she touches the negative side of R_4, she reads zero volts. What fault is indicated?

Voltage and Current Dividers

Chapter objectives

After reading this chapter and answering the questions and problems, you should be able to:

- Given any three values for the equation $a/b = c/d$, correctly solve for the unknown.
- Given statements containing ratios and proportions, correctly set up the equation in the $a/b = c/d$ format.
- Given the voltage and resistance of one series resistor and either the voltage or resistance of another, calculate the missing quantity (voltage or resistance) using a voltage divider.
- Given the current and resistance of one parallel resistor and either the current or resistance of another, calculate the missing quantity (resistance or current) using a current divider.
- Describe the difference between unloaded and loaded voltage dividers.

In this chapter we discuss unloaded voltage dividers and current dividers. The names may seen ominous; however, a voltage divider is nothing more than a series circuit of the same general type as the series circuits studied earlier. A current divider is another name for a parallel circuit. We examine some different ways to view these circuits, ways that will make direct calculations of voltage easier on a series circuit and make direct calculations of current easier for parallel circuits. It is important that you master these ways of viewing series and parallel circuits, as all advanced textbooks use them.

GETTING STARTED: UNDERSTANDING RATIOS

The major idea behind voltage dividers is that voltage and resistance are directly proportional. To fully understand voltage dividers, it will be necessary for you to have a good understanding of ratios and proportions. In the following section we demonstrate and explain their meaning.

Arithmetic The major portion of this chapter will require that you know how to work problems that are written using this arithmetic format:

$$\frac{a}{b} = \frac{c}{d}$$

Three of these variables will be known quantities. One will be an unknown. You will need to find the unknown. Several examples follow. Each will demonstrate the steps to follow when the unknown is located in the a, b, c, or d position.

EXAMPLE 8-1

Find a for the following equation:

$$\frac{a}{10} = \frac{12}{6}$$

Solution: First examine the right side of the equation:

$$\frac{12}{6} = 2$$

so that you can replace the 12/6 in the original equation with the number 2:

$$\frac{a}{10} = 2$$

Next notice that the *a* is being divided by 10. To remove the 10 from the left side of the equation it is necessary to multiply both sides by 10:

$$\frac{a}{10} \times 10 = 2 \times 10$$

The 10's on the left side of the equation cancel, which gives

$$a = 20$$

as the final answer. ∎

In Example 8-1 notice that $12/6 \times 10$ is also 20. A shortcut to solving this problem is to cross-multiply. *Cross-multiply* refers to multiplying the denominator (bottom number) of the left-side fraction and the right-side fraction. Most proportional problems can be solved in this manner.

EXAMPLE 8-2

Find *b* in the following problem:

$$\frac{20}{b} = \frac{7}{14}$$

Solution: Any time that you find the unknown used as the denominator of the left side of the equation, begin by inverting both sides of the equation.

$$\frac{b}{20} = \frac{14}{7}$$

This is the format that was demonstrated in Example 8-1. The solution follows.

$$\frac{20}{b} = \frac{7}{14} \qquad \text{Inverting}$$

$$\frac{b}{20} = \frac{14}{7} \qquad \text{Cross-multiplying}$$

$$b = \frac{14}{7} \times 20$$

$$= 40 \qquad \blacksquare$$

When the unknown is on the right side of the equation, simply exchange positions of the right and left sides of the equation. So the equation

$$\frac{12}{24} = \frac{20}{x}$$

becomes

$$\frac{20}{x} = \frac{12}{24}$$

The unknown is now in the *b* position and can be solved in the same manner as Example 8-2.

GETTING STARTED: UNDERSTANDING RATIOS

SELF-TEST

Solve for the unknown.

1. $\dfrac{w}{22} = \dfrac{7}{2}$
2. $\dfrac{b}{81} = \dfrac{5}{9}$
3. $\dfrac{g}{12} = \dfrac{42}{72}$
4. $\dfrac{k}{20} = \dfrac{25}{100}$
5. $\dfrac{h}{6} = \dfrac{14}{12}$
6. $\dfrac{a}{100} = \dfrac{3}{4}$
7. $\dfrac{20}{50} = \dfrac{h}{5}$
8. $\dfrac{60}{200} = \dfrac{q}{35}$
9. $\dfrac{50}{n} = \dfrac{75}{100}$
10. $\dfrac{25}{e} = \dfrac{52}{80}$
11. $\dfrac{78}{d} = \dfrac{20}{415}$
12. $\dfrac{35}{200} = \dfrac{80}{u}$

ANSWERS TO SELF-TEST

1. 77
2. 45
3. 7
4. 5
5. 7
6. 75
7. 2
8. 10.5
9. 66.67
10. 38.46
11. 1619
12. 457.1

Ratios A *ratio* is a numerical representation of how one thing relates to another. They are usually written as factions. A proportion is an equality between ratios. In math classes you have probably had to calculate rate problems where you were given a distance and a time and asked to find the distance after a second time.

EXAMPLE 8-3

Tom is driving down the road. After 2 hours, he notices that he has traveled 85 miles. If he continues to travel at the same rate, how far will he have traveled in 5 hours?

Solution: Examine the problem and find which two numbers are related to each other. In this case, the 2 hours and 85 miles are related. Eight-five miles is the distance traveled in 2 hours. The key to solving any rate or ratio problem is to say to yourself: "2 hours is to 85 miles as 5 hours is to X miles." The "is to" in this statement means "divide." The "as" means "equal." The equation would be

$$\frac{2}{85} = \frac{5}{X}$$

Any statement that uses "one number is to another as a third is to a fourth" will work. The trick to solving these ratios is to make the statement in such a manner that the unknown quantity is first. Rewording the first statement to say "X miles is to 5 hours as 85 miles is to 2 hours" is an example of this. The equation for this statement would be

$$\frac{X}{5} = \frac{85}{2}$$

Notice that in this equation the X (unknown) is placed on top (numerator) of the fractional ratio. This is the easiest manner to solve for the unknown. Cross multiplication is then used to find X.

$$\frac{X}{5} = \frac{85}{2}$$

$$X = 5\left(\frac{85}{2}\right)$$

$$= 212.5 \text{ miles}$$ ∎

Ratios of this type are referred to as *directly proportional* statements. With directly proportional relationships it is necessary to have the related quantities on the same plane (either as numerators or denominators). As long as they are on the same plane, the mathematic statement (equation) is directly proportional in nature.

SELF-TEST

1. A trencher digs 12 meters in 4 hours. How long will it take to dig 50 meters?
2. Pete gets two vacation days for every 28 days worked. If Pete has worked 310 days since his last vacation, how many vacation days are owed him?
3. A recipe calls for cooking a chicken 6 minutes in a microwave oven for every 1.5 lb. How long should a 2-lb chicken cook?
4. If a car uses 12 gallons in 420 miles of travel, how far could it travel on 1 gallon of gas?

ANSWERS TO SELF-TEST

1. 16 hours and 40 minutes
2. 22 days
3. 8 minutes
4. 35 miles

UNLOADED VOLTAGE DIVISION

Examine the circuit shown in Figure 8-1.

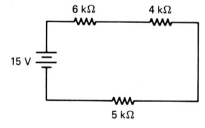

FIGURE 8-1

Circuit You should be able to look at these resistances and the supply voltage and know intuitively the voltage drops across each resistor. The 6-kΩ resistor has a 6-V drop, the 4-kΩ resistor has a 4-V drop, and the 5-kΩ resistor drops 5 V. How did you know this? Glancing at this circuit showed that total resistance was 15 kΩ. Intuition indicated that the individual voltage drops must be equal to the kilo-ohm value of the resistor in question. Remember one of the Ohm's law concepts was that resistance and voltage were directly proportional (when current was held to a constant value). A series circuit has one constant current value, so on a series circuit, voltage must be directly proportional to resistance. Mathematically, this would be

$$\frac{V_T}{V_1} = \frac{R_T}{R_1}$$

Related Formulas The 1's in this equation could be substituted by any of the individual resistor component numbers. Also, this equation can take any form, which keeps individual levels on the same plane. That is, the total voltage and the individual voltage must be one of the following:

On the same side of the equation:

$$\frac{V_T}{V_1} = \frac{R_T}{R_1}$$

or

$$\frac{V_1}{V_T} = \frac{R_1}{R_T}$$

The top (numerator) of the equation:

$$\frac{V_T}{R_T} = \frac{V_1}{R_1}$$

The bottom (denominator) of the equation:

$$\frac{R_T}{V_T} = \frac{R_1}{V_1}$$

Direct Proportion Any time the words *directly proportional* appear, you can make statements that keep individual component levels on the same plane. Voltage being directly proportional to resistance means that voltage divided by resistance will give us the same answer as long as the voltage and resistance are on the same level (for the same component). The ratio of voltage to resistance is current. Also, in a series circuit the resistance-to-resistance ratio must be the same as the corresponding voltage-to-voltage ratio. The following are some examples of how this can be applied.

Applications

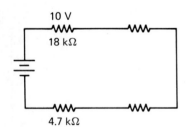

FIGURE 8-2

EXAMPLE 8-4

Find the voltage drop across the 4.7-kΩ resistor in the circuit shown in Figure 8-2.

Solution: Examine the circuit and set the equation up in such a manner that the missing parameter (voltage) is on top of the formula (expressed as a numerator). That would be in this manner:

Voltage on top:

$$\frac{V}{4700} = \frac{10}{18,000}$$

$$= 5.556 \times 10^{-4}$$

$$V = 5.556 \times 10^{-4} \times 4700$$

$$= 2.611 \text{ V}$$

or voltage on one side:

$$\frac{V}{10} = \frac{4700}{18,000}$$

$$= 0.2611$$

$$V = 0.2611 \times 10$$

$$= 2.611 \text{ V}$$

The original formulas indicate that it must be the total resistance that is used. This is not true. The formula will work as long as the direct relationship of component to component value is maintained.

EXAMPLE 8-5

Find the total resistance of the circuit shown in Figure 8-3.

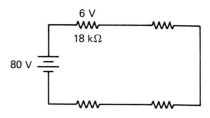

FIGURE 8-3

Solution: You know the voltage drop across a particular resistor and the supply voltage, so you can find the total resistance of the circuit by using this direct proportionality between resistance and voltage. Set the equations up in a manner where the missing resistance is on top (a numerator). That would be of this nature:

Resistance on top:

$$\frac{R}{80} = \frac{18,000}{6}$$

$$= 3000$$

$$R = 3000 \times 80$$

$$= 240 \text{ k}\Omega$$

or resistance on one side of the equation:

$$\frac{R}{18,000} = \frac{80}{6}$$

$$= 13.33$$

$$R = 13.33 \times 18,000$$

$$= 240 \text{ k}\Omega$$

SELF-TEST

1. In a series circuit a 2.7-kΩ resistor has a voltage drop of 15 V. One of the other resistors in the circuit has a voltage drop of 48 V. What is the resistance of the second resistor?
2. The supply voltage for a five-resistor series circuit is 50 V. A 68-kΩ resistor has a voltage drop of 12.6 V. What is the total resistance of the circuit?
3. A 4.7-kΩ resistor and a 2.2-kΩ resistor are connected in series. The voltage drop across the 2.2-kΩ resistor is 25 V. What is the voltage drop across the 4.7-kΩ resistor?
4. The voltage drop a across a 470-kΩ resistor is 33 V. If this resistor is connected in series with a 150-kΩ resistor, what is the voltage drop across the 150-kΩ resistor?

ANSWERS TO SELF-TEST

1. 8.64 kΩ
2. 269.8 kΩ
3. 53.41 V
4. 10.53 V

Standard Voltage Divider Formula When current held at one constant value, as in a series circuit, voltage is directly proportional to resistance, any relation (equation) that maintains this proportionally is true and can be used to find any missing value. The most commonly used formula is

$$V = \frac{R}{R_T} \times V_T$$

This form requires fewer steps. Instead of setting up the ratio of resistance to voltage, this equation sets up how to find the answer. A demonstration of the use of this equation is given in Example 8-6.

EXAMPLE 8-6

Use the voltage divider formula to find the voltage across the 4.7-kΩ resistor in Figure 8-4.

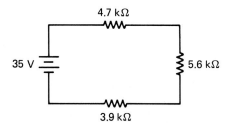

FIGURE 8-4

Solution: Find the total resistance of the circuit.

$$4700 + 5600 + 3900 = 14.2 \text{ k}\Omega$$

The supply voltage is 35 V. Insert these values into the formula:

$$V = \frac{R}{R_T} \times V_T$$

$$= \frac{4700}{14{,}200} \times 35$$

$$= 11.58 \text{ V}$$

SELF-TEST

1. In a series circuit having a total resistance of 12 kΩ, find the voltage drop across a 2.7-kΩ resistor, if the source voltage is 48 V.
2. The total resistance of a series circuit is 56 kΩ. The supply voltage is 20 V. What would be the voltage drop across a 12-kΩ resistor in this circuit?
3. The supply voltage for a five-resistor series circuit is 50 V. Find the voltage drop across the 68-kΩ resistor if the other four resistors are 120, 39, 180, and 27 kΩ.
4. A 4.7-kΩ resistor and a 2.2-kΩ resistor are connected in series. The supply voltage is 30 V. What is the voltage drop across the 4.7-kΩ resistor?

ANSWERS TO SELF-TEST

1. 10.8 V
2. 4.286 V
3. 7.834 V
4. 20.43 V

CURRENT DIVISION

Basic Formulas Ohm's law also indicates that current and resistance are inversely proportional when voltage is held at a constant value. In a parallel circuit, voltage is the same across all the branches. It is important to realize what *inversely proportional* means. One thing that it means is that any increase in resistance will cause a decrease in current. Mathematically it means that when you set up an equation, you must use

$$I_1 \times R_1 = I_T \times R_T$$

as a starting point. This can be manipulated to give the basic current division formulas

$$\frac{R_T}{R_1} = \frac{I_1}{I_T}$$

$$\frac{I_T}{R_1} = \frac{I_1}{R_T}$$

$$\frac{R_T}{R_1} = \frac{I_1}{I_T}$$

$$\frac{R_1}{I_T} = \frac{R_T}{I_1}$$

Developing an understanding of what *inversely proportional* means will help immensely with the setup of any parallel circuit analysis. An important thing to notice about the previous formulas is that (1) when the resistance increases the current will decrease (2) equal parallel voltage is the starting point for each of these formulas. Several examples follow that demonstrate these formulas.

Applications

EXAMPLE 8-7

Use the inverse proportionality of current and resistance to find the current through the 15-kΩ resistor in Figure 8-5.

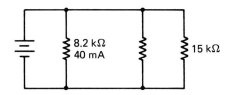

FIGURE 8-5

Solution: Begin by writing an equation so that the parallel voltages are equal.

$$R_1 \times I_1 = R_2 \times I_2$$
$$15000 \times I_2 = 8200 \times 0.04$$
$$I_2 = \frac{8200 \times 0.04}{15000}$$
$$= 21.87 \text{ mA} \qquad \blacksquare$$

Notice that the formulas listed in the basic formula section of current dividers imply their use on circuits only if the total values are known. This is not the case. They will work on any parallel circuit as long as the equality of parallel voltage is maintained. This was demonstrated in Example 8-7.

EXAMPLE 8-8

Find the total resistance of the circuit shown in Figure 8-6.

Solution: The circuit shown in Figure 8-6 shows a total current of 0.2A. The current through the 5.6-kΩ resistor is 50 mA. Again set the equation up so that the parallel voltages are equal.

$$I_T \times R_T = I_1 \times R_1$$
$$0.2 \times R_T = 0.05 \times 5600$$
$$R_T = \frac{0.05 \times 5600}{0.2}$$
$$= 1.4 \text{ K}\Omega \qquad \blacksquare$$

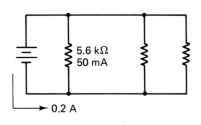

FIGURE 8-6

The ratios shown in both Examples 8-7 and 8-8 are used most often when there are only two resistors in parallel. They can be manipulated to form an equation. That formula is

$$I = \frac{R_T}{R} \times I_T$$

Further examination of this will show that the ratio of total resistance to individual resistance is equal to the opposite resistor divided by the sum of the two resistors. That is,

$$I_1 = \frac{R_2}{R_1 + R_2} \times I_T$$

Remember when using this formula that the resistance of the other resistor (not the one whose current you are finding) is the numerator of the ratio. In other words, when solving for the current through R_1, use the resistor value for R_2, not R_1, as the numerator. This formula is used in the analysis of more advanced circuitry and applies only to parallel circuits that have two resistors. An example of this application follows.

EXAMPLE 8-9

Referring to Figure 8-7, find the current through the 4.7-kΩ resistor.

Solution: Remember that you want to know the current through the 4.7-kΩ

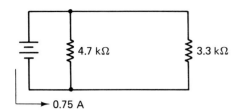

FIGURE 8-7

resistor. You must use the value of the other resistor (3.3 kΩ) as R_2 in the formula. The formula is straightforward and would appear as

$$I_1 = \frac{R_2}{R_1 + R_2} \times I_T$$
$$= \frac{3300}{4700 + 3300} \times 0.75$$
$$= \frac{3300}{8000} \times 0.75$$
$$= 0.3094 \text{ A} \qquad \blacksquare$$

The concept of current division will help in the speed and ease of parallel circuit analysis. Also forms of this concept are used in the theoretical proofs of advanced electronic formulas. As such, it should be mastered.

SELF-TEST

1. A 8.2-kΩ resistor, a 6.8-kΩ resistor, and a 3.9-kΩ resistor are connected in parallel. The current through the 3.9-kΩ resistor is 4.8 mA. Find the current through the other two resistors.
2. A 100-kΩ resistor is in parallel with a 270-kΩ resistor. The total current through the pair is 32 μA. Find the current through the 270-kΩ resistor.
3. A 400-Ω resistor has a current of 0.3 A flowing through it. A resistor connected in parallel with this resistor has a current of 0.2 A flowing through it. What is its resistance?
4. The total resistance of a parallel circuit is 2.3-kΩ. A single resistor in this circuit has a resistance of 10-kΩ and a current of 13 mA. What is the total current of the circuit?
5. A parallel circuit is constructed from a 120-Ω, a 560-Ω, and a 470-Ω resistor. The total current for this circuit is 125 mA. Find the current through the 560-Ω resistor.

ANSWERS TO SELF-TEST

1. I_{8200} = 2.283 mA and I_{6800} = 2.753 mA
2. 8.649 μA   3. 600 Ω 4. 56.52 mA   5. 18.23 mA

LOADED VOLTAGE DIVISION

Electronic circuits sometimes require less than the total amount of voltage available. A volume control on a radio is an example of this kind of requirement. One does not listen to a radio at full volume at all times. The volume control is a voltage divider, similar to the ones discussed previously. Consider the circuit shown in Figure 8-8.

Most circuits will operate differently when they have a load to drive. In general, any load will pull the normal operating voltage down. That is why some circuits cannot be treated like the voltage dividers you have seen previously. The following examples calculate the voltage across points A and B in Figure 8-8 with and without the load connected.

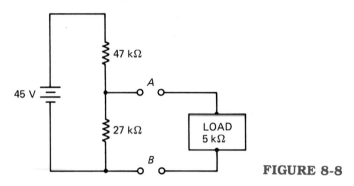

FIGURE 8-8

EXAMPLE 8-10

Find the unloaded voltage drop (the voltage drop when the 5-kΩ load is removed) from point A to point B in Figure 8-8.

Solution: Using the unloaded voltage formulas, you can find the unloaded voltage drop from point A to point B.

$$\frac{V_1}{V_T} = \frac{R_1}{R_T}$$ Inserting values

$$\frac{V_1}{45} = \frac{27{,}000}{27{,}000 + 47{,}000}$$ Cross multiplying

$$V_1 = \frac{27{,}000}{27{,}000 + 47{,}000} \times 45$$

$$= 16.42 \text{ V}$$ ∎

The voltage across points A and B would be 16.42 V. This would be the voltage before the 5-kΩ load was attached. After the load is attached, this voltage will drop drastically.

EXAMPLE 8-11

Find the loaded (the voltage drop when the 5-kΩ parallel resistor is connected to the circuit) voltage drop from point A to point B in the circuit shown in Figure 8-8.

Solution: In analyzing this circuit you must consider the 5-kΩ load in parallel with the 27-kΩ resistor. This parallel resistance would be

$$R_1 \| R_2 = \frac{1}{1/R_1 + 1/R_2}$$

$$= \frac{1}{1/5000 + 1/27{,}000}$$

$$= 4219 \ \Omega$$

and the loaded voltage across points A and B would be

$$\frac{V_1}{V_T} = \frac{R_1}{R_T} \qquad \text{Inserting values}$$

$$\frac{V_1}{45} = \frac{4219}{4219 + 47{,}000} \qquad \text{Cross-multiplying}$$

$$V_1 = \frac{4219}{4219 + 47{,}000} \times 45$$

$$= 3.707 \text{ V} \qquad\qquad\qquad\qquad ■$$

Notice that the loaded and unloaded voltages are not the same. It is possible for the load to pull the voltage down to a level that will not allow the load to operate. This is sometimes seen when using a low-resistance (less than 20,000 Ω/V) voltmeter.

SUMMARY

In summary, series circuits are sometimes called voltage dividers. They divide the supply voltage between the resistors as IR drops. The amount of the IR drop is directly proportional to the resistance of the resistor. Typically, the following formula is used to predict the voltage dropped across a resistor:

$$V = \frac{R}{R_T} \times V_T$$

Parallel circuits are sometimes referred to as current dividers. They divide the main-line current between the resistors as branch currents. The size of each branch current is inversely proportional to the resistance of that branch resistor. Typically, the following formula is used to predict the current flowing through the resistor:

$$I = \frac{R_T}{R} \times I_T$$

QUESTIONS

8-1. Describe the concept of directly proportional.

8-2. To what does the term *voltage divider* refer?

8-3. What advantages do the voltage division formulas provide?

8-4. Discuss the concept of inversely proportional.

8-5. To what does the term *current divider* refer?

8-6. What advantages do the current divider formulas provide?

8-7. The current divider formula is sometimes written as $I = G/G_T \times I_T$. Discuss how this is equivalent to the formula $I = R_T/R \times I_T$.

8-8. Explain how the voltage divider formulas could be used to find the voltage across two resistors. Then refer to Figure 8-4 and find the voltage across the combination of the 5.6-kΩ resistor and the 3.9-kΩ resistor.

8-9. In a two-resistor parallel circuit, it is the opposite resistor that is used to calculate the resistor current. Explain why. Remember that current is inversely proportional.

8-10. In a series circuit, V/R is used as a ratio. In a parallel circuit, I/G is used as a ratio. Relate these to the series and parallel circuit concepts studied earlier.

PRACTICE PROBLEMS

The voltage divider formulas are used most often to find the voltage of one resistor in a two-resistor series circuit. This applies primarily to solving transistor circuits. It will be beneficial for you to learn to apply this voltage divider formula. Here are some problems that will give you the necessary practice. All of the problems refer to a two-resistor series circuit.

8-1. $V_T = 20$ V
$R_1 = 4.7$ kΩ
$R_2 = 33$ kΩ
$V_1 = $ _____

8-2. $V_T = 30$ V
$R_1 = 5.6$ kΩ
$R_2 = 8.2$ kΩ
$V_1 = $ _____

8-3. $V_T = 20$ V
$R_1 = 4.7$ kΩ
$R_2 = 2.7$ kΩ
$V_1 = $ _____

8-4. $V_T = 50$ V
$R_1 = 4.7$ kΩ
$R_2 = 8.2$ kΩ
$V_2 = $ _____

8-5. $V_T = 40$ V
$R_1 = 47$ kΩ
$R_2 = 33$ kΩ
$V_2 = $ _____

8-6. $V_T = 30$ V
$R_1 = 390$ kΩ
$R_2 = 470$ kΩ
$V_2 = $ _____

8-7. $V_T = 25$ V
$R_1 = 2.7$ kΩ
$R_2 = 3.9$ kΩ
$V_2 = $ _____

8-8. $V_T = 45$ V
$R_1 = 1.2$ kΩ
$R_2 = 3.9$ kΩ
$V_2 = $ _____

8-9. $V_T = 20$ V
$R_1 = 6.8$ kΩ
$R_2 = 1.2$ kΩ
$V_1 = $ _____

8-10. Prove that for a two-resistor parallel circuit

$$I_1 = \frac{R_T}{R_1} \times I_T$$

is equivalent to

$$I_1 = \frac{R_2}{R_1 + R_2} \times I_T$$

The two-resistor current divider formula is also used a great deal when working with transistors. For that reason it is important for you to be familiar with its use. Here are some problems to give you the necessary practice. All of these circuits refer to a two-resistor parallel circuit similar to the one shown in Figure 8-7.

8-11. $I_T = 120$ mA
$R_1 = 8.2$ kΩ
$R_2 = 3.9$ kΩ
$I_1 = $ _____

8-12. $I_T = 22.5$ mA
$R_1 = 120$ kΩ
$R_2 = 270$ kΩ
$I_1 = $ _____

8-13. $I_T = 60$ mA
$R_1 = 330\ \Omega$
$R_2 = 560\ \Omega$
$I_2 = $ _____

8-14. $I_T = 100$ mA
$R_1 = 5.6$ kΩ
$R_2 = 1.2$ kΩ
$I_1 = $ _____

8-15. $I_T = 45$ mA
$R_1 = 27$ kΩ
$R_2 = 3.9$ kΩ
$I_2 = $ _____

8-16. $I_T = 1.25$ A
$R_1 = 20\ \Omega$
$R_2 = 5.6\ \Omega$
$I_2 = $ _____

8-17. $I_T = 73$ mA
$R_1 = 2.7$ kΩ
$R_2 = 3.9$ kΩ
$I_1 = $ _____

8-18. $I_T = 22.5\ \mu$A
$R_1 = 560$ kΩ
$R_2 = 330$ kΩ
$I_2 = $ _____

8-19. $I_T = 120$ nA
$R_1 = 1.2$ MΩ
$R_2 = 0.39$ MΩ
$I_2 = $ _____

8-20. $I_T = 2.5$ A
$R_1 = 5.6\ \Omega$
$R_2 = 4.7\ \Omega$
$I_1 = $ _____

PRACTICE PROBLEMS

9

Network Theorems

Chapter objectives

After reading this chapter and answering the questions and problems, you should be able to:

- State Thévenin's rule, Norton's rule, the superposition theorem, and Millman's theorem, together with advantages and disadvantages of each.
- Apply Thévenin's rule, Norton's rule, the superposition theorem, and Millman's theorem to any circuit.
- Make conversions between Norton and Thévenin equivalent circuits.
- Explain load matching as it pertains to source and load impedance.
- Describe load matching as it relates to voltage, current, and power transfer.
- Describe delta and wye circuits, and list their applications.
- Given any delta or wye network calculate values for the alternate circuit.

The vast majority of electronic circuits are really combinations of simpler circuits, or networks. In this chapter we discuss the major ways of viewing these circuit networks. Several of these theorems are used in the analysis of semiconductor circuits and, as such, need special attention. They will be noted in the discussion. Most of these theorems pertain to the combination of ac and dc signals when they occur at the same time in a circuit, and their respective composite result. It is important that you understand that although the circuits we are examining contain only dc voltage sources, these theorems can and will be applied to ac (alternating-current) circuits, as well as to hybrid combinations of ac and dc voltages in future courses.

SUPERPOSITION

Superposition is an important physical law that generally relates to the combining of water waves. The principle indicates that when two waves are moving in the same direction, their combination produces a larger wave. When they are moving in opposite directions, their result is a smaller wave.

In electronics, *superposition* describes the total as the sum of the individual results of each of the power sources. Consider the circuit shown in Figure 9-1.

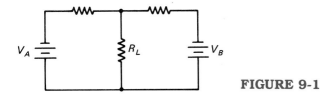

FIGURE 9-1

The superposition law indicates that the actual voltage across R_L is equal to the voltage across R_L if V_B were removed from the circuit and a wire inserted in its place, plus the voltage across R_L if V_B were then reinstalled into the circuit and V_A were removed and replaced with a conductor. To apply the superposition law, examine the effect that V_A (by itself) has on the circuit. Then examine the effect that V_B (by itself) has on the circuit. The actual circuit would have the sum of the effects produced by V_A and the effects produced by V_B. That is, the total voltage across R_L is a combination of the effects produced by V_A and V_B. One thing to remember when analyzing this circuit is to pay particular attention to the polarity of the voltage generated across R_L. In some cases these voltages will not have the same polarity. Example 9-1 shows this type of circuit.

Superposition Example with Opposing Voltages

EXAMPLE 9-1

Find the potential difference between points A and B of Figure 9-2 by using the superposition law.

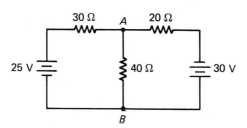

FIGURE 9-2

Solution: Superposition indicates that if you find the voltage produced by the 25-V battery (discounting any effect of the 30-V battery) and combine this voltage with the voltage produced by the 30-V battery (discounting any effect of the 25-V battery), the result will be the potential across points A and B.

Begin by redrawing the circuit, leaving out the 30-V battery (this battery is shorted out; see Figure 9-3). Analyze this circuit as any other series–parallel circuit. Sketch in arrows. This shows which components are in series and which are in parallel, but it also shows the polarity of the voltage. (Remember, current flows into the negative side of the component.) Find R_T.

$$20 \| 40 = 13.33 \ \Omega$$
$$R_T = 13.33 + 30$$
$$= 43.33 \ \Omega$$

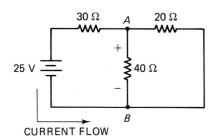

FIGURE 9-3

Solve for the voltage across points A and B.

$$\frac{V}{25} = \frac{13.33}{43.33}$$

$$V = \frac{13.33}{43.33} \times 25$$

$$= 7.692 \ V$$

Therefore, the voltage across points A and B is +7.692 V with respect to point B. (The current enters point B. Negative grounds have positive voltages.)

Redraw the original circuit again, leaving out the 25-V battery this time (Figure 9-4). Sketch in arrows and find R_T.

$$30 \| 40 = 17.14 \ \Omega$$
$$R_T = 17.14 + 20$$
$$= 37.14 \ \Omega$$

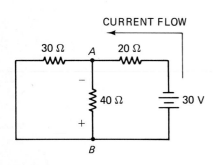

FIGURE 9-4

Solve for the voltage across points A and B.

$$\frac{V}{30} = \frac{17.14}{37.14}$$

$$V = \frac{17.14}{37.14} \times 30$$

$$= 13.85 \text{ V}$$

Therefore, the voltage produced by this battery is −13.85 V with respect to point B. (The current enters point A, making point B positive. Positive grounds have negative voltages.)

Combining the results of the two sets of calculations provides

$$+7.692 + (-13.85) = -6.158 \text{ V}$$

This is the voltage with respect to point B, which means that the combined voltage has the same polarity as that of the single voltage in Figure 9-4, the second sketch. ∎

Required Superposition Steps Superposition can be used to find the total voltage or current effect any time there is more than one voltage source present in a network. The basic steps are:

1. Remove all except one voltage source from the circuit being analyzed. Replace the others with a short (good conductor).
2. Find the effect of this one source.
3. Repeat this procedure until the single effects of each source are known.
4. Combine these single effects to find the result of all the sources.

Superposition is especially useful in the analysis of communication circuits, where there are both ac and dc components to analyze. Superposition allows us to view the ac (alternating current is discussed in a later chapter) portion separately from the dc portion and when needed, combine these effects to see the total effect of the circuit. Although this is not the fastest way to analyze dc circuits containing more than one voltage source, it uses the simplest concepts, concepts that have been discussed in earlier chapters.

SELF-TEST

1. A dual-voltage circuit of the type shown in Figure 9-1, has three branches. The first has a 4.7-kΩ resistor in series with 50-V supply. This supply has a positive ground. The second branch has a 3.9-kΩ resistor and a 8.2-kΩ resistor in series with each other. The third branch has a 2.7-kΩ resistor and a 25-V battery in series. This voltage also has a positive ground. Use superposition to find the voltage across the 3.9-kΩ resistor and the 8.2-KΩ resistor. Include a sketch of the circuit, including the polarity of these voltages.
2. Refer to the circuit described in problem 1. Reverse the polarity of the 50-V supply. Again find the voltage across the 3.9-kΩ resistor and the 8.2-kΩ resistor. Include a sketch of the circuit, showing the polarity of these voltages.
3. A 470-kΩ resistor and a 390-kΩ resistor are connected in series. A dual-voltage source powers these two resistors. The positive terminal of the 15-V source is connected to the 470-kΩ resistor. The negative terminal is connected to ground. The negative terminal of a 10-V source is connected to the 390-kΩ resistor. The positive terminal is connected to ground. Use superposition to

find the voltage from the connection between these two resistors to ground. Include a sketch of this circuit and show the polarity to ground.

4. Refer to the circuit described in problem 3. Reverse the polarity of the 15-V source. Use superposition to find the potential from the connection between the two resistors and ground. Include a sketch of the circuit which shows the polarity of this voltage.

5. A dual-voltage circuit of the type shown in Figure 9-1 has four branches. The first contains a 100-V battery (positive ground) in series with a 20-kΩ resistor. The second and third contain a 47-kΩ resistor and a 56-kΩ resistor, respectively. The fourth is made up of a series connection of a 27-kΩ resistor and a 150-V battery (positive ground). Use superposition to find the voltage across the 47-kΩ resistor.

ANSWERS TO SELF-TEST

1. Combined −29.89 V
2. 2.071 V
3. 1.337 V
4. −12.27 V
5. −83.65 V

THÉVENIN'S RULE

Thévenin's rule simply stated is: *Any complex network can be reduced to a single voltage source* (V_{Th}), *a current-limiting resistor* (R_{Th}), *and the load*. Thévenin is used in the analysis of semiconductor circuits, where this rule is often combined with the superposition theorem to form an ac model of a transistor circuit. This is especially useful when testing the performance capabilities of particular loads (speakers and antennas). It allows us to view the performance of several different loads in a particular circuit and choose the best load for our purposes without having built the actual circuit.

Procedure There are four basic steps that allow us to Thévenize any circuit.

1. Open the circuit by removing the load (the component around which you are Thévenizing).
2. Find the Thévenin voltage (the voltage across the opened portion of the circuit).
3. Short all of the voltage sources (examine the circuit as if all of the voltage sources were removed and replaced with good conductors).
4. Find the Thévenin resistance (the resistance as seen by the load).

Several examples of this process follow. Also several helpful hints are discussed which are meant to provide you with knowledge of common errors made in Thévenizing circuits. Consider the series–parallel circuit shown in Figure 9-5.

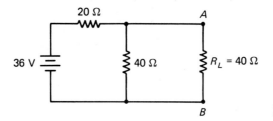

FIGURE 9-5

Simple Thévenin Analysis

EXAMPLE 9-2

Thévenize the circuit shown in Figure 9-5.

Solution:

1. *Remove the load and redraw the circuit.* The load in this case is the 40-Ω resistor located between points A and B on the schematic. This is shown in Figure 9-6.

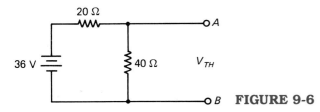

FIGURE 9-6

2. *Find the voltage across the open.* This will require analyzing a circuit; however, the circuit that will be analyzed at this step will be simpler than the original circuit. Any method can be used to find the voltage between points A and B. In this example the voltage divider formula will be used to find the voltage.

$$V_{Th} = \frac{40}{20 + 40} \times 36$$
$$= 24 \text{ V}$$

3. *Redraw the circuit one more time.* This time all the voltage sources are to be shorted out or simply removed from the circuit. This is shown in Figure 9-7.

4. *Find the Thévenin resistance.* Of all of the steps, this is the one that causes the most confusion. The key point to remember about finding the Thévenin resistance is that you are not interested in how much resistance the source would see. You are interested in the amount of resistance the load would see. You are pretending that the load is supplying the current for the circuit in this step. Looking back into the circuit from points A and B, you see the 20-Ω resistor in parallel with the 40-Ω resistor. The Thévenin resistance would be

$$R_{Th} = 20 \parallel 40$$
$$= 13.33 \text{ Ω}$$

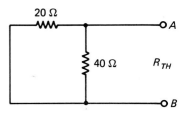

FIGURE 9-7

The Thévenin model for this circuit is shown in Figure 9-8. ∎

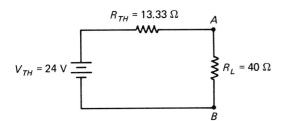

FIGURE 9-8

The 40-Ω load in Example 9-2 can be replaced by any other load and in minutes the performance of the two loads can be compared. The Thévenin equivalent circuit contains the circuit parameters as seen by the load. The maximum voltage that the load can ever expect is the Thévenin voltage. The maximum current the load can ever expect is the Thévenin voltage divided by the Thévenin resistance. These two parameters (voltage and current) are the only two factors that the circuit itself has any control of or effect upon.

When you examine the circuit and find V_{Th} and R_{Th}, you are determining these two factors. V_{Th} is also called the open-circuit voltage. *Open-circuit voltage* is the maximum voltage that can ever be generated between two points. R_{Th} is a current-limiting device. Any battery or power supply has a limit to the amount of current that it can provide. This current is limited by the series resistance of the circuit. In this case the series resistance is the Thévenin resistance of the circuit.

Two more examples follow. Each demonstrates how to avoid a particular mistake that is often made. The first demonstrates the effect of resistance in series with the load. Notice that they do not have any effect on the Thévenin voltage. They are seen only when calculating the Thévenin resistance.

Resistances in Series with the Load Do Not Affect the Thévenin Voltage Beginning students have trouble with resistances in series with the load. These resistances have no effect on the Thévenin voltage. It is difficult for students to realize this fact. Example 9-3 will demonstrate the concept.

EXAMPLE 9-3

Thévenize the circuit shown in Figure 9-9.

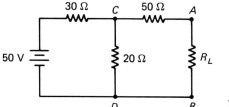

FIGURE 9-9

Solution:

1. *Redraw the circuit of Figure 9-9 with the load removed.* This is shown in Figure 9-10.

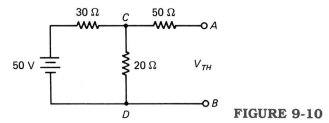

FIGURE 9-10

2. *Find the Thévenin voltage.* The Thévenin voltage of this circuit is rather hard for most students to understand at first. Since you have opened the circuit at points A and B, there is no current through the 50-Ω resistor and no resultant voltage drop. In other words, the voltage felt at points A and B is the same as the amount felt at points C and D. The Thévenin voltage would be

$$V_{Th} = \frac{20}{20 + 30} \times 50$$

$$= 20 \text{ V}$$

3. *Redraw the circuit with the voltage source shorted.* This is shown in Figure 9-11.

4. *Find the Thévenin resistance.* Looking back into the circuit from points A and B, you see the 50-Ω resistor in series with the parallel combination of the 30-Ω and 20-Ω resistors, and R_{Th} is

$$30 \parallel 20 = 12 \ \Omega$$

$$R_{Th} = 12 + 50 = 62 \ \Omega$$

The Thévenin equivalent circuit would be as shown in Figure 9-12. ∎

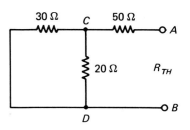

FIGURE 9-11

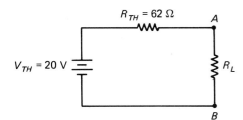

FIGURE 9-12

Resistances in Parallel with the Battery Do Not Affect the Thévenin Resistance The final example of Thévenin is shown in Figure 9-13 and an explanation of it follows. It demonstrates the effect of resistances in parallel with the battery on Thévenin's rule. Notice that these parallel resistances have no effect on the Thévenin resistance.

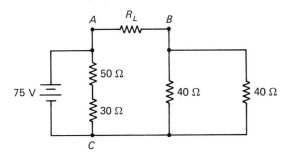

FIGURE 9-13

EXAMPLE 9-4

Thévenize the circuit shown in Figure 9-13.

Solution:

1. *Redraw the circuit without the load.* This is shown in Figure 9-14.

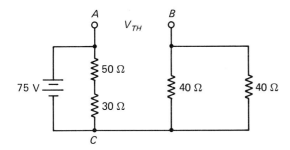

FIGURE 9-14

2. *Find the Thévenin voltage.* This appears to be quite difficult at first; however, since the circuit is open at points A and B, the two 40-Ω resistors do not drop any voltage. (There is no current through these resistors. Without current there is no voltage drop.) Therefore, the Thévenin voltage is 75 V. It

will take some thought to fully understand this. Remember that an open-series circuit will drop a voltage equal to the supply voltage.

3. *Redraw the circuit with the battery shorted.* This is shown in Figure 9-15.

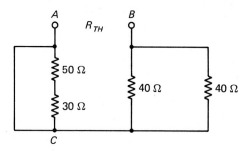

FIGURE 9-15

4. *Find the Thévenin resistance.* Again this seems to be quite difficult, but a careful examination of the circuit reveals that when the battery was shorted, it also shorted out the resistance path from A to C. In other words, the 50-Ω and 30-Ω resistors do not affect the Thévenin resistance. R_{Th} is just the parallel combination of the two 40-Ω resistors, or 20-Ω. The Thévenin equivalent circuit is shown in Figure 9-16. ∎

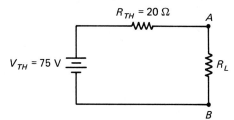

FIGURE 9-16

SELF-TEST

1. Thévenize the circuit shown in Figure 9-17 using R_3 as the load resistor. Draw the equivalent circuit and find the load voltage and current when R_3 is 12, 40, and 100 Ω.

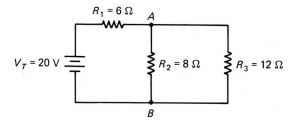

FIGURE 9-17

2. Thévenize the circuit shown in Figure 9-18 using R_2 as the load resistor. Draw the equivalent circuit and find the load voltage and current.

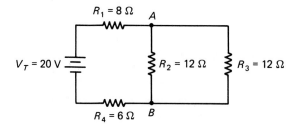

FIGURE 9-18

3. Thévenize the circuit shown in Figure 9-19 using R_4 as the load resistor. Draw the equivalent circuit and find the load voltage and current.

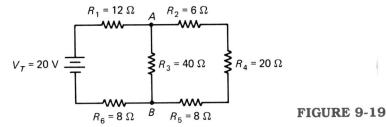

FIGURE 9-19

4. Thévenize the circuit shown in Figure 9-20 using R_2 as the load resistor. Draw the equivalent circuit and find the load voltage and current.

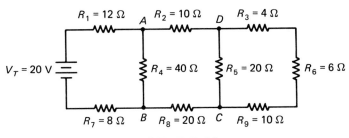

FIGURE 9-20

5. Thévenize the circuit shown in Figure 9-2 using the 40-Ω resistor as the load. Draw the equivalent circuit and find the load voltage and current.

ANSWERS TO SELF-TEST

1. $V_{Th} = 11.43$ V, $R_{Th} = 3.429$ Ω, $I_{L12} = 740.7$ mA, $V_{L12} = 8.889$ V, $I_{L40} = 263.2$ mA, $V_{L40} = 10.53$ V, $I_{L100} = 110.5$ mA, $V_{L100} = 11.05$ V
2. $V_{Th} = 9.231$ V, $R_{Th} = 6.462$ Ω, $I_L = 500$ mA, $V_L = 6$ V
3. $V_{Th} = 13.33$ V, $R_{Th} = 27.33$ Ω, $I_L = 281.7$ mA, $V_L = 5.634$ V
4. $V_{Th} = 13.33$ V, $R_{Th} = 43.33$ Ω, $I_L = 250$ mA, $V_L = 2.5$ V
5. $V_{Th} = 8$ V, $R_{Th} = 12$ Ω, $I_L = 153.8$ mA, $V_L = 6.154$ V

NORTON'S RULE

Like Thévenin's rule, Norton's rule is used primarily in the analysis of semiconductor circuits. Norton can either be used instead of, or in conjunction with, Thévenin. They are equivalent methods for accomplishing the same tasks. Where Thévenin views everything from the viewpoint of the voltage provided the load, Norton views things with respect to the current provided to the load.

Simply stated, Norton's rule says: *Any complex network can be reduced to a single current source (I_N) in parallel with a current-limiting resistor (R_N) and a load (R_L).*

Procedure Like Thévenin, Norton has four steps that allow us to Nortonize any circuit.

1. Short the circuit by removing the load (the component around which you are Nortonizing) and replacing it with a good conductor,

2. Find the Norton current (the current through the shorted portion of the circuit).
3. Short all the voltage sources (examine the circuit as if all the voltage sources were removed and replaced with good conductors).
4. Find the Norton resistance (the resistance as seen by the load).

Simple Norton Analysis

EXAMPLE 9-5

Nortonize the circuit shown in Figure 9-21.

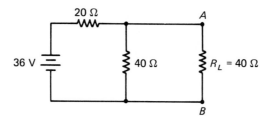

FIGURE 9-21

Solution

1. *Remove the load and redraw the circuit with the load replaced by a good conductor* (Figure 9-22).

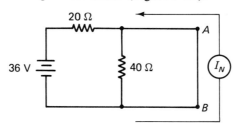

FIGURE 9-22

2. *Find the current flow through the short*. This current is the Norton current. Examining the circuit you will see that the 40-Ω resistor has been shorted out by the shorting of the load resistance, leaving the 36-V battery and the 20-Ω resistor in control of the current provided.

$$I_N = \frac{36}{20} = 1.8 \text{ A}$$

3. *Redraw the circuit one more time* (Figure 9-23). This time all of the voltage sources are to be shorted out or simply removed from the circuit. The third and forth steps are the same whether you are Thévenizing or Nortonizing a circuit.

4. *Look back into the circuit from the load's point of view to find the amount of resistance seen by the load*. Looking into points A and B, we see the 20-Ω resistor in parallel with the 40-Ω resistor.

$$R_N = 20 \parallel 40$$
$$= 13.33 \text{ Ω}$$

The Norton equivalent circuit would look like Figure 9-24. ■

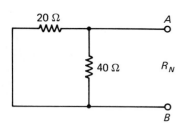

FIGURE 9-23

Thévenin and Norton Comparison Comparing the Norton equivalent to the Thévenin equivalent of the same circuit leads to two very interesting points.

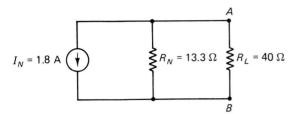

FIGURE 9-24

First, the Thévenin and Norton resistances are identical. They are always identical because they are found in the same manner.

$$R_{Th} = R_N$$

Second, Ohm's law can be used to find the Thévenin voltage or Norton current if the opposite rule was applied, That is,

$$I_N \times R_{Th} = V_{Th}$$
$$13.33 \times 1.8 = 24 \text{ V}$$

The Norton current times the Norton resistance will provide the Thévenin voltage. From Thévenin's rule, this would be that the Thévenin voltage divided by the Thévenin resistance yields the Norton current.

$$I_N = \frac{V_{Th}}{R_{Th}}$$

This is important in that it will allow you to choose the easier of the two methods to use for simplification. Which is easier? That depends on the circuit. There is no clearcut method to determine which is easier. In part, the person performing the task will have a personal preference. In general, Thévenin will be easier to perform on a circuit that is primarily series, while Norton is faster on circuits that are mostly parallel branches. Consider the circuit shown in Figure 9-25.

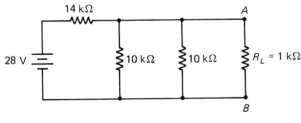

FIGURE 9-25

Norton Is More Effective on a Parallel Circuit

EXAMPLE 9-6

Use Norton's rule or Thévenin's rule to simplify the circuit shown in Figure 9-25.

Solution: This is primarily a parallel circuit. It will be easier to Nortonize this circuit. Prove it to yourself. Try to Thévenize it. Then examine the Nortonization that follows.

1. *Redraw the circuit with the load shorted* (Figure 9-26).

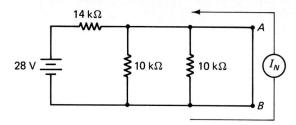

FIGURE 9-26

2. *Find the current through the short.* A careful examination of this circuit will reveal that the two 10-kΩ resistors are being shorted by the short across points A and B, so

$$I_N = \frac{28}{14{,}000} = 2 \text{ mA}$$

3. *Redraw the circuit with the voltage sources shorted* (Figure 9-27).

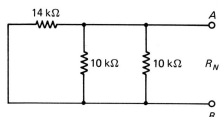

FIGURE 9-27

4. *Find the Norton resistance.* Look back into the circuit from points A and B. The two 10-kΩ resistors are in parallel with the 14-kΩ resistor. The Norton resistance would be

$$R_N = 10{,}000 \parallel 10{,}000 \parallel 14{,}000$$
$$= 3684 \text{ } \Omega$$
∎

In summary, Norton and Thévenin are equivalent ways of viewing circuits. They are also interchangeable and by using Ohm's law, it is possible to use a Nortonization as a Thévenin circuit, and vice versa. In general, if a circuit is predominantly series, use Thévenin; if a circuit is primarily parallel, use Norton.

Implications of Thévenin include the concept of low source resistance being preferred when dealing with a voltage source, while Norton implies that a high source resistance is needed when working with a current source. Both of these rules are used extensively in circuits that contain semiconductors.

SELF-TEST

1. Nortonize the circuit shown in Figure 9-17 using R_3 as the load resistor. Draw the equivalent circuit and find the load voltage and current when R_3 is 12, 40, and 100 Ω.

2. Nortonize the circuit shown in Figure 9-18 using R_2 as the load resistor. Draw the equivalent circuit and find the load voltage and current.

3. Nortonize the circuit shown in Figure 9-19 using R_4 as the load resistor. Draw the equivalent circuit and find the load voltage and current.

4. Nortonize the circuit shown in Figure 9-20 using R_2 as the load resistor. Draw the equivalent circuit and find the load voltage and current.

5. Nortonize the circuit shown in Figure 9-2 using the 40-Ω resistor as the load. Draw the equivalent circuit and find the load voltage and current.

ANSWERS TO SELF-TEST

1. $I_N = 3.333$ A, $R_N = 3.429$ Ω, $I_{L12} = 740.7$ mA, $V_{L12} = 8.889$ V, $I_{L40} = 263.2$ mA, $V_{L40} = 10.53$ V, $I_{L100} = 110.5$ mA, $V_{L100} = 11.05$ V
2. $I_N = 1.429$ A, $R_N = 6.462$ Ω, $I_L = 500$ mA, $V_L = 6$ V
3. $I_N = 487.7$ mA, $R_N = 27.33$ Ω, $I_L = 281.7$ mA, $V_L = 5.634$ V
4. $I_N = 307.6$ mA, $R_N = 43.33$ Ω, $I_L = 250$ mA, $V_L = 2.5$ V
5. $I_N = 666.7$ mA, $R_N = 12$ Ω, $I_L = 153.8$ mA, $V_L = 6.154$ V

SOURCE IMPEDANCE AND LOAD MATCHING

As suggested by Thévenin's and Norton's rule, it is the relative size of the source impedance (resistance) when compared to the load resistance that determines how well current, voltage, and power are transferred from the source to the load. In general, a supply will transmit voltage to the load better when the source resistance is low compared to the load resistance. This type of source is called a voltage source. Example 9-7 is a voltage source.

EXAMPLE 9-7

A 10-V source has a 100-Ω source impedance. Find the load voltage for each of the following load resistances: 10, 100, and 1000 Ω. Which has the largest voltage transfer?

Solution: The circuit above is describing a series circuit containing a 10-V battery, a 100-Ω resistor, and the load. The voltage divider formula can be used to find the load voltage for each load.

10-Ω load:
$$V_L = \frac{10}{100 + 10} \times 10$$
$$= 0.9091 \text{ V}$$

100-Ω load:
$$V_L = \frac{100}{100 + 100} \times 10$$
$$= 5. \text{ V}$$

1000-Ω load:
$$V_L = \frac{1000}{100 + 1000} \times 10$$
$$= 9.091 \text{ V}$$

As can be seen by the calculations, when the source resistance (100 Ω) was small compared to the load resistance (1000 Ω), most of the source voltage was transmitted to the load. When the source resistance (100 Ω) was large compared to the load resistance (10 Ω), very little of the source voltage was transmitted to the load. ∎

When the source resistance is large compared to the load resistance, more of the source current is transmitted to the load. That is, when you want most of the source current to flow through the load, you need to have a load resistance that is small compared to the source resistance. This type of source is called a *current source*. Example 9-8 is a current source.

EXAMPLE 9-8

A 10-V battery has a source resistance of 100 Ω and can produce a maximum of 0.1 A. Find the load current for each of the following loads: 10, 100, and 1000 Ω. Which had the largest current transfer?

Solution: Notice that the maximum current occurs when the battery and the source are alone in the circuit.

$$\frac{10}{100} = 0.1 \text{ A}$$

To find the load current for each of the loads treat the circuit as a series circuit consisting of the battery, source resistance and the load. Then:

10-Ω load:

$$I_L = \frac{10}{100 + 10}$$

$$= 0.09091 \text{ A}$$

100-Ω load:

$$I_L = \frac{10}{100 + 100}$$

$$= 0.05 \text{ A}$$

1000-Ω load:

$$I_L = \frac{10}{100 + 1000}$$

$$= 0.009091 \text{ A}$$

Notice that the largest current transfer occurred when the source resistance (100 Ω) was large when compared to the load resistance (10 Ω). ∎

Power is best transmitted when the source and the load resistance are equal. Consider Table 9-1. It demonstrates when voltage, current, and power are best transmitted. There will be times when you are working with electronic circuitry that it will be necessary for you to transmit voltage. At other times it will be necessary for you to transmit current or power. It is important that you remember how source and load resistances compare for each different transfer. In general, if you do not know whether a voltage or current transfer is preferred, use a power transfer. A power transfer provides the best combination of voltage and current transfer.

TABLE 9-1 Most Favorable Transmission Points

	Maximum	10 Ω	100 Ω	1000 Ω
Current (A)	0.1	0.09	0.05	0.009
Voltage (V)	10	0.9	5	9
Power (W)	1	0.081	0.25	0.081

Also, for good voltage or current transfer, the load and source should differ by at least a factor of 10. This means that if the source has a resistance of 250 Ω, the load should have a resistance of at least 2.5 kΩ for good voltage transfer. For good current transfer, the load should have a maximum of 25 Ω of resistance for the same 250 Ω of source resistance.

SELF-TEST

1. An audio technician has a choice of a 300-, 500-, and 1000-Ω microphone. Which would give the best voltage transfer? (The microphone is a source.)
2. An amplifier has an input resistance of 500 Ω. Which of the microphones listed in question 1 would provide the best power transfer? (The amplifier is a load.)
3. An amplifier has an output resistance of 8 Ω. If a 4-Ω speaker was used, would there be an increase in output current, voltage, or power? (The amplifier is the source and the speakers are the load.)
4. A current amplifier would have a high or a low input resistance. (The amplifier is a load.)
5. A voltage amplifier would have a high or a low output impedance. (The amplifier is a source.)
6. A microphone has specifications of maximum signal strength = 100 mV (V_{Th}) and output impedance = 1500 Ω. (R_{Th}). If this microphone were driving an amplifier whose input impedance is 10 kΩ (R_L), what is the maximum signal (V_L) that the amplifier will receive?

ANSWERS TO SELF-TEST

1. The 300-Ω microphone
2. *The* 500-Ω microphone
3. The current would increase.
4. A current amplifier would have low input resistance. (The amplifier is a load.)
5. A voltage amplifier would have low output impedance. (The amplifier is a source.)
6. V_L = 86.96 mV

MILLMAN'S THEOREM

Probably the least used of all the network theorems is Millman's. As with several of the nontraditional theorems, if there wasn't an immediate use for them, they were overlooked. In electronics today you must be concerned with two needs. One is speed. The other is adaptability to computers. With the rising cost of technical labor, it is necessary for the technician to get answers fast. Also in many cases a computer can be programmed to give answers as correct as the data that is fed to it. Millman's theorem provides both of these needs. It gives a quick answer and at the same time it can be applied to a program that will allow a computer to solve the circuit for you. This section is devoted to an explanation of Millman's theorem.

Millman's theorem was originally applied to the dual-voltage source circuits viewed earlier. Millman views each branch of such a circuit as a current generator. The sum of the currents provided by each current generator is then divided by the sum of the conductance (reciprocals of the source resistance) of each of the current generators. In other words, Millman's method is done in four steps.

1. Find the total current generated by each branch.
2. Find the total current produced by the generators.
3. Find the total conductance.
4. Divide the total current by the total conductance.

Millman Applied to a Dual-Voltage Circuit

EXAMPLE 9-9

Consider the circuit shown in Figure 9-28. Use Millman's theorem to find V_{AB}.

Solution: When you Millmanize a circuit you must begin by picking the points about which you are going to apply the theorem. In the circuit shown in Figure 9-28, you are going to work about points A and B. Examining point A, you will see that there are three branches or paths to point B. One is through the 25-V battery and the 30-Ω resistor. Another is through the 40-Ω resistor in the center. The last is through the 30-V battery and the 20-Ω resistor. These three paths are viewed as current generators. You must find the current generated by each as well as the conductance of each.

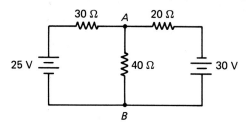

FIGURE 9-28

Path 1:

$$I_{\text{generator 1}} = \frac{25}{30}$$

$$= 0.8333 \text{ A}$$

$$G_{\text{generator 1}} = \frac{1}{30}$$

$$= 0.03333 \text{ ℧}$$

Path 2:

$$I_{\text{generator 2}} = \frac{0}{40}$$

$$= 0 \text{ A}$$

Note: Since there was no battery in this branch there was no current generated.

$$G_{\text{generator 2}} = \frac{1}{40}$$

$$= 0.025 \text{ ℧}$$

Path 3:

$$I_{\text{generator 3}} = -\frac{30}{20}$$

$$= -1.5 \text{ A}$$

Note: Since this voltage has a positive ground, it is negative.

$$G_{\text{generator 3}} = \frac{1}{20}$$

$$= 0.05 \, \mho$$

The total generator current is found as well as the total conductance. The total current is divided by total conductance to get the voltage across points A and B.

Totals:

$$I_{\text{total generator}} = 0.8333 + 0 - 1.5$$

$$= -0.6667 \, \text{A}$$

$$G_{\text{total generator}} = 0.3333 + 0.05 + 0.025$$

$$= 0.1083 \, \mho$$

$$V_{AB} = \frac{-0.6667}{0.1083}$$

$$= -6.154 \, \text{V}$$

This is the voltage across points A and B. ∎

A formula that demonstrates this process mathematically would be

$$V_{AB} = \frac{V_1/R_1 + V_2/R_2 + V_3/R_3}{1/R_1 + 1/R_2 + 1/R_3}$$

This formula can be applied to any circuit. The only requirement is imagination on the part of the operator. Several examples follow, together with an explanation of each.

Millman Applied to Other Circuits

Form 1

EXAMPLE 9-10

Use Millman's theorem to find the voltage across points A and B in Figure 9-17.

Solution: This is one of the easier applications of Millman's theorem. There are three current generators to examine. The 6 Ω of R_1 and the 20 V of the supply form one generator. The other two will generate no current. They are the 8 Ω of R_2 and the 12 Ω of R_3, respectively. Using the formula for Millman's theorem provides the following result.

$$V_{AB} = \frac{20/6 + 0/8 + 0/12}{1/6 + 1/8 + 1/12}$$

$$= 8.889 \, \text{V}$$

∎

Form 2

EXAMPLE 9-11

Use Millman's theorem to find the voltage across points *A* and *B* in Figure 9-18.

Solution: In this application it is necessary to first combine the resistances of R_1 and R_4. They are in series. Here again, there are three current generators, but only one actually produces current. That one is the combination of the 8 Ω of R_1 and the 6 Ω of R_4, which is driven by the source voltage of 20 V. Using Millman's formula provides

$$V_{AB} = \frac{20/(6+8) + 0/12 + 0/12}{1/(6+8) + 1/12 + 1/12}$$

$$= 6 \text{ V} \qquad \blacksquare$$

Form 3

EXAMPLE 9-12

Use Millman's theorem to find the voltage across points *A* and *B* in Figure 9-19.

Solution: This application requires the prior combination of R_1 and R_6. These resistors are in series. Also it is necessary to combine R_2, R_4, and R_5. They are in series also. This circuit also has three current generators, and the use of Millman's equations should be obvious.

$$V_{AB} = \frac{20/(12+8) + 0/40 + 0/(6+20+8)}{1/(12+8) + 1/40 + 1/(6+20+8)}$$

$$= 9.577 \text{ V} \qquad \blacksquare$$

In conclusion, Millman is a powerful theorem. It has been overlooked in the past. Using the examples will help you to realize the importance of this theorem.

SELF-TEST

1. Use Millman's formula to find the voltage across the load resistor in Figure 9-21.
2. Use Millman's formula to find the voltage across the 1-kΩ resistor in Figure 9-25.
3. Use Millman's formula to find the voltage across the 20-Ω resistor in Figure 9-9. The load resistance is 10 Ω.
4. Use Millman's formula to find the voltage across the 40-Ω resistor in Figure 9-2.
5. Use Millman's formula to find the load voltage in Figure 9-5 if the load resistance were changed to 60 Ω.

ANSWERS TO SELF-TEST

1. 18 V
2. 1.573 V
3. 16.67 V
4. −6.154 V
5. 19.63 *V*

DELTA AND WYE CIRCUITS

Delta and wye circuits are used primarily in the winding of generator and motor fields. They are used to provide multiphase ac electricity. The two circuits shown in Figure 9-29 can be equivalent, in that for every delta circuit there is an equivalent wye circuit. The values of the resistors used in the two equivalent circuits will differ, however.

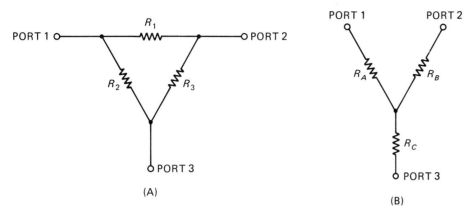

FIGURE 9-29

Delta and wye configurations are also used in the design of filters. In this case the delta configuration is usually referred to as a pi configuration while the wye is called a tee configuration. Whether the application is as a field winding or as a filter network, it is often necessary to make a conversion from a delta network to its equivalent wye network, or vice versa.

Delta and wye conversions can also be used in the simplification of bridge circuits and other complex circuitry. One such complex circuit is a box constructed of equal resistors (see Figure 9-35). Repeated delta-to-wye conversions and wye-to-delta conversions will provide a solution. There are better ways of solving these complex types of circuits. They are discussed in a later chapter.

Conversion from Delta to Wye Formulas for delta-to-wye conversions follow.

$$R_A = \frac{R_1 \times R_2}{R_1 + R_2 + R_3}$$

$$R_B = \frac{R_1 \times R_3}{R_1 + R_2 + R_3}$$

$$R_C = \frac{R_2 \times R_3}{R_1 + R_2 + R_3}$$

Notice that R_A is connected to port 1, as are R_1 and R_2. To make a conversion from delta to wye, first look to see to which port the wanted wye resistor will be connected. Then multiply the two delta resistors which also connect to this port:

$$R_1 \times R_2$$

and divide their product by the sum of all the delta resistors.

$$\frac{R_1 \times R_2}{R_1 + R_2 + R_3}$$

Conversion from Wye to Delta Formulas for wye-to-delta conversion follow.

$$R_1 = \frac{R_A \times R_B + R_B \times R_C + R_A \times R_C}{R_C}$$

$$R_2 = \frac{R_A \times R_B + R_B \times R_C + R_A \times R_C}{R_B}$$

$$R_3 = \frac{R_A \times R_B + R_B \times R_C + R_A \times R_C}{R_A}$$

Notice here that you are using the sum of the cross products in the wye as a dividend:

$$R_A \times R_B + R_B \times R_C + R_A \times R_C$$

and the opposite resistor as the divisor. These formulas are developed by the use of Kirchhoff's laws. An example of each of these conversions follows. They will be demonstrated using the pi and tee configurations.

EXAMPLE 9-13

Use delta-to-wye conversions to prove that the circuit in Figure 9-30 is equivalent to the circuit in Figure 9-31.

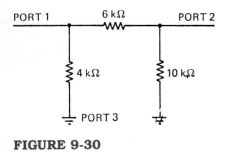

FIGURE 9-30

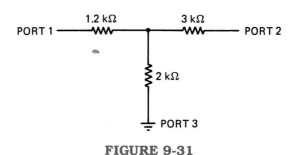

FIGURE 9-31

Solution: The circuit in Figure 9-30 is a pi network. You must convert it to an equivalent tee network.

$$R_A = \frac{4000 \times 6000}{4000 + 6000 + 10{,}000}$$

$$= 1200 \, \Omega$$

$$R_B = \frac{6000 \times 10{,}000}{4000 + 6000 + 10{,}000}$$

$$= 3000 \, \Omega$$

$$R_C = \frac{4000 \times 10{,}000}{4000 + 6000 + 10{,}000}$$

$$= 2000 \ \Omega$$

Since these are the same values as shown in Figure 9-31, the two circuits must be equivalent. ∎

EXAMPLE 9-14

Use wye-to-delta conversions to prove that the circuit in Figure 9-31 is equivalent to the circuit in Figure 9-30.

Solution: The circuit in Figure 9-31 is a tee configuration. You must convert it to an equivalent pi configuration.

$$R_1 = \frac{1200 \times 3000 + 3000 \times 2000 + 1200 \times 2000}{2000}$$

$$= 6000 \ \Omega$$

$$R_2 = \frac{1200 \times 3000 + 3000 \times 2000 + 1200 \times 2000}{3000}$$

$$= 4000 \ \Omega$$

$$R_3 = \frac{1200 \times 3000 + 3000 \times 2000 + 1200 \times 2000}{1200}$$

$$= 10{,}000 \ \Omega$$

Since these are the same values as those listed in Figure 9-30, the two circuits are equivalent. ∎

SELF-TEST

Convert the following delta configurations to their equivalent wye configurations.

1. $R_1 = 560 \ \Omega$
   $R_2 = 950 \ \Omega$
   $R_3 = 390 \ \Omega$

2. $R_1 = 1.5 \ k\Omega$
   $R_2 = 2.2 \ k\Omega$
   $R_3 = 2.7 \ k\Omega$

3. $R_1 = 33 \ k\Omega$
   $R_2 = 27 \ k\Omega$
   $R_3 = 39 \ k\Omega$

4. $R_1 = 820 \ \Omega$
   $R_2 = 680 \ \Omega$
   $R_3 = 1 \ k\Omega$

Convert the following wye configurations to their equivalent delta configurations.

5. $R_A = 270 \ \Omega$
   $R_B = 390 \ \Omega$
   $R_C = 470 \ \Omega$

6. $R_A = 1.5 \ k\Omega$
   $R_B = 2.7 \ k\Omega$
   $R_C = 3.3 \ k\Omega$

7. $R_A = 56 \ k\Omega$
   $R_B = 82 \ k\Omega$
   $R_C = 120 \ k\Omega$

8. $R_A = 1.2 \ k\Omega$
   $R_B = 5.6 \ k\Omega$
   $R_C = 3.9 \ k\Omega$

ANSWERS TO SELF-TEST

1. $R_A = 280\ \Omega$, $R_B = 114.9\ \Omega$, $R_C = 195\ \Omega$
2. $R_A = 515.6\ \Omega$, $R_B = 632.8\ \Omega$, $R_C = 928.1\ \Omega$
3. $R_A = 9000\ \Omega$, $R_B = 13\ \text{k}\Omega$, $R_C = 10.64\ \text{k}\Omega$
4. $R_A = 223\ \Omega$, $R_B = 328\ \Omega$, $R_C = 272\ \Omega$
5. $R_1 = 884\ \Omega$, $R_2 = 1.065\ \text{k}\Omega$, $R_3 = 1.539\ \text{k}\Omega$
6. $R_1 = 5.427\ \text{k}\Omega$, $R_2 = 6.633\ \text{k}\Omega$, $R_3 = 11.94\ \text{k}\Omega$
7. $R_1 = 176.3\ \text{k}\Omega$, $R_2 = 258\ \text{k}\Omega$, $R_3 = 377.7\ \text{k}\Omega$
8. $R_1 = 8.523\ \text{k}\Omega$, $R_2 = 5.936\ \text{k}\Omega$, $R_3 = 27.7\ \text{k}\Omega$

SUMMARY

In conclusion, there are many theorems that allow you to analyze network circuitry. Each of these is of special importance. Superposition allows you to separate signal levels when working with ac and dc voltages. A combination of an ac and a dc voltage is correctly termed fluctuating dc voltage. Thévenin's and Norton's rules allow you to solve a circuit looking at things from the load's point of view. Doing this will allow you to simplify the circuit down to its equivalent form where you can substitute differing loads and choose the best load for overall circuit operation. Millman's theorem is an overlooked equation that can be applied to computer programs. Delta and wye conversions must be used when dealing with voltage generation. They also play a role in circuit filtration.

QUESTIONS

9-1. List and describe the steps necessary to applying the superposition theorem.
9-2. When would you apply the superposition theorem?
9-3. Why is it necessary to understand the superposition theorem?
9-4. When applying the superposition theorem, why do we consider only one source at a time?
9-5. Why is polarity important when the superposition theorem is used?
9-6. List and describe each of the steps required to Thévenize a circuit.
9-7. List and describe each of the steps required to Nortonize a circuit.
9-8. To what does the term *open-ciruit voltage* relate?
9-9. To what does *short-circuit current* relate?
9-10. Explain in your own words, why the Thévenin resistance and the Norton resistance are equal.
9-11. Discuss the application of Thévenin's and Norton's rules. Include how to determine which would provide the simplest solution.
9-12. What can be said of resistors that are in series with the voltage source and the load? Explain.
9-13. What can be said of resistors in parallel with the voltage source? Explain.
9-14. List and describe each of the steps required to apply Millman's formula to a circuit.
9-15. Where might one find a delta configuration?

9-16. Why is a delta configuration sometimes called a pi configuration?
9-17. Describe the conversion of a delta configuration to a wye configuration.
9-18. What is meant by the term *cross product*?
9-19. Describe the conversion of a wye configuration to a delta configuration.
9-20. Define what is meant by a network circuit; a network equation.

PRACTICE PROBLEMS

9-1. The circuit shown in Figure 9-32 is constructed using 2-kΩ resistors. Find the Thévenin resistance looking into points *A* and *B*.

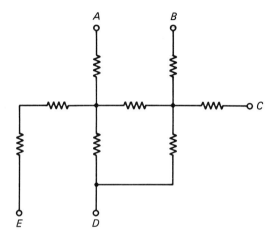

FIGURE 9-32

9-2. All of the resistors in Figure 9-32 are 5.6-kΩ resistors. Find the Thévenin resistance looking into points *D* and *E*.

9-3. All of the resistors in Figure 9-32 are 8.2-kΩ resistors. Find the Thévenin resistance looking into points *C* and *E*.

9-4. All of the resistors in Figure 9-32 are 10-kΩ resistors. Find the Thévenin resistance looking into points *C* and *D*.

9-5. Suppose that all the resistors in Figure 9-33 are 2.2-kΩ resistors. Find the Thévenin resistance looking into points *A* and *D*.

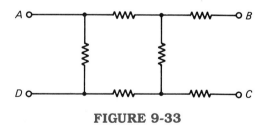

FIGURE 9-33

9-6. Repeat Problem 9-5 looking into points *B* and *C* of Figure 9-33.

9-7. Thévenize the circuit shown in Figure 9-34. Then find the load current and voltage for each of the following loads: 100 Ω, 200 Ω, 1.2 kΩ, 2.7 kΩ, 5.6 kΩ, and 10 kΩ.

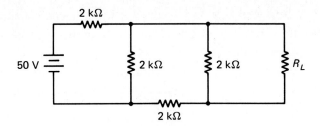

FIGURE 9-34

9-8. Nortonize the circuit shown in Figure 9-34. Then find the load current and voltage for each of the following loads: 500 Ω, 1.5 kΩ, 4.7 kΩ, and 15 kΩ.

9-9. Twelve 1-kΩ resistors are connected in such a manner that they form a box (see Figure 9-35). Use delta and wye conversions to find the total resistance between two opposing corners.

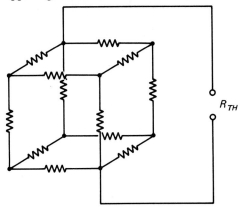

FIGURE 9-35

9-10. The input resistor of a tee configuration is 27 kΩ (port 1). The output resistor is 33 kΩ (port 2). The pole resistor is 56 kΩ (port 3). Find the values for the resistors needed to make an equivalent pi configuration.

10

Additional Methods of Analysis

Chapter objectives

After reading this chapter and answering the questions and problems, you should be able to:

- Given a set of linear equations containing two unknowns, correctly graph the straight lines represented by the equations.
- Given a system of simultaneous linear equations containing two or three unknowns, use linear combination to correctly solve for each unknown.
- State, in your own words, Kirchhoff's voltage and current laws and give an example of each.
- Use mesh analysis to write loop equations for any series–parallel circuit discussed in earlier chapters, as well as the dual-voltage circuits discussed in Chapter 9.

Electronics is a complex field and as such there are times when advanced mathematics is appropriate. This chapter is devoted to the simplest manner of writing simultaneous linear equations. Also, the text will demonstrate how to set up linear equations for electronic circuits. The use of linear combination in solving linear equations will be demonstrated. At first this method of solving equations may seem complicated, but it is the normal method of teaching how to solve linear equations. Also, it is the basis for most computer programs that solve linear equations. Anyone who plans to obtain a degree in electrical engineering needs to master linear equations because they are the basis for all engineering work.

GETTING STARTED: UNDERSTANDING LINEAR ALGEBRA

Linear algebra is used extensively in the solution of electrical circuits. The graph of a linear equation produces a straight line. That is where the name *linear* (meaning "line-like") equation comes from. Usually, a linear equation is written in the form

$$X + Y = V$$

Several examples of linear equations follow.

$$4X + 10Y = 20$$
$$-3X + 5Y = -3$$
$$10X - 25Y = 12$$

EXAMPLE 10-1

Graph the equation $4X + 10Y = 20$.

Solution: Begin by choosing a value for X and solving for Y. Usually, $X = 0$ will provide a simple solution.

$$4(0) + 10Y = 20$$
$$10Y = 20$$
$$Y = 2$$

So when $X = 0$, then $Y = 2$. Plot this point on the graph (Figure 10-1).

Then choose a value for Y and solve for X. Again, $Y = 0$ will usually provide a simple solution.

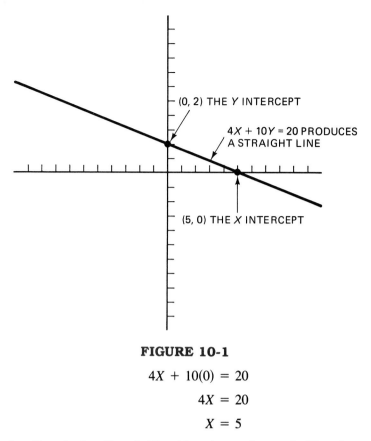

FIGURE 10-1

$$4X + 10(0) = 20$$
$$4X = 20$$
$$X = 5$$

So when $Y = 0$, then $X = 5$. Plot this point on the graph. Now draw a straight line through both of these points. ∎

Any value of X or Y on this straight line will provide a solution for this equation. When you are given a system of simultaneous linear equations, the solution for the set is the point where the lines intersect (cross). It is time consuming to use a graphic method to solve systems of linear equations, and the results are only approximations of the actual solution set. For that reason, linear algebra is most often used to find the solution of simultaneous linear equations.

Solving Linear Equations When solving simultaneous linear equations, remember that there must be at least one equation for each unknown. That is, if you are solving for X and Y (two unknowns), you must have at least two working equations that contain either X or Y, or both. There are several ways to solve simultaneous linear equations. In this book we demonstrate the use of linear combination. The process is demonstrated and explained in the the following paragraphs and examples.

EXAMPLE 10-2

Use linear combination to find the values of X and Y that will solve the following set of equations.

(1) $3X + 4Y = 20$

(2) $2X - 5Y = 12$

GETTING STARTED: UNDERSTANDING LINEAR ALGEBRA

Solution: First, examine the equations. Notice that the X terms form a column and the Y terms form a column in the equations shown above. The constants, 20 and 12, also form a column. The equations must be arranged in such a manner that each unknown forms its own column and each constant forms its own column.

Notice also that the first equation contains a $+4Y$ and the second equation contains a $-5Y$. It is important for the coefficients of these Y terms to have opposite signs. If the Y terms are both positive (or both negative), your next step must result with opposite signs for these terms. The idea is for the combination of these two equations to contain no Y terms. You can multiply or divide each equation in its entirety to provide this cancelation. We will divide. This is called *normalizing with respect to Y*.

$$(1) \quad \frac{3X}{4} + \frac{4Y}{4} = \frac{20}{4}$$

$$0.75X + Y = 5$$

Dividing the first equation by 4 gave us a single Y term in the result. That is what we want. To do the same in the second equation, we must divide by 5.

$$(2) \quad \frac{2X}{5} - \frac{5Y}{5} = \frac{12}{5}$$

$$0.4X - Y = 2.4$$

We want to have the Y terms of both equations equal and opposite in sign. The resulting set of equations is

$$(1) \quad 0.75X + Y = 5$$

$$(2) \quad 0.4X - Y = 2.4$$

Combine (add) these two equations.

$$1.15X = 7.4$$

$$X = \frac{7.4}{1.15}$$

$$= 6.435$$

The value for Y can be found in a similar manner. This time we normalize with respect to X.

$$(1) \quad 3X + 4Y = 20 \qquad (2) \quad 2X - 5Y = 12$$

$$\frac{3X}{3} + \frac{4Y}{3} = \frac{20}{3} \qquad \frac{2X}{-2} - \frac{5Y}{-2} = \frac{12}{-2}$$

$$X + 1.3333Y = 6.6667 \qquad -X + 2.5Y = -6$$

Combine these two equations to get

$$3.8333Y = 0.66667$$

$$Y = \frac{0.66667}{3.8333}$$

$$= 0.17391 \qquad \blacksquare$$

When working this type of problem, it is your choice whether to use multiplication or division to provide the cancelation. Example 10-3 demonstrates multiplication.

EXAMPLE 10-3

Find the solution for the following set of equations.

$$(1) \quad X - Y = 14$$
$$(2) \quad 10X + 5Y = -12$$

Solution: First, notice that there is no number in front of the X or Y in the first equation. Any time that X or Y is alone in an equation, the coefficient is understood to be 1. This problem would usually be solved in the following manner.

$$(1) \quad X - Y = 14 \qquad (2) \quad 10X + 5Y = -12$$
$$5(X) - 5(Y) = 5(14) \qquad 1(10X) + 1(5Y) = 1(-12)$$
$$5X - 5Y = 70 \qquad 10X + 5Y = -12$$

Combine.

$$15X = 58$$
$$X = \frac{58}{15}$$
$$= 3.8667$$

and

$$(1) \quad X - Y = 14 \qquad (2) \quad 10X + 5Y = -12$$
$$10(X) - 10(Y) = 10(14) \qquad -1(10X) + -1(5Y) = -1(-12)$$
$$10X - 10Y = 140 \qquad -10X - 5Y = +12$$

Combine.

$$-15Y = 152$$
$$= \frac{152}{-15}$$
$$= -10.133 \qquad \blacksquare$$

At times it is necessary to solve linear equations that have more than two unknowns. These are solved in a similar manner. The equations are still combined, two at a time, to provide a cancellation of one variable (unknown). This process is repeated until there is only one variable and its value is known. Remember that the set must have at least one equation per unknown. The process is demonstrated in Example 10-4.

EXAMPLE 10-4

Find the solutions for the following set of linear equations.

$$(1) \quad X + Y + 2Z = 0$$

(2) $2X - 2Y + Z = 8$

(3) $3X + 2Y + Z = 2$

Solution: Begin by combining the equations two at a time, removing the Z term from each combination.

Combination of (1) and (2):

(1) $\quad X + Y + 2Z = 0 \qquad$ (2) $\qquad 2X - 2Y + Z = 8$

$\quad -1(X) + -1(Y) + -1(2Z) = -1(0) \qquad 2(2X) - 2(2Y) + 2(Z) = 2(8)$

$\qquad\qquad -X - Y - 2Z = 0 \qquad\qquad\qquad 4X - 4Y + 2Z = 16$

Combine.

$$3X - 5Y = 16$$

Combination of (2) and (3)

(2) $\qquad 2X - 2Y + Z = 8 \qquad$ (3) $\qquad 3X + 2Y + Z = 2$

$\quad -1(2X) - -1(2Y) + -1(Z) = -1(8) \qquad 1(3X) + 1(2Y) + 1(Z) = 1(2)$

$\qquad\qquad -2X + 2Y - Z = -8 \qquad\qquad\qquad 3X + 2Y + Z = 2$

Combine.

$$X + 4Y = -6$$

These combinations are now used to solve for X and Y.

$$3X - 5Y = 16 \qquad X + 4Y = -6$$

$$\frac{3X}{5} - \frac{5Y}{5} = \frac{16}{5} \qquad \frac{X}{4} + \frac{4Y}{4} = \frac{-6}{4}$$

$$0.6X - Y = 3.2 \qquad 0.25X + Y = -1.5$$

Combine.

$$0.85X = 1.7$$

$$X = \frac{1.7}{0.85}$$

$$= 2$$

$$3X - 5Y = 16 \qquad\qquad X + 4Y = -6$$

$$\frac{3X}{-3} - \frac{5Y}{-3} = \frac{16}{-3} \qquad\qquad X + 4Y = -6$$

$$-X + 1.6667Y = -5.3333 \qquad X + 4Y = -6$$

Combine.

$$5.6667Y = -11.333$$

$$Y = \frac{-11.333}{5.6667}$$

$$= -2$$

The original equations are combined again. This time the X or the Y terms are removed. We will remove the X terms.

Combination of (1) and (2):

(1) $\quad X + Y + 2Z = 0 \qquad$ (2) $\quad 2X - 2Y + Z = 8$

$\quad -2(x) + -2(Y) + -2(Z) = 0 \qquad\qquad 2X - 2Y + Z = 8$

$\quad\quad -2X - 2Y - 4Z = 0 \qquad\qquad\quad 2X - 2Y + Z = 8$

Combine.

$$-4Y - 3Z = 8$$

Combination of (2) and (3):

(2) $\quad 2X - 2Y + Z = 8 \qquad$ (3) $\quad 3X + 2Y + Z = -2$

$\quad -3(2X) - -3(2Y) + -3(Z) = -3(8) \qquad 2(3X) + 2(2Y) + 2(Z) = 2(2)$

$\quad\quad -6X + 6Y - 3Z = -24 \qquad\qquad\quad 6X + 4Y + 2Z = 4$

Combine.

$$10Y - Z = -20$$

These combinations are now used to solve for z.

$$-4Y - 3Z = 8 \qquad 10Y - Z = -20$$

$$\frac{-4Y}{4} - \frac{3Z}{4} = \frac{8}{4} \qquad \frac{10Y}{10} - \frac{Z}{10} = \frac{-20}{10}$$

$$-Y - 0.75Z = 2 \qquad Y - 0.1Z = -2$$

Combine.

$$-0.85Z = 0$$
$$Z = 0$$

The solution set is $X = 2$, $Y = -2$, and $Z = 0$. ∎

SELF-TEST

1. $3X + 4Y = 15$
   $14X - 2Y = 32$

2. $5X - 3Y = 21$
   $X + 2Y = 10$

3. $\frac{X}{2} + \frac{Y}{4} = 2$
   $X - 2Y = 3$

4. $-2X - 3Y = 1$
   $X + 2Y = 0$

5. $2X - 3Y = -1$
   $X + 4Y = 5$

6. $3X - 4Y = -2$
   $X - 2Y = 0$

7. $3X - 4Y = -2$
   $6X + 12Y = 36$

8. $2X - 4Y = 7$
   $X + 2Y = 1$

9. $2X - 3Y = 0$
   $X - 2Y = 1$

10. $2X - 4Y = 3$
    $3X + 3Y = 3$

11. $2X - 3Y + Z = 3$
    $X - Y - 3Z = -1$
    $-X + 2Y - 3Z = -4$

12. $X + 5Y - Z = 2$
    $3X - 9Y + 3Z = 6$
    $X - 6Y + Z = 4$

GETTING STARTED: UNDERSTANDING LINEAR ALGEBRA

ANSWERS TO SELF-TEST

1. $X = 2.548, Y = 1.839$
2. $X = 5.538, Y = 2.231$
3. $X = 3.8, Y = 0.4$
4. $X = -2, Y = 1$
5. $X = 1, Y = 1$
6. $X = -2, Y = -1$
7. $X = 2, Y = 2$
8. $X = 2.25, Y = -0.625$
9. $X = -3, Y = -2$
10. $X = 1.167, Y = -0.1667$
11. $X = -6, Y = -5, Z = 0$
12. $X = 2.667, Y = -0.6667, Z = -2.667$

KIRCHHOFF'S LAWS

Kirchhoff's laws are used to implement equations whose solutions will describe the circuit operation. Kirchhoff has two basic laws, one that describes current and one that describes voltage. Both of these laws were discussed in part in an earlier chapter. They are discussed in detail here.

The Voltage Law *The algebraic sum of the voltages around any closed path is zero.* Here the term *algebraic* seems to cause the most confusion. Algebra is a form of mathematics that uses positive and negative numbers. When you sum the voltages around a closed path, it is necessary for you to designate the polarity of each voltage. From a component point of view, when current flows into the negative terminal and out of the positive terminal (the normal direction of current flow), the voltage is said to be *positive*. Understand that from a voltage source viewpoint, this would be backwards (current flows into the positive terminal and out of the negative terminal of the battery). How to determine the polarity of a voltage is demonstrated in Example 10-5.

EXAMPLE 10-5

Write the voltage equation, using Kirchhoff's voltage law, for the circuit shown in Figure 10-2.

Solution: Label the schematic with the polarities for each component. Remember that current will enter the negative terminal of a component and exit the positive. Then pick a point of reference. This is done by sketching an arrow in the circuit. It makes no difference which direction the arrow is drawn as it is used only for a point of reference. If the arrow enters the negative side of any component that component's voltage is positive. If the arrow enters the positive side of the component then that component's voltage is negative. Using a clockwise convention and starting at the battery, the voltage equation for the circuit shown in Figure 10-2 would be

$$-3 \text{ V} -5 \text{ V} -2 \text{ V} +10 \text{ V} = 0 \text{ V}$$

Using a counter clockwise direction the equation would be

$$+2 \text{ V} +5 \text{ V} +3 \text{ V} -10 \text{ V} = 0 \text{ V}$$

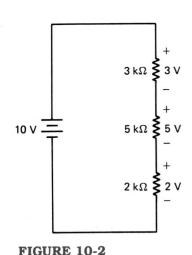

FIGURE 10-2

Notice this form of Kirchhoff's law relates to the voltage source (battery) as a component. Another form, the one used in an earlier chapter, relates to the voltage source as such. In that convention, the voltage drops are said to equal the voltage source and the equations above would be

$$-3\text{ V} -5\text{ V} -2\text{ V} = -10\text{ V}$$

for a clockwise direction and

$$+3\text{ V} +5\text{ V} +2\text{ V} = 10\text{ V}$$

for a counter clockwise direction.

This is an important concept and it is necessary for you to understand the difference.

The Current Law *The algebraic sum of the currents into and out of any point must equal zero.* Here again, the word *algebraic* refers to the use of positive and negative numbers in summing the currents. The currents flowing into a point are generally referred to as *positive* while the currents flowing out of a point are referred to as *negative*. Example 10-6 demonstrates how current equations are written.

EXAMPLE 10-6

Use Kirchhoff's current law to write equation for the circuit shown in Figure 10-3.

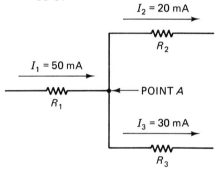

KIRCHHOFF'S CURRENT LAW   **FIGURE 10-3**

Solution: Examine the circuit. Here you will find 50 mA entering point A (the connection between R_1, R_2, and R_3). Notice that 20 mA leaves the point through R_2 and 30 mA leaves the point through R_3. The equation for this would be

$$+50\text{ mA} -20\text{ mA} -30\text{ mA} = 0\text{ mA}$$ ∎

Both of Kirchhoff's laws are used to write linear equations that describe circuit operation and will be demonstrated in the following sections.

MESH CURRENT ANALYSIS

Mesh current analysis is the most widely used method of circuit analysis. Examine Figure 10-4. The X current indicated in the sketch flows through the loop consisting of the 25-V battery, the 30-Ω resistor and the 40-Ω resistor. The Y current flows through the loop consisting of the 35-V battery, the 20-Ω resistor, and the 40-Ω resistor. A loop can be defined as any closed current path. The mesh occurs at the 40-Ω resistor. Notice that here (the 40-Ω resistor) the X and Y currents mesh (oppose each other) together like one gear driving another.

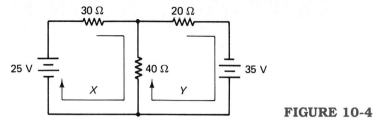

FIGURE 10-4

Procedure Mesh current analysis is done in four steps.

1. Sketch in the arrows.
2. Examine the circuit.
3. Write the equations.
4. Solve the equations.

Example 10-7 details these steps.

EXAMPLE 10-7

Use mesh current analysis to find the mesh current for the circuit shown in Figure 10-4.

Solution:

1. *Sketch in the arrows.* Mesh current analysis is started by drawing arrows in the circuit. Label one X and the other Y. Both arrows are to be drawn in the same direction (both clockwise or both counterclockwise). That way they will be going in opposite directions through the middle resistor (one up the other down). These arrows are thought of as currents and an equation is written for each.

2, 3. *Examine the circuit and write the equations.* The general form for the equations set is

$$X - Y = V$$
$$-X + Y = V$$

The X-path equation. Examine the circuit by considering only the portion of the circuit through which the X current flows. This portion consists of the 25-V battery, the 30-Ω resistor, and the 40-Ω resistor. The X current flows through both the 30- and 40-Ω resistors → $(30 + 40)X$. Of these three components the Y current flows through only the 40-Ω resistor → $40Y$. For this circuit the X current flows from the 25-V battery. The X current is also flowing against normal current flow (from negative to positive) for the 25-V battery. So when you use this voltage in your equation, it is -25 V. The voltage equation for the X current is

$$(30 + 40)X - 40Y = -25$$

Notice that you are multiplying resistance by current. Ohm's law indicates that this is a voltage. So you are applying Kirchhoff's voltage law.

The Y-path equation. Examine the circuit by considering only the portion of the circuit through which the Y current flows. This consists of the 35-V battery, the 20-Ω resistor, and the 40-Ω resistor. The Y current flows through both the 20- and 40-Ω resistors → $(20 + 40)Y$. Of these three components the X flows through only the 40-Ω resistor → $40X$. The Y current is going against the direction of normal current flow for the 35-V battery → −35. The equation for the Y current would be

$$-40X + (20 + 40)Y = -35$$

The set of equations would be

(1) $70X - 40Y = -25$

(2) $-40X + 60Y = -35$

4. *Solve the equations.* The X current is

(1) $70X - 40Y = -25$ (2) $-40X + 60Y = -35$

$$\frac{70X}{40} - \frac{40Y}{40} = -\frac{25}{40} \qquad -\frac{40X}{60} + \frac{60Y}{60} = -\frac{35}{60}$$

$$1.75X - Y = -0.625 \qquad -0.66667X + Y = -0.58333$$

Combine.

$$1.0833X = -1.2083$$

$$X = -\frac{1.2083}{1.0833}$$

$$= -1.1154 \text{ A}$$

This is the current through the 30-Ω resistor. Notice that linear combination produced a negative value for this current. The only thing the negative value indicates is that the current is actualy going in the direction opposite to the arrow labeled X. In the original schematic the X current was drawn in the clockwise direction; however, in the actual circuit current would really be flowing in a counterclockwise direction. This is indicated by the negative sign.

The Y current is

(1) $70X - 40Y = -25$ (2) $-40X + 60Y = -35$

$$\frac{70X}{70} - \frac{40Y}{70} = -\frac{25}{70} \qquad -\frac{40X}{40} + \frac{60Y}{40} = -\frac{35}{40}$$

$$X - 0.57143Y = -0.35714 \qquad -X + 1.5Y = -0.875$$

Combine.

$$0.92857Y = -1.2321$$

$$Y = -\frac{1.2321}{0.92857}$$

$$= -1.3269 \text{ A}$$

This is the current through the 20-Ω resistor. Here, once more, the negative sign indicates that clockwise is the wrong direction for this current.

To find the current through the middle resistor, simply subtract these two currents. Remember to pay close attention to the sign when you do so. Using the X portion as a reference, you have

$$-1.1154 - (-1.3269) = 0.21154 \text{ A}$$

This is the current through the 40-Ω resistor. ∎

There are three main things to remember about mesh current analysis.

1. The arrows are drawn in the same direction.
2. The general form for the equation set is

$$X - Y = V$$
$$-X + Y = V$$

3. When an arrow is going against current flow, the voltage is negative.

Also, you are not limited to performing this technique only on circuits containing two or more voltage sources. They work equally well on single-supply circuits. Examples of several circuit types and their corresponding equations follow.

Applications

Form 1

EXAMPLE 10-8

Use mesh current analysis to find the current through each of the resistors in Figure 10-5.

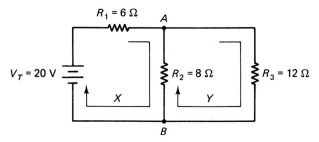

FIGURE 10-5

Solution: Begin by finding the equation set

(1) $14X - 8Y = -20$

(2) $-8X + 20Y = 0$

Then solve for X.

(1) $14X - 8Y = -20$ (2) $-8X + 20Y = 0$

$$\frac{14X}{8} - \frac{8Y}{8} = -\frac{20}{8} \qquad -\frac{8X}{20} + \frac{20Y}{20} = 0$$

$$1.75X - Y = -2.5 \qquad -0.4X + Y = 0$$

Combine.

$$1.35X = -2.5$$

$$X = -\frac{2.5}{1.35}$$

$$= -1.8519 \text{ A}$$

Now solve for Y.

$$\begin{array}{ll}(1) \quad 14X - 8Y = -20 & (2) \quad -8X + 20Y = 0 \\ \dfrac{14X}{14} - \dfrac{8Y}{14} = -\dfrac{20}{14} & -\dfrac{8X}{8} + \dfrac{20Y}{8} = 0 \\ X - 0.57143Y = -1.4286 & -X + 2.5Y = 0\end{array}$$

Combine.

$$1.9286Y = -1.4286$$
$$Y = -\dfrac{1.4286}{1.9286}$$
$$= -0.74074 \text{ A}$$

$$I_1 = X \qquad I_2 = X - Y \qquad I_3 = Y \qquad \blacksquare$$

Form 2

EXAMPLE 10-9

Use mesh current analysis to find the current through each of the resistors in Figure 10-6.

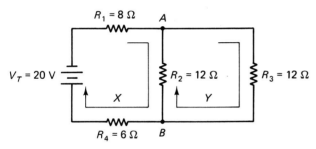

FIGURE 10-6

Solution: Begin by finding the equation set.

$$(1) \quad 26X - 12Y = 20$$
$$(2) \quad -12X + 24Y = 0$$

Then solve for X.

$$\begin{array}{ll}(1) \quad 26X - 12Y = 20 & (2) \quad -12X + 24Y = 0 \\ \dfrac{26X}{12} - \dfrac{12Y}{12} = \dfrac{20}{12} & -\dfrac{12X}{24} + \dfrac{24Y}{24} = 0 \\ 2.1667X - Y = 1.6667 & -0.5X + Y = 0\end{array}$$

Combine.

$$1.6667X = 1.6667$$
$$X = 1 \text{ A}$$

Now solve for Y.

$$(1) \quad 26X - 12Y = 20 \qquad (2) \quad -12X + 24Y = 0$$

$$\frac{26X}{26} - \frac{12Y}{26} = \frac{20}{26} \qquad -\frac{12X}{12} + \frac{24Y}{12} = 0$$

$$X - 0.46154Y = 0.76923 \qquad -X + 2Y = 0$$

Combine.

$$1.5385Y = 0.76923$$

$$Y = \frac{0.76923}{1.5385}$$

$$= 0.5 \text{ A}$$

$$I_1 = I_4 = X \qquad I_2 = X - Y \qquad I_3 = Y \qquad \blacksquare$$

Form 3

EXAMPLE 10-10

Use mesh current analysis to find the current through each of the resistors in Figure 10-7.

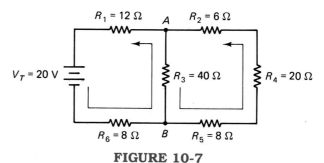

FIGURE 10-7

Solution: Begin by finding the equation set

$$(1) \quad 60X - 40Y = 20$$
$$(2) \quad -40X + 74Y = 0$$

Then solve for X.

$$(1) \quad 60X - 40Y = 20 \qquad (2) \quad -40X + 74Y = 0$$

$$\frac{60X}{40} - \frac{40Y}{40} = \frac{20}{40} \qquad -\frac{40X}{74} + \frac{74Y}{74} = 0$$

$$1.5X - Y = 0.5 \qquad -0.54054X + Y = 0$$

Combine.

$$0.95946X = 0.5$$

$$X = \frac{0.5}{0.95946}$$

$$= 0.52113 \text{ A}$$

Now solve for Y.

$$(1) \quad 60X - 40Y = 20 \qquad (2) \quad -40X + 74Y = 0$$

$$\frac{60X}{60} - \frac{40Y}{60} = \frac{20}{60} \qquad -\frac{40X}{40} + \frac{74Y}{40} = 0$$

$$X - 0.66667Y = 0.33333 \qquad -X + 1.85Y = 0$$

Combine.

$$1.1833Y = 0.33333$$

$$Y = \frac{0.33333}{1.1833}$$

$$= 0.28169 \text{ A}$$

$$I_1 = I_6 = X \qquad I_3 = X - Y \qquad I_2 = I_4 = I_5 = Y \qquad \blacksquare$$

Form 4

EXAMPLE 10-11

Use mesh current analysis to find the current through each of the resistors in Figure 10-8.

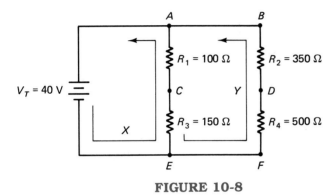

FIGURE 10-8

Solution: The equation set is

$$250X - 250Y = 40$$
$$-250X + 1100Y = 0$$

Individual currents are

$$I_1 = I_3 = X + Y$$
$$I_2 = I_4 = Y$$

The calculations are left for you. $\blacksquare$

Form 5

EXAMPLE 10-12

Use mesh current analysis to find the current through each of the resistors in Figure 10-9.

Solution: Begin by finding the equation set.

$$(1) \quad 250X - 100Y - 150Z = 40$$

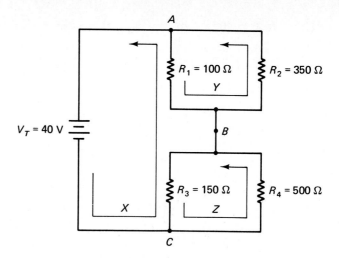

FIGURE 10-9

$$(2) \quad -100X + 450Y - 0Z = 0$$
$$(3) \quad -150X - 0Y + 650Z = 0$$

Then combine (1) and (3) to remove the Z current.

(1) $\quad 250X - 100Y - 150Z = 40 \qquad$ (3) $\quad -150X - 0Y + 650Z = 0$

$$\frac{250X}{150} - \frac{100Y}{150} - \frac{150Z}{150} = \frac{40}{150} \qquad -\frac{150X}{650} - 0Y + \frac{650Z}{650} = 0$$

$$1.6667X - .66667Y - Z = .26667 \qquad -0.23077X - 0Y + Z = 0$$

Combining gives

$$1.4359X - 0.66667Y = 0.26667$$

Use this with (2) to find X.

$1.4359X - 0.66667Y = 0.26667 \qquad$ (2) $\quad -100X + 450Y = 0$

$$\frac{1.4359X}{0.66667} - \frac{0.66667Y}{0.66667} = \frac{0.26667}{0.66667} \qquad -\frac{100X}{450} + \frac{450Y}{450} = 0$$

$$2.1538X - Y = 0.4 \qquad -0.22222X + Y = 0$$

Combined,

$$1.9316X = 0.4$$
$$X = \frac{0.4}{1.9316}$$
$$= 0.20708 \text{ A}$$

Then find Y.

$1.4359X - .66667Y = 0.26667 \qquad$ (2) $\quad -100X + 450Y = 0$

$$\frac{1.4359X}{1.4359} - \frac{0.66667Y}{1.4359} = \frac{0.26667}{1.4359} \qquad -\frac{100X}{100} + \frac{450Y}{100} = 0$$

$$X - .46429Y = 0.18571 \qquad -X + 4.5Y = 0$$

Combined,

$$4.0357Y = 0.18571$$

$$Y = \frac{0.18571}{4.0357}$$

$$= 0.046018 \text{ A}$$

Now combine (1) and (2) to remove the Y current.

(1) $\quad 250X - 100Y - 150Z = 40 \qquad$ (2) $\quad -100X + 450Y - 0Z = 0$

$\qquad \dfrac{250X}{100} - \dfrac{100Y}{100} - \dfrac{150Z}{100} = \dfrac{40}{100} \qquad\qquad -\dfrac{100X}{450} + \dfrac{450Y}{450} - 0Z = 0$

$\qquad\qquad 2.5X - Y - 1.5Z = 0.4 \qquad\qquad\qquad -0.22222X + Y - 0Z = 0$

Combined,

$$2.2778X - 1.5Z = 0.4$$

Combine this with (3) to find Z.

$\quad 2.2778X - 1.5Z = 0.4 \qquad\qquad$ (3) $\quad -150X - 0Y + 650Z = 0$

$\quad \dfrac{2.2778X}{2.2778} - \dfrac{1.5Z}{2.2778} = \dfrac{0.4}{2.2778} \qquad\qquad -\dfrac{150X}{150} + \dfrac{650Z}{150} = 0$

$\qquad\quad X - 0.65854Z = 0.175616 \qquad\qquad\quad -X + 4.3333Z = 0$

Combined,

$$3.6748Z = 0.17561$$

$$Z = \frac{0.17561}{3.6748}$$

$$= 0.047788$$

$$I_1 = X - Y \qquad I_2 = Y \qquad I_3 = X - Z \qquad I_4 = Z \qquad \blacksquare$$

Form 6

EXAMPLE 10-13

Use mesh current analysis to find the current through each of the resistors in Figure 10-10.

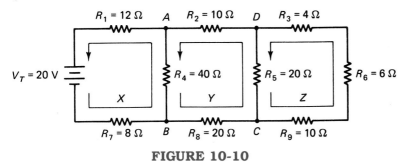

FIGURE 10-10

Solution: The equation set

(1) $\quad 60X - 40Y - 0Z = 20$

(2) $\quad -40X + 90Y - 20Z = 0$

$$(3) \quad -0X - 20Y + 40Z = 0$$

Begin by combining (2) and (3) to remove the Z current.

$$(2) \quad -40X + 90Y - 20Z = 0 \qquad (3) \quad -0X - 20Y + 40Z = 0$$

$$-\frac{40X}{20} + \frac{90Y}{20} - \frac{20Z}{20} = 0 \qquad\qquad -\frac{20Y}{40} + \frac{40Z}{40} = 0$$

$$-2X + 4.5Y - Z = 0 \qquad\qquad -0.5Y + Z = 0$$

Combined,

$$-2X + 4Y = 0$$

Use this with (1) to find X.

$$-2X + 4Y = 0 \qquad (1) \quad 60X - 40Y - 0Z = 20$$

$$-\frac{2X}{4} + \frac{4Y}{4} = 0 \qquad\qquad \frac{60X}{40} - \frac{40Y}{40} = \frac{20}{40}$$

$$-0.5X + Y = 0 \qquad\qquad 1.5X - Y = 0.5$$

Combined,

$$X = 0.5 \text{ A}$$

Then find Y.

$$-2X + 4Y = 0 \qquad (1) \quad 60X - 40Y - 0Z = 20$$

$$-\frac{2X}{2} + \frac{4Y}{2} = 0 \qquad\qquad \frac{60X}{60} - \frac{40Y}{60} = \frac{20}{60}$$

$$-X + 2Y = 0 \qquad\qquad X - 0.66667Y = 0.33333$$

Combined,

$$1.3333Y = 0.33333$$

$$Y = \frac{0.33333}{1.3333}$$

$$= 0.25 \text{ A}$$

Now combine (1) and (2) to remove X.

$$(1) \quad 60X - 40Y - 0Z = 20 \qquad (2) \quad -40X + 90Y - 20Z = 0$$

$$\frac{60X}{60} - \frac{40Y}{60} - 0Z = \frac{20}{60} \qquad\qquad -\frac{40X}{40} + \frac{90Y}{40} - \frac{20Z}{40} = 0$$

$$X - 0.66667Y - 0Z = 0.33333 \qquad\qquad -X + 2.25Y - 0.5Z = 0$$

Combined,

$$1.5833Y - 0.5Z = 0.33333$$

Use this with (3) to find the Z current.

$$1.5833Y - 0.5Z = 0.33333 \qquad (3) \quad -0X - 20Y + 40Z = 0$$

$$\frac{1.5833Y}{1.5833} - \frac{0.5Z}{1.5833} = \frac{0.33333}{1.5833} \qquad\qquad -\frac{20Y}{20} + \frac{40Z}{20} = 0$$

$$Y - 0.31580Z = 0.21053 \qquad\qquad -Y + 2Z = 0$$

Combined,

$$1.6842Z = 0.21053$$
$$Z = 0.125 \text{ A}$$

$$I_1 = I_7 = X \qquad I_4 = X - Y \qquad I_2 = I_8 = Y$$
$$I_5 = Y - Z \qquad I_3 = I_6 = I_9 = Z$$

∎

As has been demonstrated, mesh current analysis can be used to evaluate any of the circuits discussed so far. In fact, the rules (resistance additive in a series circuit; conductance additive in a parallel circuit; etc.) learned earlier were developed by using mesh current analysis.

SUMMARY

In summary, Kirchhoff's voltage and current laws are applied when writing linear equations. Linear equations are the basic tool of electrical engineers. Mesh current analysis is the method most used to write linear equations for electrical circuits. Mesh current analysis can be used to solve any circuit for voltage and current values. Some other important topics discussed in this chapter follow.

Currents can be added algebraically at any point in a circuit. Current going into the point is referred to as positive (current coming out of the point is referred as negative). The algebraic sum (positive and negative) of the currents into and out of a point must equal zero. This is Kirchhoff's current law.

Voltages can be added algebraically around any closed path in the circuit. The algebraic sum of these voltages must equal zero. This is Kirchhoff's voltage law.

Any closed current path can be referred to as a current loop. When loops are connected so that the currents mesh (like gears), we say they provide a mesh current.

Kirchhoff's voltage and current laws can be used to write linear equations for electrical circuits. Each mesh will provide a separate equation.

A linear equation, when graphed, produces a straight line.

When simultaneous (more than one occurring at the same time) linear equations are graphed, the intersection of the straight lines occurs at the solution values.

To find the solution values for simultaneous linear equations, you must have at least one equation for each unknown. Also, each equation must be unique (not just a multiple of another).

Linear combination (addition or subtraction of linear equations) is used to find the answers to simultaneous linear equations.

QUESTIONS

10-1. In your own words, define linear equation.

10-2. Briefly explain the use of linear combination in solving sets of linear equations.

10-3. Describe Kirchhoff's voltage law in your own words.

10-4. Refer to Figure 10-2. Use algebra to demonstrate that 3 V + 5 V + 2 V = 10 V (the IR drops equal the applied voltage) is the equivalent to 3 V +

5 V + 2 V − 10 V = 0 V (the algebraic sum of the voltages must equal zero).

10-5. Describe how an arrow can be used to reference polarity in a circuit.

10-6. In your own words, describe Kirchhoff's current law.

10-7. When using KCL, how does one determine which point to use?

10-8. Describe the manner the arrows are drawn in the circuit for mesh current analysis.

10-9. Why is the mesh current method of circuit analysis an application of Kirchhoff's voltage law?

10-10. Why do you think we must write separate equations for the X current and the Y current when performing current analysis?

10-11. When the solution of a circuit produces a negative answer, what is indicated?

10-12. Describe the method used to find other circuit currents after the X and Y currents have been found.

10-13. List the three points to remember when performing mesh current analysis.

10-14. What do you do when there is no voltage source in a current loop? How does this affect the loop equation?

10-15. After graphing two simultaneous linear equations, how do you find the solution set?

PRACTICE PROBLEMS

Refer to Figure 10-11. Use mesh current analysis to solve for circuit currents and voltages.

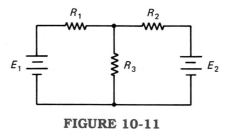

FIGURE 10-11

10-1. $E_1 = 20$ V
$E_2 = 14$ V
$R_1 = 20\ \Omega$
$R_2 = 33\ \Omega$
$R_3 = 18\ \Omega$

10-2. $E_1 = 25$ V
$E_2 = 10$ V
$R_1 = 2.2$ kΩ
$R_2 = 4.7$ kΩ
$R_3 = 5.6$ kΩ

10-3. $E_1 = 100$ V
$E_2 = 80$ V
$R_1 = 39$ kΩ
$R_2 = 47$ kΩ
$R_3 = 22$ kΩ

10-4. $E_1 = 15$ V
$E_2 = 20$ V
$R_1 = 100$ kΩ
$R_2 = 150$ kΩ
$R_3 = 220$ kΩ

Refer to Figure 10-12. Use mesh current to solve for circuit currents and voltages.

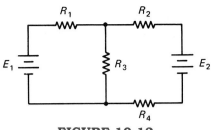
FIGURE 10-12

10-5. $E_1 = 15$ V
$E_2 = 20$ V
$R_1 = 39\ \Omega$
$R_2 = 10\ \Omega$
$R_3 = 18\ \Omega$
$R_4 = 22\ \Omega$

10-6. $E_1 = 25$ V
$E_2 = 20$ V
$R_1 = 5.6$ kΩ
$R_2 = 2.7$ kΩ
$R_3 = 1.5$ kΩ
$R_4 = 2.2$ kΩ

10-7. $E_1 = 10$ V
$E_2 = 8$ V
$R_1 = 39$ kΩ
$R_2 = 15$ kΩ
$R_3 = 22$ kΩ
$R_4 = 17$ kΩ

10-8. $E_1 = 25$ V
$E_2 = 20$ V
$R_1 = 330$ kΩ
$R_2 = 150$ kΩ
$R_3 = 220$ kΩ
$R_4 = 180$ kΩ

Refer to Figure 10-13. Use mesh current to solve for circuit currents and voltages.

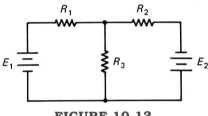
FIGURE 10-13

10-9. $E_1 = 20$ V
$E_2 = 14$ V
$R_1 = 20\ \Omega$
$R_2 = 33\ \Omega$
$R_3 = 18\ \Omega$

10-10. $E_1 = 25$ V
$E_2 = 10$ V
$R_1 = 2.2$ kΩ
$R_2 = 4.7$ kΩ
$R_3 = 5.6$ kΩ

10-11. $E_1 = 100$ V
$E_2 = 80$ V
$R_1 = 39$ kΩ
$R_2 = 47$ kΩ
$R_3 = 22$ kΩ

10-12. $E_1 = 15$ V
$E_2 = 20$ V
$R_1 = 100$ kΩ
$R_2 = 150$ kΩ
$R_3 = 220$ kΩ

Refer to Figure 10-14. Use mesh current to solve for circuit currents and voltages.

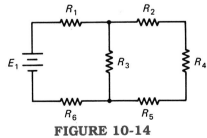
FIGURE 10-14

PRACTICE PROBLEMS

10-13. $E_1 = 20$ V
$R_1 = 20\ \Omega$
$R_2 = 33\ \Omega$
$R_3 = 82\ \Omega$
$R_4 = 27\ \Omega$
$R_5 = 20\ \Omega$
$R_6 = 15\ \Omega$

10-14. $E_1 = 25$ V
$R_1 = 2.2$ kΩ
$R_2 = 0.5$ kΩ
$R_3 = 5.6$ kΩ
$R_4 = 2.7$ kΩ
$R_5 = 2.7$ kΩ
$R_6 = 1.8$ kΩ

10-15. $E_1 = 100$ V
$R_1 = 39$ kΩ
$R_2 = 47$ kΩ
$R_3 = 68$ kΩ
$R_4 = 4.7$ kΩ
$R_5 = 3.3$ kΩ
$R_6 = 10$ kΩ

10-16. $E_1 = 15$ V
$R_1 = 100$ kΩ
$R_2 = 47$ kΩ
$R_3 = 220$ kΩ
$R_4 = 100$ kΩ
$R_5 = 27$ kΩ
$R_6 = 12$ kΩ

Refer to Figure 10-15. Use mesh current to solve for circuit currents and voltages.

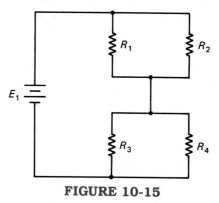

FIGURE 10-15

10-17. $E_1 = 20$ V
$R_1 = 20\ \Omega$
$R_2 = 33\ \Omega$
$R_3 = 18\ \Omega$
$R_4 = 18\ \Omega$

10-18. $E_1 = 25$ V
$R_1 = 2.2$ kΩ
$R_2 = 4.7$ kΩ
$R_3 = 5.6$ kΩ
$R_4 = 6.8$ kΩ

10-19. $E_1 = 100$ V
$R_1 = 39$ kΩ
$R_2 = 47$ kΩ
$R_3 = 22$ kΩ
$R_4 = 56$ kΩ

10-20. $E_1 = 15$ V
$R_1 = 100$ kΩ
$R_2 = 150$ kΩ
$R_3 = 220$ kΩ
$R_4 = 150$ kΩ

Refer to Figure 10-16. Use mesh current to solve for circuit currents and voltages.

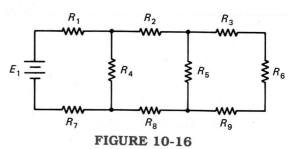

FIGURE 10-16

10-21. $E_1 = 20$ V
$R_1 = 200\ \Omega$
$R_2 = 330\ \Omega$
$R_3 = 180\ \Omega$
$R_4 = 270\ \Omega$
$R_5 = 560\ \Omega$
$R_6 = 220\ \Omega$
$R_7 = 150\ \Omega$
$R_8 = 180\ \Omega$
$R_9 = 200\ \Omega$

10-22. $E_1 = 25$ V
$R_1 = 2.2$ kΩ
$R_2 = 4.7$ kΩ
$R_3 = 5.6$ kΩ
$R_4 = 3.3$ kΩ
$R_5 = 3.9$ kΩ
$R_6 = 2.7$ kΩ
$R_7 = 1.8$ kΩ
$R_8 = 3.3$ kΩ
$R_9 = 4.7$ kΩ

10-23. $E_1 = 100$ V
$R_1 = 39$ kΩ
$R_2 = 47$ kΩ
$R_3 = 22$ kΩ
$R_4 = 56$ kΩ
$R_5 = 27$ kΩ
$R_6 = 10$ kΩ
$R_7 = 15$ kΩ
$R_8 = 22$ kΩ
$R_9 = 33$ kΩ

10-24. $E_1 = 15$ V
$R_1 = 100$ kΩ
$R_2 = 150$ kΩ
$R_3 = 220$ kΩ
$R_4 = 560$ kΩ
$R_5 = 120$ kΩ
$R_6 = 220$ kΩ
$R_7 = 330$ kΩ
$R_8 = 100$ kΩ
$R_9 = 220$ kΩ

PRACTICE PROBLEMS

11

Circuit Conductors and Controlling Devices

Chapter objectives

After reading this chapter and answering the questions and problems, you should be able to:

- List the advantages and disadvantages of each of the following metals as conductors: copper, silver, gold, aluminum, and Moleculoy.
- Define *resistivity*, *wire gauge*, *annealed*, *throw*, and *pole of switch*.
- Given the dimensions and resistivity (in Ω/cm or Ω cmils/ft), calculate the resistance of a material.
- Use tables to determine the resistivity, gauge diameter, resistance, and temperature coefficient of various metals.
- Identify large- or small-diameter wires by comparing wire gauge.
- State the meaning of positive-, negative-, and zero-temperature coefficients of metals.
- List several types of wire construction and the applications of each.
- List several types of switch construction and the operation of each.
- Describe the method used to rate fuses, circuit breakers, and fusible links.
- Describe various methods of testing fuses, circuit breakers, and fusible links.

Every circuit discussed so far has considered the circuit connectors as perfect, having no resistance. Unfortunately that is not the case. Every wire has resistance. In this chapter we discuss how to decide if this resistance is large enough to affect the overall circuit operation. Wire resistance is found by reading charts, so in this chapter we provide you with practice at reading charts. Also, circuits are not designed to be operational at all times. Switches are used to turn a portion, or all, of a circuit off or on. The various types of switches and their construction and use are discussed in this section. Last, but not least, since circuits do malfunction from time to time, we discuss the use and ratings of circuit protectors (fuses, circuit breakers, and fusible links) which protect the circuit in the event of a short circuit.

CONDUCTORS

There are several different metals that are used as conductors. They all have one thing in common, resistance. They all have a very small amount of resistance, unless specifically designed otherwise. Copper, gold, aluminum, and silver are all valued for their low resistance.

Copper. Copper is the material most often used as a conductor. This is due to its low resistance and relative low cost. Copper has become an industry standard, so much so that other materials are sometimes rated by their performance compared to that of copper. There are two types of copper: hard and soft. Soft copper is technically referred to as *annealed* copper.

Annealed copper has gone through a tempering process, in which it is heated and allowed to cool slowly. This leaves the copper softer and less brittle; it also has a brighter appearance and conducts electricity better. Annealed copper is used as a base when comparing the conductivity of two metals. For example, hard copper has only 97% of the conducting capabilities of annealed (soft) copper. Again, copper is a popular conductor because of its low resistance, second only to silver, and because of its cheapness compared to other metals.

Silver. Silver has the least resistance of all metals. It is used when it is absolutely necessary that resistance be kept at a minimum. Although silver has the lowest resistance, its resistivity is only slightly lower than that of copper. Nontempered silver has 98% of the conducting capabilities of annealed copper, while tempered silver has 102.8% of annealed copper's conducting abilities. In most instances this slight improvement does not warrant the extra expense of silver. In some of his early experiments, Thomas Edison used silver as a conductor. Airplanes sometimes use silver as a conductor. Finer stereo and television sets use silver. For the most part silver is "a money is no object" type of conductor.

Gold. Down through the ages, gold has been the most precious metal. One reason for this is that unlike other materials, gold does not tarnish. It is not subject to oxidation or rust. Gold is used as a conductor when corrosion must be minimized. The tempering process has no effect on gold. Gold does not have nearly the conducting ability of copper, in fact, only 73.4%. Its only advantage is that it will not be affected by other chemicals and as such will not loose its conducting abilities because of corrosion, which can be caused by circuit operation.

Aluminum. For many years aluminum was too expensive to make its use as a conductor worthwhile. In the last 20 years chemical processes for refining aluminum have made the use of aluminum more and more appealing. While aluminum has only approximately 60% of the conducting power of annealed copper, it also has only 30% of the weight of copper. So when a copper and aluminum wire of the same diameter are compared, the copper wire has 40% less resistance, but the aluminum wire has 70% less weight. The larger the diameter of a wire, the less resistance that wire has. Thus it is possible for a larger aluminum wire to have less resistance than a smaller copper wire and also for that same aluminum wire to weigh less. When weight is a factor aluminum is a good choice.

Moleculoy. Some metals are prized for their resistance. They are used to manufacture resistors and as resistance wires. Since the resistance of materials changes with temperature, it is necessary to use metals that have a stable resistance, one that is not affected by changes in temperature. Moleculoy is one such metal. It has a chemical composition of 20% chromium, 4% aluminum, 1% silicon, 1 to 5% manganese, and the balance nickel. This alloy has a resistance that changes only 5 to 10 parts per million per degree Celsius. It has almost 80 times the resistance of annealed copper, which changes 6800 parts per million per degree Celsius. In comparison, Moleculoy has considerably more resistance, which remains practically constant at all temperatures.

There are many, many metals and alloys that are used as conductors. Most are listed in the tables included in this chapter. The tables that follow and their explanations are meant to demonstrate the various scientific methods used to determine the resistivity, temperature coefficients, and size of wires.

RESISTIVITY

Resistivity is the amount of resistance in a solid (or wire). Resistivity is determined by the number of ohms contained in a cross section of the area of the solid. The specific resistance (another name for resistivity) is found by multiplying the resistance of the wire by its cross-sectional area and dividing this product by the wire's length (see Figure 11-1). As a result, the resistivity is given in units of ohm-centimeter (Ω-cm). The symbol for resistivity is ρ (the Greek lowercase letter rho). A list of popular metals and their respective resistivities is shown in Table 11-1. A careful inspection of the table will demonstrate that silver has the lowest resistivity, only slightly lower than that of copper. Aluminum and gold have about 50% more resistance than silver and copper. Most conductors are of a wire construction. Although the resistivity of wire is measured in a similar manner, there are some differences.

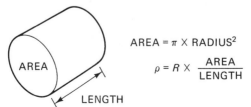

FIGURE 11-1

TABLE 11-1 Resistivity of Common Metals

Element	Resistivity (μΩ-cm)	Element	Resistivity (μΩ-cm)
Advance	48	Magnesium	4.6
Aluminum	2.828	Manganin	44
Brass	7.00	Molybdenum	5.7
Climax	87	Monel metal	42
Constantan	44.1	Nichrome	150
Copper	1.724	Nickel	7.8
Eureka	47	Platinum	10
Excello	92	Silver	1.629
German silver	33	Tantalum	15.5
Gold	2.44	Tin	11.5
Iron	10	Tungsten	5.51
Lead	22	Zinc	5.75

SELF-TEST

Use Table 11-1 to find the resistivity of each of the following.

1. Brass
2. Nickel
3. Iron
4. Aluminum
5. Copper
6. Silver
7. Gold
8. Nichrome
9. Manganin
10. Excello

ANSWERS TO SELF-TEST

1. 7.00
2. 7.80
3. 10
4. 2.828
5. 1.724
6. 1.629
7. 2.44
8. 150
9. 44
10. 92

RESISTIVITY OF WIRE

The diameter of a wire is measured in thousandths of an inch. Each thousandth of an inch is referred to as a *mil*. In other words, 1 mil = 0.001 inch. Naturally, wires are round. Rather than calculate the cross-sectional area of a wire in units of square inches, another unit was developed. That unit is a circular mil. A 0.001-in.-diameter wire has a cross-sectional area of 0.00000078539 in., but its circular mil area is 1 cmil. The circular mil area is figured by squaring the mil diameter.

EXAMPLE 11-1

Find the cmil area of a wire having a diameter of 5 mils.

Solution: The circular mil area is found by squaring the mil diameter of the wire.

$$5^2 = 25$$

This wire would have an area of 25 cmils.

As demonstrated in Example 11-1, this is an easy way of calculating the cross-sectional area of a piece of wire. The resistivity of a wire is then given in the units ohm cmils per foot. The conversion factor between Ω-cmils/ft and $\mu\Omega$-cm is 6.018, on average.

EXAMPLE 11-2

Given that a metal has a resistivity of 800 Ω-cmils/ft, find the $\mu\Omega$-cm resistivity of this metal.

Solution: To convert from Ω cmils/ft to $\mu\Omega$-cm, divide by 6.018.

$$\rho = \frac{800}{6.018}$$

$$= 133 \ \mu\Omega\text{-cm}$$

SELF-TEST

Find the resistivity (in Ω-cmils/ft) for each of the following materials.

1. Advance
2. Platinum
3. Lead
4. Copper
5. Manganin
6. Aluminum

ANSWERS TO SELF-TEST

1. 288.9 Ω-cmils/ft
2. 60.18 Ω-cmils/ft
3. 132.4 Ω-cmils/ft
4. 10.38 Ω-cmils/ft
5. 264.8 Ω-cmils/ft
6. 17.02 Ω-cmils/ft

In conductors resistance is generally thought to be bad; however, there are certain instances where resistance in wire can have its advantage. One such application is in making wirewound resistors. Wirewound resistors can dissipate more power than can the conventional carbon-film resistors. The majority of metals used as resistance wires are alloys, mixtures of several metal elements. Table 11-2 lists the most common metal alloys used as resistance wires, their composition, and resistivity in both $\mu\Omega$-cm and Ω cmils/ft.

TABLE 11-2 Resistance Wires

Material	Composition	Specifie Resistance $\mu\Omega$-cm	Ω-cmils/ft
Moleculoy	20 Cr, 4 Al, 1 Si, 1–5 Mn, bal Ni	133	800
Electroly	60 Ni, 15 Cr, bal Fe	112	675
Protoloy	80 Ni, 20 Cr	108	650
35–20 Ni chromium	35 Ni, 20 Cr, bal Fe	101.5	610
Neutroloy	55 Cu, 45 Ni	50.5	300
Manganin	12 Mn, 4 Ni, bal Cu	48	290
18% Nickel silver	55 Cu, 27 Zn, 18 Ni	31	190
180 Alloy	23 Ni, bal Cu	30	180
Pelcoloy	70 Ni, 30 Fe	19.9	120
90 Alloy	11 Ni, bal Cu	14.9	90
60 Alloy	6 Ni, bal Cu	10	60
Nickel 200	99.5 Ni	9.5	57
Nickel 270	99.9 Ni	7.5	45
30 Alloy	2 Ni, bal Cu	5	30

SELF-TEST

Refer to Table 11-2. Find the resistivity (in the specified unit) and composition for each of the following.

1. Manganin resistivity, in $\mu\Omega$-cm.
2. 35–20 nickel–chromium resistivity, in Ω-cmils/ft.
3. Moleculoy resistivity, in $\mu\Omega$-cm.
4. 18% nickel–silver resistivity, in Ω-cmils/ft.

ANSWERS TO SELF-TEST

1. 48 $\mu\Omega$-cm
2. 610 Ω-cmils/ft
3. 133 $\mu\Omega$-cm
4. 190 Ω-cmils/ft

The diameters of wires have been standardized into groupings called *gauge size*. Although in the past there have been numerous manufacturers, each with their own specifications, Brown and Sharpe has become what is termed American Wire Gauge, which is the standard for rating wire diameter. This system uses numbers to indicate a wire's diameter. The smaller the number, the larger the diameter of the wire. This inverse relationship seems to cause the most confusion about gauge size versus wire diameter. Table 11-3 shows the diameter and resistance of copper wire.

You will notice that the word "annealed" is used in this table. Copper has two forms, soft and hard. Annealed copper has gone through a special heating and cooling process, which has tempered it. It is the softer form of copper and can usually be identified by its bright, shiny appearance. Most electrical wire is made of annealed copper, because annealed copper has less resistance.

EXAMPLE 11-3

Use Table 11-3 to find the cmil area and the resistance of 460 ft of 30-gauge copper wire at 20°C.

Solution: Look down the column marked "gauge no." until you find the row listed as 30. Look across this row until you find the column marked "circular mil area." The cmil area is listed as 100.5 cmils.

Continue looking down this row until you find the column marked "resistance at 20°C" (68°F). That listing is 103.2 Ω. Notice that this is the resistance of 1000 ft of 30-gauge wire. Set up a ratio to find the resistance of 460 ft of this wire.

$$\frac{R}{460} = \frac{103.2}{1000}$$

$$R = 460 \times \frac{103.2}{1000}$$

$$= 47.47\ \Omega$$

The resistance of 460 ft of 30-gauge copper wire is 47.47 Ω.

TABLE 11-3 Wire Table, Standard Annealed Copper (Brown and Sharpe, American Wire Gauge)

Gauge no.	Diameter (mils at 20°C)	Circular Mil Area	Ohms per 1000 ft			
			0°C 32°F	20°C 68°F	50°C 122°F	70°C 167°F
0000	460.0	211600	0.04516	0.04901	0.05479	0.05961
000	409.6	167800	0.05696	0.06180	0.06909	0.07516
00	364.8	133100	0.07181	0.07793	0.08712	0.09478
0	324.9	105500	0.09055	0.09827	0.1099	0.1195
1	289.3	83690	0.1142	0.1239	0.1385	0.1507
2	257.6	66370	0.1140	0.1563	0.1747	0.1900
3	229.4	52640	0.1816	0.1970	0.2203	0.2396
4	204.3	41740	0.2289	0.2485	0.2778	0.3022
5	181.9	33100	0.2887	0.3133	0.3502	0.3810
6	162.0	26250	0.3640	0.3951	0.4416	0.4805
7	144.3	20820	0.4590	0.4982	0.5569	0.6059
8	128.5	16510	0.5788	0.6282	0.7023	0.7640
9	114.4	13090	0.7299	0.7921	0.8855	0.9633
10	101.9	10380	0.9203	0.9989	1.117	1.215
11	90.74	8234	1.161	1.260	1.408	1.532
12	80.81	6530	1.463	1.588	1.775	1.931
13	71.96	5178	1.845	2.003	2.239	2.436
14	64.08	4107	2.327	2.525	2.823	3.071
15	57.07	3257	2.934	3.184	3.560	3.873
16	50.82	2583	3.700	4.016	4.489	4.884
17	45.26	2048	4.666	5.064	5.660	6.159
18	40.30	1624	5.883	6.385	7.138	7.765
19	35.89	1288	7.418	8.051	9.001	9.792
20	31.96	1022	9.355	10.15	11.35	12.35
21	28.45	810.1	11.80	12.80	14.31	15.57
22	25.35	642.4	14.87	16.14	18.05	19.63
23	22.57	509.5	18.76	20.36	22.76	24.76
24	20.10	404.0	23.65	25.67	28.70	31.22
25	17.90	320.4	29.82	32.37	36.18	39.36
26	15.94	254.1	37.61	40.81	45.63	49.64
27	14.24	201.5	47.42	51.47	57.53	62.59
28	12.64	159.8	59.80	64.90	72.55	78.93
29	11.26	126.7	75.40	81.83	91.48	99.52
30	10.03	100.5	95.08	103.2	115.4	125.5
31	8.928	79.70	119.9	130.1	145.5	158.2
32	7.950	63.21	151.2	164.1	183.4	199.5
33	7.080	50.13	190.6	206.9	231.3	251.6
34	6.305	39.75	240.4	260.9	291.7	317.3
35	5.615	31.52	303.1	329.0	367.8	400.1
36	5.000	25.00	382.2	414.8	463.7	504.5
37	4.453	19.83	482.0	523.1	584.8	636.2
38	3.965	15.72	607.8	659.6	737.4	802.2
39	3.531	12.47	766.4	831.8	929.8	1012
40	3.145	9.888	966.5	1049	1173	1276
41	2.800	7.840	1219	1323	1479	1583
42	2.500	6.250	1529	1659	1854	1985
43	2.200	4.840	1975	2143	2396	2564
44	2.000	4.000	2389	2593	2899	3102
45	1.800	3.240	2950	3201	3578	3830
46	1.600	2.560	3733	4051	4528	4847
47	1.400	1.960	4875	5291	5914	6330
48	1.200	1.440	6636	7202	8051	8616
49	1.100	1.210	7898	8571	9581	10254
50	1.000	1.000	9555	10370	11592	12407

SELF-TEST

1. Find the cmil area of a 40-gauge copper wire.
2. What is the resistance of a 1000-ft length of 32-gauge copper wire at 122°F?
3. What is the diameter of a 2-gauge copper wire?
4. What is the resistance of a 500-ft length of 20-gauge copper wire at 20°C?
5. What is the resistance of a 2000-ft length of 14-gauge copper wire at 0°C?

ANSWERS TO SELF-TEST

1. 9.888 cmils
2. 183.4 Ω
3. 257.6 mils
4. 5.057 Ω
5. 4.654 Ω

The industry standard for conductors is copper. For this reason, instead of listing the resistance or resistivity of a conductor, manufacturers give a comparison of their product to copper. An example of this is shown in Table 11-4.

TABLE 11-4 Electrical Conductor Wire

Material	Percent of Copper Conductance	
	Hard	Soft
Silver	98	102.8
Copper	97	100
Gold	N.A.	73.4
Cadmium copper	85	90
Zirconium copper	93	N.A.
1350 Aluminum	61	61.8
1100 Aluminum	57	59
5005 Aluminum	52	52

WIRE RESISTANCE

The resistance for any wire can be calculated if you know the diameter of the wire (you can calculate the area if you know the diameter), the length of the wire, and the resistivity of the wire material, by using the following formula:

$$R = \rho \times \frac{\text{length}}{\text{area}}$$

EXAMPLE 11-4

Suppose that you wanted to know the resistance of a 2000-ft length of 20-gauge aluminum wire.

Solution: You can use Table 11-3 to calculate this. First find the cmil area of the 20-gauge wire using the AWG standard. The table indicates this as 1022 cmils. Next find the resistivity of aluminum. Table 11-1 shows this as 2.828 $\mu\Omega$-cm. It is necessary to convert this value from $\mu\Omega$-cm to Ω-cmils/ft. This can be done by using the conversion factor 6.018.

$$\rho = 2.828 \times 6.018$$
$$= 17.02 \ \Omega\text{-cmils/ft}$$

The resistance can then be calculated by using the formula

$$R = \rho \times \frac{\text{length}}{\text{area}}$$

$$= \frac{17.02 \times 2000}{1022}$$

$$= 33.31 \ \Omega \qquad \blacksquare$$

SELF-TEST

1. Use the resistance formula to find the resistance of a 50-ft length of 30-gauge gold wire.
2. Calculate the Ω-cmil/ft resistivity of molybdenum.
3. How much resistance would a 30-ft section of 20-gauge platinum wire have?
4. Find the resistance of 200 ft of brass wire. The wire has a diameter of 0.03 in.
5. A wire is measured and found to be 0.04 in. in diameter and 1000 ft long. Its resistance measures 10 Ω. What is the resistivity of the material used to manufacture the wire?

ANSWERS TO SELF-TEST

1. 7.305 Ω
2. 34.30 Ω-cmils/ft
3. 1.767 Ω
4. 9.361 Ω
5. 16 Ω-cmil/ft

Table 11-5 lists the resistance, various sizes, and material used in wires. It can be used to calculate the resistance of the connectors used in a circuit. All of the tables in this chapter are given with respect to a temperature of 20°C unless otherwise indicated.

TABLE 11-5 Resistances of Wires (Ω/ft)

Size (AWG)	Advance	Aluminum	Brass	Climax	Constantan	Copper	Eureka	Excello
10	0.0278	0.00164	0.00406	0.0504	0.0255	0.000999	0.0272	0.0533
12	0.0442	0.00260	0.00645	0.0801	0.0406	0.00159	0.0433	0.0847
14	0.0703	0.00414	0.0103	0.127	0.0646	0.00253	0.0688	0.135
16	0.112	0.00658	0.0163	0.203	0.103	0.00401	0.109	0.214
18	0.178	0.0105	0.0259	0.322	0.163	0.00638	0.174	0.341
20	0.283	0.0167	0.0412	0.512	0.260	0.0102	0.277	0.542
22	0.449	0.0265	0.0655	0.815	0.413	0.0161	0.440	0.861
24	0.715	0.0421	0.104	1.30	0.657	0.0257	0.700	1.37
26	1.14	0.0669	0.166	2.06	1.04	0.0408	1.11	2.18
28	1.81	0.106	0.263	3.27	1.66	0.0649	1.77	3.466
30	2.87	0.169	0.419	5.21	2.64	0.103	2.81	5.51
32	4.57	0.269	0.666	8.28	4.20	0.164	4.47	8.75
34	7.26	0.428	1.06	13.2	6.67	0.261	7.11	13.9
36	11.5	0.680	1.68	20.9	10.6	0.415	11.3	22.1
40	29.2	1.72	4.26	52.9	26.8	1.05	28.6	56.0

TABLE 11-5 (cont.)

Size (AWG)	German silver	Gold	Iron	Lead	Magnesium	Manganin	Monel metal	Nichrome
10	0.0191	0.00141	0.00579	0.0127	0.00267	0.0255	0.0243	0.06488
12	0.0304	0.00225	0.00921	0.0203	0.00424	0.0405	0.0387	0.1029
14	0.0483	0.00357	0.0146	0.0322	0.00674	0.0644	0.0615	0.1648
16	0.0768	0.00568	0.0233	0.0512	0.0107	0.102	0.0978	0.2595
18	0.122	0.00904	0.0370	0.0815	0.0170	0.163	0.156	0.4219
20	0.194	0.0144	0.0589	0.130	0.0271	0.259	0.247	0.6592
22	0.309	0.0228	0.0936	0.206	0.0431	0.412	0.393	1.055
24	0.491	0.0363	0.149	0.328	0.0685	0.655	0.625	1.671
26	0.781	0.0577	0.237	0.521	0.109	1.04	0.994	2.670
28	1.24	0.0918	0.376	0.828	0.173	1.66	1.58	6.750
30	1.97	0.146	0.598	1.32	0.275	2.63	2.51	10.55
32	3.14	0.232	0.952	2.09	0.438	4.19	4.00	17.00
34	4.99	0.369	1.51	3.33	0.696	6.66	6.36	27.00
36	7.94	0.587	2.41	5.29	1.11	10.6	10.1	42.19
40	20.1	1.48	6.08	13.4	2.80	26.8	25.6	70.24

Size (AWG)	Molybdenum	Nickel	Platinum	Silver	Steel piano wire	Steel invar	Tantalum	Tin
10	0.00330	0.00452	0.00579	0.000944	0.00684	0.0469	0.00898	0.00666
12	0.00525	0.00718	0.00921	0.00150	0.0109	0.0746	0.0143	0.0106
14	0.00835	0.0114	0.0146	0.00239	0.0173	0.119	0.0227	0.0168
16	0.0133	0.0182	0.0233	0.00379	0.0275	0.189	0.0361	0.0268
18	0.0211	0.0289	0.0370	0.00603	0.0437	0.300	0.0574	0.0426
20	0.0336	0.0459	0.0589	0.00959	0.0695	0.477	0.0913	0.0677
22	0.0534	0.0730	0.0936	0.0153	0.110	0.758	0.145	0.108
24	0.0849	0.116	0.149	0.0243	0.176	1.21	0.231	0.171
26	0.135	0.185	0.237	0.0386	0.279	1.92	0.367	0.272
28	0.215	0.294	0.376	0.0613	0.444	3.05	0.583	0.433
30	0.341	0.467	0.598	0.0975	0.706	4.85	0.928	0.688
32	0.542	0.742	0.952	0.155	1.12	7.71	1.47	1.09
34	0.863	1.18	1.51	0.247	1.76	12.3	2.35	1.74
36	1.37	1.88	2.41	0.392	2.84	19.5	3.73	2.77
40	3.47	4.75	6.08	0.991	7.18	49.3	9.43	7.00

Size (AWG)	Tungsten	Zinc
10	0.00319	0.00333
12	0.00508	0.00530
14	0.00807	0.00842
16	0.0128	0.0134
18	0.0204	0.0213
20	0.0324	0.0339
22	0.0516	0.0538
24	0.0820	0.0856
26	0.130	0.136
28	0.207	0.216
30	0.330	0.344
32	0.524	0.547
34	0.834	0.870
36	1.33	1.38
40	3.35	3.50

Another important consideration when choosing wire is the amount of current that it can safely handle. There are several standards that can be used to determine this. The safest is generally the 500-cmil/A rule. Different geographic locations, applications, and codes, may have differing requirements. It is always necessary to check the governing policy. Table 11-6 demonstrates the various requirements of several agencies.

TABLE 11-6 Allowable Current Carrying Capacities of Conductors

| Size (AWG) | Military Specifications | | | | National Electrical Code | Underwriter's Laboratory | | American Insurance Association | 500 cmils/A |
| | Copper | | Aluminum | | | | | | |
	Single wire	Wire bundled	Single wire	Wire bundled		60°C	80°C		
30	N.A.	N.A.	N.A.	N.A.	N.A.	0.2	0.4	N.A.	0.20
28	N.A.	N.A.	N.A.	N.A.	N.A.	0.4	0.6	N.A.	0.32
26	N.A.	N.A.	N.A.	N.A.	N.A.	0.6	1.0	N.A.	0.51
24	N.A.	N.A.	N.A.	N.A.	N.A.	1.0	1.6	N.A.	0.81
22	9	5	N.A.	N.A.	N.A.	1.6	2.5	N.A.	1.28
20	11	7.5	N.A.	N.A.	N.A.	2.5	4.0	3	2.04
18	16	10	N.A.	N.A.	6	4.0	6.0	5	3.24
16	22	13	N.A.	N.A.	10	6.0	10.	7	5.16
14	32	17	N.A.	N.A.	20	10.	16.	15	8.22
12	41	23	N.A.	N.A.	30	16.	26.	20	13.05
10	55	33	N.A.	N.A.	35	N.A.	N.A.	25	20.8
8	73	46	58	36	50	N.A.	N.A.	35	33.0
6	101	60	86	51	70	N.A.	N.A.	50	52.6
4	135	80	108	64	90	N.A.	N.A.	70	83.4
2	181	100	149	82	125	N.A.	N.A.	90	132.8
1	211	125	177	105	150	N.A.	N.A.	100	167.5
0	245	150	204	125	200	N.A.	N.A.	125	212.0
00	283	175	237	146	225	N.A.	N.A.	150	266.0
000	328	200	N.A.	N.A.	275	N.A.	N.A.	175	336.0
0000	380	225	N.A.	N.A.	325	N.A.	N.A.	225	424.0

TEMPERATURE COEFFICIENT OF RESISTANCE

As stated earlier, any change in temperature also effects the resistance of the material. The symbol for the temperature coefficient is α (the Greek lowercase letter alpha). There are three designations for α:

1. *Positive α*. The resistance increases as the temperature increases.
2. *Negative α*. The resistance decreases as the temperature increases.
3. *Zero α*. The temperature has virtually no effect on resistance.

Most materials have a positive α. Their resistances change almost linearly with temperature. A few materials have a negative α. They are generally in the class of semiconductors. That is, transistors, diodes, and many IC chips are made of materials that have a negative α. No materials have a truly zero α; however, when a material changes only, say, ±15 parts per million (0.000015 or 0.0015%), we consider this negligible or zero. A formula for showing how temperature effects resistance is:

$$R_{temp} = R_{base} + R_{base} \times [\alpha \times (temp_2 - temp_1)]$$

EXAMPLE 11-5

Given that the temperature coefficient for copper is 0.0039312 and the resistance of 1000 ft of 14-gauge wire is 2.575 Ω at 25°C, how much resistance will this wire have at 70°C?

Solution: Use the R_{temp} formula to find this resistance.

$$R_{70} = R_{25} + R_{25} \times [0.0039312 \times (70 - 25)]$$
$$= 2.575 + 2.575 \times (0.0039312 \times 45)$$
$$= 2.575 + 0.4555278$$
$$= 3.031 \, \Omega$$

Table 11-7 lists the values of temperature coefficient of resistance for several materials. You will notice that almost all of those materials have positive α. Constantan, manganin, and Therlo are the only materials that exhibit zero α. Carbon, silicon, germanium, and other semiconductor materials have a negative α.

TABLE 11-7 Temperature Coefficient of Conductors at 20°C

Metal	α	Melting Point °C	Metal	α	Melting Point °C
Aluminum	0.0039	659	Magnesium	0.004	651
Antimony	0.0036	630	Manganin	0.00001	910
Arsenic	0.0042		Mercury	0.00089	−38.9
Bismuth	0.004	271	Molybdenum	0.004	2500
Brass	0.002	900	Monel metal	0.0020	1300
Cadmium	0.0038	321	Nichrome	0.0004	1500
Cilmax	0.0007	1250	Nickel	0.006	1452
Cobalt	0.0033	1480	Palladium	0.0033	1550
Constantan	0.00001	1190	Phosphor bronze	0.0018	750
Copper			Platinum	0.003	1755
Annealed	0.00393	1083	Silver	0.0038	960
Hard-drawn	0.00382		Steel, Siemens–Martin	0.003	1510
Excello	0.00016	1500			
Gas carbon	−0.0005	3500	Tantalum	0.0031	2850
German silver 18% nickel	0.004	1100	Therlo	0.00001	
			Tin	0.0042	232
Gold	0.0034	1063	Tungsten	0.0045	3400
Iron	0.005	1530	Zinc	0.0037	419
Lead	0.0039	327			

SELF-TEST

1. A length of aluminum wire has 5 Ω of resistance at 20°C. What will its resistance be at 50°C and at 150°C?
2. The resistance of a length of nickel wire is 25 Ω at 120°C. Find its resistance at 20°C.
3. Find the resistance of a length of platinum wire if its resistance was 6.5 Ω at 20°C and the present temperature is 70°C.
4. Which would experience more change in resistance, aluminum or gold, if the temperature were to increase?
5. A length of constantan wire has 100 Ω of resistance at 20°C. Calculate its resistance at 150°C.
6. A length of annealed copper wire has 100 Ω of resistance at 20°C. Calculate its resistance at 150°C.
7. Compare the answers for problems 5 and 6. What can be said about the two metals?

ANSWERS TO SELF-TEST

1. $R_{50} = 5.585 \ \Omega$, $R_{150} = 7.535 \ \Omega$
2. $R_{20} = 10 \ \Omega$
3. $R_{70} = 7.475 \ \Omega$
4. Aluminum
5. $R_{150} = 100.1 \ \Omega$
6. $R_{150} = 151.1 \ \Omega$
7. Constantan is not affected by temperature change, whereas copper is.

WIRE CONSTRUCTION

There are several ways in which wire is manufactured, each with its own special purpose. The most common of these are shown in Figure 11-2.

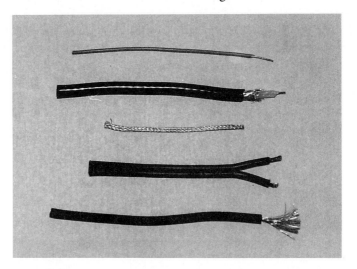

FIGURE 11-2 Types of Wire. Top to bottom solid, coaxial cable, braided, twin lead, stranded.

Solid-core wires are best used in applications where movement is kept at a minimum. Solid wires have the most tensile strength; however, when used on circuits that are continually moved (doors and windows or meter leads, for instance), they have a tendency to break. Stranded wires serve this purpose best.

Grounding cables and other conductors that need very low resistance are sometimes *braided*. Braided wires allow both flexibility and, since the majority of surface area is in constant contact, low resistance.

Television and radio communication requires wire that has a specific amount of ac circuit resistance called *impedance*. Usually, this is either 300 or 75 Ω. The *twin lead* used on TV antennas has 300 Ω of resistance, and coaxial cables have 75 Ω of resistance. For differing reasons, 300-Ω twin lead will generally provide the user with reception at a greater distance than will 75-Ω coaxial lead. The 75-Ω coaxial cable, on the other hand, provides a signal that has much less radio-frequency interference.

There are also differing types of extension and power cord wire. For 120-V ac, lights and polarized appliances generally use a two-wire power cord. Equipment, particularly hand-held or operated equipment, will usually need a three-wire or grounded power cord. It is important that the required power cord be used for replacement purposes. This is done for safety reasons. These reasons and requirements are discussed in detail in a later chapter.

Of course, using a bare wire should not be done. Even when a wire is used only for grounding the circuit, it should have a protective covering. This covering is an insulator (generally plastic). It is a good practice to examine this insulator for cracks and areas that show excessive heat. In either case, wires having these defects need to be replaced.

Transformer windings, armature windings, and field windings must also have an insulating covering. This covering is usually a shellac or varnish material. Cracks and visible hot spots again are reasons for these components to be replaced.

SWITCHES

Most circuits are not designed to be operational at all times. These and other circuits require controlling devices that will turn them on and off at the operator's discretion. This section is devoted to an explanation of these components. In Figure 11-3 you will find the schematic symbols for several popular types of switches. Notice that the dot indicates the throw, while the arrow indicates the pole. The upper symbols show two switches. You will notice that each of these has only one arrow. The bottom set of symbols, on the other hand, each contain two arrows—hence the names *single* (one) *pole* and *double* (two) *pole* used in the description of these types of switches.

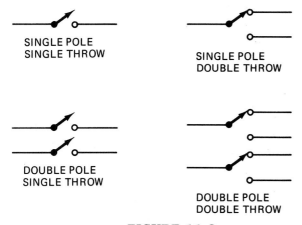

FIGURE 11-3

The *throw* is the number of operational positions available to each pole. The switches on the left have only one position for each pole. The right-hand switches have two positions for each pole.

The switch symbols shown in Figure 11-4 are of two other popular switch types, *rotary* and *slide*. These switches offer more possible contact positions than do the SPST, STDP, or any of the aforementioned switch configurations. The rotary

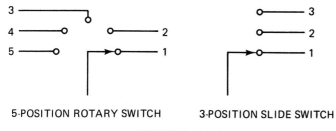

FIGURE 11-4

and slide switches are both available in multipole packages. Some of these are shown in Figure 11-5.

FIGURE 11-5 Assorted Switches. SPST Toggle–upper left, SPDT DIP–lower left, SPDT Slide–upper center, DP 3 position Slide–lower center, SP 5 position Rotary–right.

There are several different types of pushbutton switches. Momentary contact, positive action, and push on/push off are some of the more popular names. Each of these types of pushbutton switches is rated by how far the button must be depressed before the switching contact actually takes place. A special group of pushbutton switches requires very little movement for the switching action to occur. They are sometimes called *microswitches*. See Figure 11-6 for the schematic symbol of a pushbutton switch. Pushbutton switches are also listed by their normal operational position: *normally closed* (on) and *normally open* (off). Figure 11-6 shows a normally open switch. The button must be pressed to turn this switch on. When the button is released, the switch will turn off.

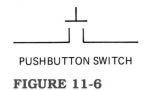

PUSHBUTTON SWITCH

FIGURE 11-6

CIRCUIT PROTECTION

There are three main devices used for circuit protection: fuses, circuit breakers, and fusible links. In this section we discuss the rating, testing, and theory of operation for each of these.

Fuse Types At the component level, fuses are the most used protective device. There are two basic types of fuses used by manufacturers. One is the *slow-blow fuse*. A slow-blow fuse is used on any device that requires more current on initial operation than on normal operation. Devices that have motors typically fall into this category. When a motored device is turned on, a large amount of current is required to start the motor turning. After the motor picks up speed (rpm), the current draw is lessened. A slow-blow fuse will allow a large amount of current to pass during initial operations but will open in the event that an overcurrent condition is present. A slow-blow fuse can usually be identified by a small internal spring and sometimes, an internal resistor (see Figure 11-7).

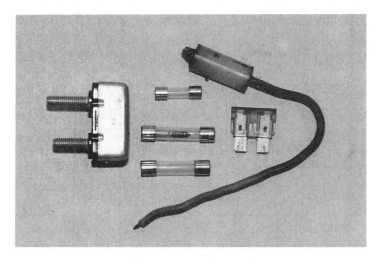

FIGURE 11-7 Circuit Protectors. Circuit breaker–left, Quick Blow Mini—top center, Slow Blow AGC–center, Quick Blow AGC–lower center, Fusible link–right.

A *fast-blow fuse* is the other common type of fuse. This type of fuse is used on any circuit that requires immediate shutdown in the event of an overcurrent condition. These are the most common of the fuse types (see Figure 11-7).

Current Rating No matter which type of fuse construction is used, every fuse has a specific current and voltage rating that cannot be exceeded. The current rating is the amount of current that a fuse can handle without opening. This is a prime consideration when fuse replacement is required. The current rating of a replaced fuse should never be more than the original fuse rating. Using such a fuse will endanger the other circuit components.

Voltage Rating The voltage rating of a fuse indicates the maximum supply voltage that the fuse can protect. If this rating is less than the original rating, the circuit may be subjected to voltage flash or arcing. That is, when the fuse opens, the voltage source may be capable of forcing a spark across the open portion of the fuse. This is similar to the way a spark is forced across the gap of a spark plug on a gasoline engine. The voltage rating can be safely exceeded because a larger voltage rating will provide greater circuit protection. The voltage rating should never be reduced.

Testing a Fuse The best way to test for an opened fuse is with a ohmmeter. The fuse must be removed from the circuit to perform this test. The resistance of the fuse is measured. A good fuse will have little or no resistance. A faulty fuse will show an open. Never rely on a visual inspection of the fuse. Too many times a faulty fuse will appear to be good. A voltage test is sometimes used to find the faulty fuse on a live circuit. A voltmeter is placed across the fuse when this test is performed. A good fuse should show little or no voltage. A faulty one will show full supply voltage.

Circuit Breakers A circuit breaker is the most common type of circuit protection. Most homes are equipped with circuit breakers. The circuit breaker is a current-sensitive switch. In the event that an overcurrent condition occurs, the switch portion of the breaker will open. Turning the breaker off and then on again will reinstate the power to the circuit. Circuit breakers have the advantage of not needing replacement in the event of a short, while almost any short will require the replacement of a fuse (see Figure 11-7).

Fusible Links A fusible link is a weak portion of the circuit. It is a piece of wire that has a good amount of resistance. As current passes through this wire, the wire will heat up. In the event of an overcurrent condition, the wire will burn open, stopping the current flow. Fusible links are used on circuits that normally require large amounts of current (50 A or more). Automobiles are generally equipped with fusible links that protect the unfused portions of their electrical system (see Figure 11-7).

SUMMARY

Wires and other connectors have resistance. In general, that resistance is not large enough to affect a circuit's performance. However, in instances where large currents are flowing, even small resistances can cause a substantial voltage drop. Under these circumstances it is necessary to take the wire resistance into consideration.

The resistance of a wire is affected by three factors. One is the diameter or gauge size of the wire. A large-diameter wire has a lower amount of resistance than that of a smaller wire. A second consideration is the material of which the wire is made. Different materials show differing resistances. The final consideration is the physical temperature of the wire. Most substances will have a higher resistance at a higher temperature.

The temperature coefficient indicates the rate at which temperature affects resistance. A positive value of α indicates that temperature and resistance vary directly. A negative value of α indicates that temperature and resistance vary inversely. Zero α shows that the temperature and resistance of a material are independent of each other.

There are a wide variety of switches that can be used to control the operation of any circuit. Every circuit also needs some sort of current protection. Fuses, circuit breakers, and fusible links provide this needed current protection.

QUESTIONS

11-1. List the five most used conductive materials. Describe their respective advantages and disadvantages.

11-2. What does the term *annealed* mean?

11-3. What is the best overall conductor? Why?

11-4. In your own words, describe what is meant by resistivity.

11-5. What is a mil?

11-6. What is a circular mil? For what is this measurement used?

11-7. Why would one want a material to have a large amount of resistance?

11-8. Which has a larger diameter, a 40-gauge wire or a 20-gauge wire?

11-9. Which has more resistance, a 20-ft length of 35-gauge copper wire or a 20-ft length of 10-gauge copper wire?

11-10. Which has more resistance, a 20-ft length of 35-gauge copper wire or a 10-ft length of 20-gauge copper wire?

11-11. A 100-ft length of copper wire has 200 Ω of resistance. How much resistance would a similar length of cadmium copper wire have? Both wires have the same diameter. (*Hint:* Use Table 11-4.)

11-12. A 210-ft length of copper wire has 50 Ω of resistance. How much resistance would a similar length of gold wire have? Both wires have the same diameter. (*Hint:* Use Table 11-4.)

11-13. Why is it important to know the maximum current-carrying capability of a conductor?

11-14. Who sets the standards for the maximum current that a conductor can safely handle?

11-15. Describe positive, negative, and zero α. Include an example of each.

11-16. List each of the five different types of wire construction. Describe and include an application of each.

11-17. What type of cord should be used when replacing a power cord on an appliance or a piece of hand-held equipment?

11-18. Briefly describe each of the various switch types. Include a schematic symbol of each.

11-19. In your own words, describe how you would determine how a switch operates: for instance, which position is off, which position is on, and so on.

11-20. What is a momentary-contact, normally open microswitch?

11-21. List each of the various circuit protection devices. Describe each briefly.

11-22. What is the difference between the operation of a slow-blow fuse and that of a fast-blow fuse.

11-23. When examining a fuse, how can one determine if it is a fast- or a slow-blow fuse?

11-24. In your own words, describe what is meant by the *current rating* of a fuse.

11-25. Why is it important that the current rating not be exceeded when replacing a fuse?

11-26. Describe in your own words what the voltage rating of a fuse indicates.

11-27. Why can the voltage rating be exceeded without worrying? Is it dangerous to lessen this value when replacing a fuse?

11-28. Describe how to use an ohmmeter to test a fuse.

11-29. Describe how to use a voltmeter to test a fuse.

11-30. In the event that a circuit breaker opens (throws), what should be done?

PRACTICE PROBLEMS

11-1. A 120-V source (the ac power line) provides 25 A of current to a household circuit. The circuit contains 250 ft of 10-gauge copper wire. Calculate the line voltage drop to find the available voltage to the household appliances.

11-2. Repeat Problem 11-1 using 12-gauge and 14-gauge copper wire.

11-3. Use Table 11-5 to find the resistance of 500 ft of 22-gauge Eureka wire.

11-4. Use Table 11-5 to find the resistance of 750 ft of 30-gauge Monel metal wire.

11-5. Find the resistance of a coil made by winding 2500 turns of 26-gauge copper wire around a 2-in.-diameter coil form. (Circumference = $\pi \times$ diameter.)

11-6. A technician discovers an open fuse. The fuse has 1.5 A and 25 V as specifications. Could the technician replace this fuse with a 1.5-A 120-V fuse without worry? Why?

11-7. A material has 400 Ω of resistance. Its cross-sectional area is 20 cm^2 and its length is 100 cm. What is the resistivity of this material?

11-8. A piece of molybdenum measures 1 cm × 0.5 cm and is 2000 m long. How much resistance does this block of metal have?

11-9. A wire measures 0.032 in. in diameter and is 330 ft long. It is made of copper. Use Table 11-3 to find its gauge size, circular mil area, and its resistance at 0, 20, and 50°C.

11-10. A circuit uses a rotary switch to select a 2-, 5-, or 10-V power source. The circuit consists of a 10-kΩ resistor in series with a parallel combination of 22 and 18 kΩ. Sketch the circuit and find the current and voltages for each switch position.

VOM Meter Design and Considerations

12

Chapter objectives

After reading this chapter and answering the questions and problems, you should be able to:

- Describe the effects of meter loading.
- Define *sensitivity*, *cascaded multiplier*, *cascaded shunt*, *individual multiplier*, and *individual shunt* as they apply to a VOM meter.
- Determine the value of R_M given I_M and maximum power supply voltage.
- Design cascaded and individual multiplier voltmeters.
- Design cascaded and individual shunt current meters.
- Design ohmmeters.
- Extend the range of a voltmeter or a current meter.

The most used single piece of test equipment is the VOM (volt-ohm-milliampere) meter. In this chapter we remove some of the mysteries that shroud the design and construction of this meter. A VOM meter is a combination of four meters, all housed in the same meter case. The four meters are an ac voltmeter, a dc voltmeter, a dc milliampere meter, and an ohmmeter. In this chapter we discuss the design and construction of the dc portion of the VOM meter. With the exception of diodes (semiconductors that change alternating current to direct current), the ac and dc portions are the same. When using a VOM meter it is important to know how the meter itself will affect the circuit; for that reason, meter loading and voltmeter sensitivity will be explained. At one time or another every technician is placed in a situation where the meter needed is not available. This chapter also includes a section devoted to meter range extension. This can prove invaluable under certain circumstances.

METER LOADING

The sensitivity is a rating that pertains to the voltmeter section of a VOM meter. It does not necessarily indicate how accurate the ohmmeter or current meter section will be; however, since voltage readings are the readings that are most used, sensitivity is generally considered to be a good indication of the meter's accuracy. Almost all meters have this rating printed on their meter face.

When it comes to choosing which meter will work the best for a particular application, a prime consideration is the meter's sensitivity. Most meters used in electronics have a sensitivity of 20 kΩ/V. This rating indicates how much internal resistance the meter has on any voltage setting.

EXAMPLE 12-1

Find the internal resistance of a meter that has a sensitivity of 20 kΩ/V when it is placed on the 2.5-V range.

Solution: The sensitivity is multiplied by the voltage range setting to find the internal resistance.

$$R_{int} = 2.5 \times 20{,}000 = 50 \text{ k}\Omega$$

The internal resistance is the amount of resistance that will be placed in parallel with the circuit any time the meter is used on that particular range. ∎

The circuit, in Figure 12-1, shows a VOM meter being used to measure the voltage across a 100-kΩ resistor. This meter will load down the circuit because the meter has an internal resistance that is a lot smaller than the 100-kΩ resistor that is being be measured across.

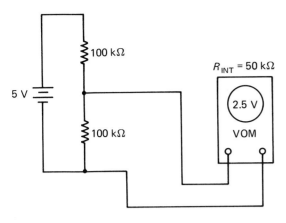

FIGURE 12-1 The internal resistance of the meter can pull down the circuit voltage.

EXAMPLE 12-2

Refer to Figure 12-1. Calculate the amount of voltage that the voltmeter indicates.

Solution: When you consider that the meter is placed in parallel with the resistor, you will find that the circuit has been changed. The resistance of the lower resistor has been pulled down to

$$100 \text{ k}\Omega \parallel 50 \text{ k}\Omega = 33.33 \text{ k}\Omega$$

This change in resistance is due to the meter, not to a circuit malfunction. However, this change in resistance also causes an accompanying change in voltage. The meter would read

$$5 \times \frac{33{,}330}{133{,}300} = 1.25 \text{ V}$$

not the 2.5 V that we know is present. ■

In general, it is necessary for the VOM meter to have an internal resistance higher than (10 times) the resistance that is being measured. Also, the higher the sensitivity, the better (more accurate) the VOM voltmeter will be. A meter that has a sensitivity of 50 kΩ/V would be more accurate than one that has a sensitivity of 20 kΩ/V.

Digital meters, transistorized meters, and vacuum-tube voltmeters all have an internal resistance that is fixed. This means that their internal resistance does not change as the range is switched. Usually, they have a large amount of resistance at all times. It is not unusual for such a meter to have an internal resistance of 10 or 11 MΩ. Thus they are less likely to load down the circuit. It is best to use a meter of this type when there is a possibility of circuit loading.

METER MOVEMENTS AND THEIR CONSTRUCTION

One area of confusion about the construction of various meters is the meter movement. It seems difficult, at first, to differentiate between a meter and a meter movement. The movement is the portion of the meter that consists of the pointer assembly, which indicates the quantity being measured. Sometimes the meter movement is called a panel meter. Most meter movements are of moving-coil (D'Arsonval) construction (see Figure 12-2).

FIGURE 12-2

Such meter movements consist of four basic parts, a magnet, an armature & windings, a pointer, and a return spring. A permanent magnet is placed around the outside of the movable armature. The armature is made up of several windings (coils of wire) around a movable iron core. A pointer is attached to the armature. Any time that current flows through the armature coil, the armature turns and its movement is indicated by the pointer. Moving the armature coil places tension on a coil return spring, which returns the pointer and armature to their original position when current is removed.

The meter movement has two specifications, I_M and R_M. The meter movement current, I_M, is the amount of current that is needed to move the pointer to the full-scale position. Panel meters come with a variety of movement currents, ranging from 10 μA or so to several milliamperes. VOM meters typically use 20- or 50-μA movements.

The wire that makes up the coil has resistance. Ideally, this resistance is small, but it is also an important meter design consideration. This resistance is called the meter movement resistance, R_M. Panel meters also have a variety of R_M values from which to choose. These range from as little as 2 Ω to as much as several thousands of ohms. VOM meters typically use meter movements that have 1200 or 4500 Ω as the value for R_M.

The I_M and R_M of the meter should be listed in the manufacturer's specifications for that movement. There are some manufacturers that do not list the R_M value for their movement. When they are encountered, the circuit shown in Figure 12-3 can be used to find the resistance of the movement.

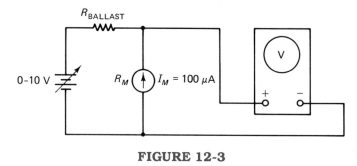

FIGURE 12-3

EXAMPLE 12-3

Refer to Figure 12-3. Find the value of R_M for a 100-μA movement. Given that the adjustable supply has a maximum voltage of 10 V.

Solution: Begin by calculating the necessary value for R_{ballast}. This resistor is used to limit the current output of the supply and to increase the control of the voltage adjustment. Do your calculations at the half-voltage point of the supply. Then

$$R_{ballast} = \frac{V_{half\,supply}}{I_M}$$

$$= \frac{5}{0.0001}$$

$$= 50\,k\Omega$$

The circuit is then constructed as shown. The supply voltage is slowly adjusted from its lowest point to the point where the current through the meter movement is equal to I_M, the full-scale current. At this point the voltmeter indication is read. Suppose, for this example, that the reading was found to be 0.25 V. Then

$$R_M = \frac{V_{reading}}{I_M}$$

$$= \frac{0.25}{0.0001}$$

$$= 2.5\,k\Omega \quad \blacksquare$$

The meter movement itself has only one adjustment. That adjustment is used to zero the meter. As a rule, any problem that requires more than this adjustment necessitates the replacement of the entire meter movement assembly.

SELF-TEST

1. A technician is using a circuit such as that shown in Figure 12-3. She is using a 15-V power supply to find the R_M of a 50-μA meter movement. What resistance should she use as a ballast? If when testing she finds that the movement, at full scale, corresponds to a 0.15-V reading on the voltmeter, what is the R_M of the meter movement?

2. Given: $V_{supply} = 30$ V   Find: $R_{ballast} = $ _____
   $I_M = 20\,\mu A$ $R_M = $ _____
   $V_{reading} = 0.3$ V

3. Given: $V_{supply} = 25$ V   Find: $R_{ballast} = $ _____
   $I_M = 28\,\mu A$ $R_M = $ _____
   $V_{reading} = 0.15$ V

ANSWERS TO SELF-TEST

1. $R_{ballast} = 150\,k\Omega$, $R_M = 3\,k\Omega$ 2. $R_{ballast} = 750\,k\Omega$, $R_M = 15\,k\Omega$
3. $R_{ballast} = 446\,k\Omega$, $R_M = 5.357\,k\Omega$

CONSTRUCTION AND DESIGN OF VOLTMETERS

The desired sensitivity is a major consideration when designing a voltmeter. The sensitivity of the voltmeter, S, is determined by the full-scale current capabilities of the meter movement. This maximum current rating is symbolized as I_M. The sensitivity of a voltmeter is always the reciprocal of this I_M figure.

$$S = \frac{1}{I_M}$$

EXAMPLE 12-4

Find the sensitivity of a voltage meter that uses a meter movement having an $I_M = 1$ mA.

Solution: The sensitivity is the reciprocal of I_M.

$$S = \frac{1}{I_M}$$

$$= \frac{1}{0.001} \text{ A} = 1000 \text{ }\Omega/\text{V}$$ ∎

Similarly, an I_M of 50 μA results in a sensitivity of 20 kΩ/V and an I_M of 20 μA results is a sensitivity of 50 kΩ/V. It is important choose a meter movement with an I_M that will correspond to the desired sensitivity, as the sensitivity is inherent to the I_M of the meter movement.

The voltmeter is constructed by placing current-limiting resistors in series with the meter movement. These resistors are called multiplier resistors and are symbolized as R_{mult}. These multipliers can be arranged in one of two fashions, a cascaded arrangement and an individual arrangement. An explanation and demonstration of these two methods follow.

Individual Multiplier Voltmeter

EXAMPLE 12-5

Refer to Figure 12-4. Calculate the multiplier resistors for each range.

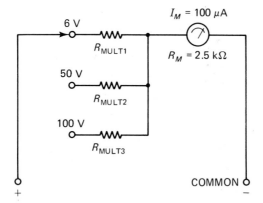

INDIVIDUAL MULTIPLIER VOLTMETER   FIGURE 12-4

Solution: The design of an individual multiplier voltmeter is simple. Use the sensitivity of the meter, the full-scale voltage value for the desired range, and the small amount of internal resistance of the meter movement (R_M) to calculate the multiplier resistance needed. The process follows (refer to Figure 12-4).

6-V range:

$$S = 1/100 \, \mu A$$
$$= 10 \text{ k}\Omega/\text{V}$$
$$R_{int} = S \times V_{range}$$
$$= 10 \text{ k}\Omega/\text{V} \times 6 \text{ V}$$
$$= 60 \text{ k}\Omega$$
$$R_{mult1} = R_{int} - R_M$$
$$= 60 \text{ k}\Omega - 2.5 \text{ k}\Omega$$
$$= 57.5 \text{ k}\Omega$$

The basic idea here is first to find the internal resistance of the meter. This is done by multiplying the sensitivity by the voltage of the range being calculated. This resistance is the total amount as seen from the two meter leads. The meter movement also has a bit of resistance. This resistance is symbolized as R_M. The meter resistance (R_M) is subtracted from the internal resistance to leave only the amount of resistance needed by the multiplier. Calculation for the other ranges follow.

50-V range:
$$R_{mult2} = S \times V_{range} - R_M$$
$$= 10{,}000 \times 50 - 2500$$
$$= 497{,}500 \, \Omega$$

100-V range:
$$R_{mult3} = S \times V_{range} - R_M$$
$$= 10{,}000 \times 100 - 2500$$
$$= 997{,}500 \, \Omega \quad \blacksquare$$

In practice it is usually not necessary to find or build exact resistance values. As a rule, the closest standard value resistor will provide a surprisingly accurate result. Manufacturers use resistors that have a tolerance of either 1% or 2% as a matter of course. If a high degree of accuracy is needed, it is possible to use a trim pot for the multiplier resistors. This also provides a method of adjustment as the meter and resistors age. The individual multiplier also has an added advantage in that if one range should be damaged for some reason, the meter is still operable on the other ranges.

Cascaded Multiplier Voltmeter

EXAMPLE 12-6

Refer to Figure 12-5. Calculate the multiplier resistors for each range.

Solution: The design of the cascaded multiplier is even simpler than that of the individual multiplier. Notice that the multipliers are placed in series with each other (refer to Figure 12-5). This allows the resistors to combine to form the internal resistance. This means that the multiplier resistor's ohm value is smaller for a cascaded design. The procedure follows.

6-V range:

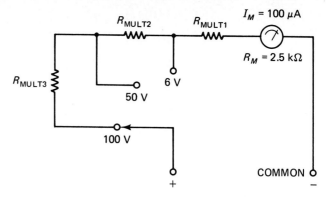

CASCADED MULTIPLIER VOLTMETER

FIGURE 12-5

$$S = 1/100 \ \mu A$$
$$= 10 \ k\Omega/V$$
$$R_{int} = S \times V_{range}$$
$$= 10 \ k\Omega/V \times 6 \ V$$
$$= 60 \ k\Omega$$
$$R_{mult1} = R_{int} - R_M$$
$$= 60 \ k\Omega - 2.5 \ k\Omega$$
$$= 57.5 \ k\Omega$$

The process is exactly that of the individual multiplier. The other two (higher) ranges are calculated in a slightly different manner. It is only necessary to use the sensitivity and the voltage change (ΔV) between the ranges in order to calculate the remaining multipliers. This method is demonstrated in the calculations that follow.

50-V range:

$$\Delta V = \text{desired range} - \text{preceding range}$$
$$= 50 - 6$$
$$= 44 \ V$$
$$R_{mult2} = \Delta V \times S$$
$$= 44 \times 10,000$$
$$= 440 \ k\Omega$$

100-V range:

$$\Delta V = \text{desired range} - \text{preceding range}$$
$$= 100 - 50$$
$$= 50 \ V$$
$$R_{mult3} = \Delta V \times S$$
$$= 50 \times 10,000$$
$$= 500 \ k\Omega$$

A trim pot is often used as the multiplier resistor R_{mult1}. This provides a method of meter adjustment as the resistors age. The cascaded multiplier is most often used by manufacturers because it requires only one nonstandard resistor value.

Choose resistor values that will provide a uniform voltage change. That is, in Example 12-6, each 10-V step in the voltage range will need a 100-kΩ step in resistance, so a 30-V step will need a 300-kΩ resistance step. Also, the voltage steps can be chosen according to the available standard resistor value.

CONSTRUCTION AND DESIGN OF CURRENT METERS

A current meter is built by placing resistors in parallel with the meter movement. These resistors are called *shunt resistors*. In general, any resistor that is installed in parallel with the circuit is called a shunt. For a meter to measure large currents it is necessary for the shunt resistance to be considerably smaller than the resistance of the meter movement (R_M). Thus a determining factor for a current meter is the R_M of the meter movement. The design of almost any current meter will require a resistor value less than 1Ω. Manufacturers use a resistance wire (at first glance, it appears to be a piece of coat hanger) for this resistor (see Figure 12-6). As with the voltmeter multiplier resistor, the shunts can be arranged in two fashions, a cascaded shunt and an individual shunt. The cascaded shunt is also referred to as an *Ayrton shunt*.

FIGURE 12-6 Internal VOM Circuitry. Low resistance, high power, wire resistor is indicated.

Individual Shunt Current Meter The design of an individual shunt current meter (Figure 12-7) is more of a commonsense application of Ohm's law than a sophisticated design problem. As learned in earlier chapters, parallel voltages are equal. The voltage value can be calculated by multiplying the R_M and I_M of the meter movement.

Current is additive, in parallel circuits, so the shunt current can be found by subtracting the current value of the meter movement from the current value for the particular range.

EXAMPLE 12-7

Refer to Figure 12-7. Calculate the values of shunt resistance needed for each current range.

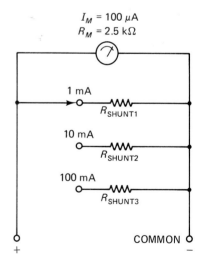

INDIVIDUAL SHUNT CURRENT METER

FIGURE 12-7

Solution:

1-mA range:

$$V_{\text{meter}} = I_M \times R_M$$
$$= 0.0001 \times 2500$$
$$= 0.25 \text{ V}$$

$$I_{\text{shunt}} = I_{\text{range}} - I_M$$
$$= 0.001 - 0.0001$$
$$= 900 \ \mu A$$

$$R_{\text{shunt1}} = \frac{0.25}{0.0009}$$
$$= 278 \ \Omega$$

This is the value of the shunt resistance to be used on the 1-mA range. The other resistor values are calculated in a similar manner. These calculations follow.

10-mA range:

$$I_{\text{shunt}} = 0.01 - 0.0001$$
$$= 0.0099 \text{ A}$$

$$R_{\text{shunt2}} = \frac{0.25}{0.0099}$$
$$= 25.25 \ \Omega$$

100-mA range:

$$I_{\text{shunt}} = 0.1 - 0.0001$$
$$= 0.0999 \text{ A}$$

$$R_{\text{shunt3}} = \frac{0.25}{0.0999}$$
$$= 2.503 \ \Omega$$ ∎

Again the value of the particular resistor is not horribly crucial; however, it is important that the power requirements be met.

The individual shunt current meter has one major disadvantage. It is mandatory that this meter be disconnected before the ranges are changed. Switching the meter while it is connected to the circuit forces all of the current through the meter movement. For one short instant during the switching, there will be no shunt in the meter circuitry. Manufacturers that choose this design incorporate a make-before-brake switch to overcome this problem.

Cascaded (Ayrton) Shunt Current Meter The design of an Ayrton shunt current meter is much more complex. It requires several calculations and a particular viewpoint. The calculations begin in the same manner as that of the individual shunt; however, this first set of calculations provides the total shunt resistance, R_{ST}.

The design of this meter allows the parallel shunt to be constructed in segments (see Figure 12-8). As the range is changed, the switch will remove one resistor from the shunt and add that value to the meter movement resistance. That is,

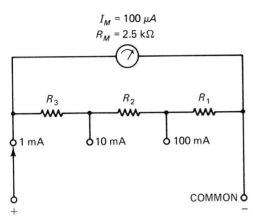

CASCADED (AYRTON) SHUNT CURRENT METER   **FIGURE 12-8**

when the meter is turned to the 1-mA position, all three of the resistors are used as the shunt resistor. When the meter is turned to the 10-mA position, the resistance of R_3 is removed from the shunt and placed in series with the meter movement.

The calculations require you to look at the circuit as one loop. The total resistance in that loop is the resistance value of R_{ST}, the total shunt resistance, plus that of the meter movement. Viewing the circuit in this manner will allow you to apply the current divider formula to the design. Some basic things to remember are:

1. The lowest current range is used to calculate the value of R_{ST},
2. The higher current ranges are used to calculate the values of the resistor segments of the Ayrton shunt.

That is, the highest range is used to find R_1, the next-highest range is used to calculate R_2, and so on. The procedure is outlined in Example 12-8.

EXAMPLE 12-8

Refer to Figure 12-8. Calculate the value of each shunt resistor in this cascaded shunt design.

Solution: Begin by finding the total shunt resistance. Remember that you must use the shunt current for the lowest range to calculate this value.

$$V_{\text{meter}} = I_M \times R_M$$
$$= 0.0001 \times 2500$$
$$= 0.25 \text{ V}$$

$$I_{\text{shunt lowest range}} = I_{\text{range}} - I_M$$
$$= 0.001 - 0.0001$$
$$= 0.0009$$

$$R_{ST} = \frac{V_{\text{meter}}}{I_{\text{shunt}}}$$
$$= \frac{0.25}{0.0009}$$
$$= 277.78 \text{ }\Omega$$

CONSTRUCTION AND DESIGN OF CURRENT METERS

Next find the total loop resistance. This is the combination of R_{ST} and R_M.

$$R_{loop} = R_{ST} + R_M$$
$$= 277.78 + 2500$$
$$= 2777.78 \; \Omega$$

Again the viewpoint is to look at the circuit as if it were a common current divider. That is why it is important to find the total resistance of the loop. This resistance value is used to find the resistance of the resistor segments. Begin with the largest desired current range and work toward the smallest.

100-mA range:

$$R_1 = \frac{I_M}{I_{range}} \times R_{loop}$$
$$= \frac{0.0001}{0.1} \times 2777.78$$
$$= 2.78 \; \Omega$$

10-mA range:

$$R_1 + R_2 = \frac{I_M}{I_{range}} \times R_{total}$$
$$= \frac{0.0001}{0.01} \times 2777.78$$
$$= 27.78 \; \Omega$$
$$R_2 = 27.78 - 2.78$$
$$= 25 \; \Omega$$

1-mA range:

$$R_1 + R_2 + R_3 = R_{ST}$$
$$= 277.78$$
$$R_3 = 277.78 - 27.78$$
$$= 250 \; \Omega$$

Notice that the same formula is used to calculate each of these resistor segments. There are two items that must be watched as this process is performed. First, the current range changes for each calculation. Second, the calculation will produce the amount of resistance needed in the shunt. This means that although the first calculation gives the value of R_1, the second calculation gives the resistance of both R_1 and R_2 combined. That is why R_1 must be subtracted from the second answer. There is no need to perform the calculation on the last resistor, as you already know the value of all of the resistor segments (i.e., R_{ST}). To find R_3, just subtract the resistance of R_1 and R_2 from R_{ST}. ∎

The most common design for a current meter is the cascaded version. This design has a distinct advantage over the individual shunt in that it is not necessary to disconnect the meter in order to change the range. The reason for this is that as the meter is switched from one range to another, the circuit is momentarily opened. This prevents an overcurrent condition from being applied to the meter movement.

DESIGN AND CONSTRUCTION OF RESISTANCE METERS

The voltmeter and the current meter are designed to be most accurate at the full-scale current position. All of the calculations are based on a current equal to I_M. The resistance meter differs greatly from these meters in that the center (middle) scale is used as a basis of design. This half-scale current position is referred to as midscale.

Basic Meter Design Examine the circuit shown in Figure 12-9. The ohmmeter is essentially a series circuit. The resistance of the meter movement, the zero adjustment potentiometer, and the resistance of the current-limiting resistor are all known values. The current in the circuit can be used to calculate the value of the unknown, or more correctly, the current scale has been renumbered to correspond to the value of the unknown resistor.

Figure 12-9 shows the basic circuitry for an ohmmeter. A careful inspection of this schematic will show that the circuit is series in nature. Also remember that the first step in using any ohmmeter is to calibrate the meter. This is done by holding the two probes together and turning R_{adj} until the meter is zeroed. This action sets the current flow to equal the I_M for the meter movement. When the unknown resistance is placed between the probes, the total resistance of the circuit is increased. An increase in the total resistance causes the current flow to decrease accordingly.

Think for a minute. The meter movement is at full-scale (to the right as a rule) when the resistance is zero. When the resistor to be measured is connected to the meter, the current level decreases and the current meter lowers (to the left).

The crucial value of resistance (to be measured) is a value equal to the total of R_M, R_{lim}, and R_{adj} (when adjusted). This value is called total meter resistance or R_{MT}. With a resistance equal to R_{MT} between the probes the meter movement drops to one-half. R_{MT} can be calculated easier by dividing the battery voltage by I_M. R_{adj} is usually set at 10% of R_{MT}, while R_{lim} is set at 95% of R_{MT} (the extra 5% is to allow for adjustment as the battery wears down). Although these values are typical, any combination that will allow the total of the three resistances (R_M, R_{lim}, and R_{adj}) to be adjusted from a value of less than R_{MT} to a value greater than R_{MT} will work satisfactorily. An outline of these steps is given in Example 12-9.

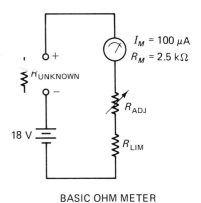

BASIC OHM METER

FIGURE 12-9

EXAMPLE 12-9

Refer to Figure 12-9. Calculate the values for R_{lim} and R_{adj} for this ohm meter circuit.

Solution: Begin by finding R_{MT}, the total meter resistance.

$$R_{MT} = \frac{V_{battery}}{I_M}$$

$$= \frac{18 \text{ V}}{100 \text{ }\mu\text{A}}$$

$$= 180 \text{ k}\Omega$$

Then find the limiting resistor, R_{lim}, which is approximately 95% of R_{MT}.

$$R_{lim} = 0.95 \times R_{MT}$$

$$= 0.95 \times 180,000$$

$$= 171 \text{ k}\Omega$$

Then find the resistance of the adjusting potentiometer, R_{adj}. It should be approximately 10% of R_{MT}.

$$R_{adj} = 0.10 \times R_{MT}$$
$$= 0.10 \times 180{,}000$$
$$= 18 \text{ k}\Omega \qquad \blacksquare$$

The calculations for R_{lim} and R_{adj} are only guidelines. For example, in Example 12-9, R_{lim} was calculated at 171,000 ohms. This is not a standard resistor value. The closest standard value that will allow meter adjustment is 150 kΩ. If the 150-kΩ resistor were used, it would be necessary for the potentiometer to be larger than the 18 kΩ that the calculations indicated. A 50-kΩ pot would allow us to adjust the meter resistance from approximately 150 kΩ to about 200 kΩ. The calculated values fall between both extremes and will provide a zero adjustment that will allow the meter to be calibrated regardless of the battery condition.

The highest range is the building block for this section of the ohmmeter. The highest range is determined by the size of the battery and the full-scale current capabilities of the meter movement. The highest range of the ohmmeter together with R_{MT} also determine the midscale for the ohmmeter.

EXAMPLE 12-10

Find the midscale and highest range for the ohmmeter designed in Example 12-9.

Solution: Midscale is essentially just R_{MT} void of any zeros.

$$R_{MT} = 180{,}000 \ \Omega$$
$$\text{Midscale} = 18$$

The highest range would be $R \times 10{,}000$, since

$$18 \times 10{,}000 = 180 \text{ k}\Omega \qquad \blacksquare$$

Ohmmeter Shunts Other ranges are constructed by placing a shunt resistor in parallel with R_{MT}. These calculations begin by choosing a lower range (usually a multiple of 10 of the highest range). This range is multiplied by the midscale to find the amount of resistance, R_{range}, needed to cause the current meter to deflect to half-scale. R_{range} is the total of R_{MT} and R_{shunt} when they are placed in parallel. A completed schematic and an outline of this procedure are demonstrated in Example 12-11.

EXAMPLE 12-11

Refer to Figure 12-10. Calculate the values for each shunt resistor in this ohmmeter.

Solution: Notice that the highest range has no shunt resistance. Notice also that the polarity of the battery appears to be reversed. It is not. To complete the series circuit it is necessary for the positive terminal of the battery to be connected to the positive probe. This requires that the battery be installed in the negative lead as shown.

$R \times 1k$ range: Begin by finding R_{range}.

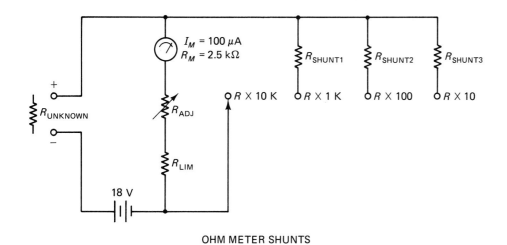

OHM METER SHUNTS

FIGURE 12-10

$$R_{\text{range}} = \text{midscale} \times \text{range}$$
$$= 18 \times 1000 \ \Omega$$
$$= 18 \ \text{k}\Omega$$

Then find R_{shunt1}.

$$R_{\text{shunt1}} = \frac{1}{1/R_{\text{range}} - 1/R_{MT}}$$
$$= \frac{1}{1/18 \ \text{k}\Omega - 1/180 \ \text{k}\Omega}$$
$$= 20 \ \text{k}\Omega$$

$R \times 100$ range:

$$R_{\text{range}} = \text{midscale} \times \text{range}$$
$$= 18 \times 100$$
$$= 1800 \ \Omega$$

$$R_{\text{shunt2}} = \frac{1}{1/R_{\text{range}} - 1/R_{MT}}$$
$$= \frac{1}{1/1800 \ \Omega - 1/180 \ \text{k}\Omega}$$
$$= 1818 \ \Omega$$

$R \times 10$ range:

$$R_{\text{range}} = 180 \ \Omega$$
$$R_{\text{shunt3}} = 180 \ \Omega$$

DESIGN AND CONSTRUCTION OF RESISTANCE METERS

Notice that the only shunt resistor which is not approximately equal to R_{range} is the one that is 10 times smaller than highest range. For all other values, the shunt can equal the value of R_{range}. That is, for the $R \times 100$ shunt we calculated 1818 Ω. The midscale times the range (18 × 100) is 1800. In terms of standard values, these two values are equal. Thus the only ranges that require calculation are those that differ from the highest range by a factor of 10. For example, if the highest range were $R \times 100\text{k}$, the only shunt resistor that requires calculation (major) is the $R \times 10\text{k}$ shunt. The shunt resistors for the other ranges can be approximated by multiplying the range times the midscale.

■

Ohmmeter Scaling The design of the ohmmeter circuitry is only a portion of the needed work. Building the circuit as outlined above provides an ohmmeter that will function; however, by itself it will not provide a readable meter. It will be necessary to scale (mark) off the meter so that you can read the ohm values directly. In other words, when you purchased the meter movement, the scale was marked off in units of the number of microamperes or milliamperes. It will be necessary for you to draw lines on the meter face that will be spaced in units of ohms (or their multiples).

This is accomplished by first thinking of the meter circuit as a simple series circuit (see Figure 12-9). Consider only the highest range components (there are no shunts on the highest range). The series circuit is comprised of R_{MT}, the battery, and a resistance equal to the point (ohm scale value) times the highest range. A formula expression would be

$$I_{\text{point}} = \frac{V_{\text{battery}}}{R_{\text{point}} + R_{MT}}$$

EXAMPLE 12-12

Suppose that you wanted to plot these remaining points on the ohmmeter scale, 1, 5, 50, and 100 of the meter described in Examples 12-9, 12-10, and 12-11. Calculate the current values that correspond to these points.

Solution: There are three points that are known at the onset. They are zero (far right), infinity (far left), and 18 (midscale). The procedure for finding other points is outlined in the following section.

Point 1:

$$R_{\text{point}} = \text{point} \times \text{highest range}$$
$$= 1 \times 10 \text{ k}\Omega$$
$$= 10 \text{ k}\Omega$$

$$I_{\text{point}} = \frac{V_{\text{battery}}}{R_{\text{point}} + R_{MT}}$$
$$= \frac{18}{10\text{k} + 180\text{k}}$$
$$= \frac{18}{190\text{k}}$$
$$= 94.74 \text{ μA}$$

Point 5:

$$R_{\text{point}} = 5 \times 10 \text{ k}\Omega = 50 \text{ k}\Omega$$

$$I_{\text{point}} = \frac{18}{50\text{k} + 180\text{k}} = 78.26 \text{ }\mu\text{A}$$

Point 50:

$$R_{\text{point}} = 50 \times 10 \text{ k}\Omega = 500 \text{ k}\Omega$$

$$I_{\text{point}} = \frac{18}{500\text{k} + 180\text{k}} = 26.47 \text{ }\mu\text{A}$$

Point 100:

$$R_{\text{point}} = 100 \times 10 \text{ k}\Omega = 1000 \text{ k}\Omega$$

$$I_{\text{point}} = \frac{18}{1000\text{k} + 180\text{k}} = 15.25 \text{ }\mu\text{A}$$ ∎

The current values found in Example 12-12 correspond to the ohmmeter points (see Figure 12-11). That is, 95 μA would correspond to point 1 on the ohmmeter. Remember that when the ohmmeter is used, it is necessary to multiply the scale reading (ohmmeter point) times the range. Similarly, 26 μA corresponds to point 50, and 15 μA corresponds to point 100.

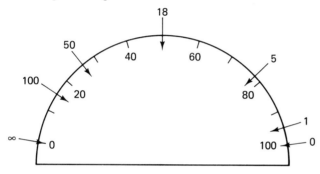

FIGURE 12-11

When designing an ohmmeter you should begin by choosing the midscale value. Typically, manufacturers use 12 or 15 for a midscale value. This value is used when deciding on a battery size. Some manufacturers go so far as to use more than one battery. One is used for low resistor values like the $R \times 1$ range. The main requirement on the battery for the lower resistance ranges is a high current capability. A second battery is used to provide the higher ranges such as $R \times 10\text{k}$. The requirement here is for the battery to have a high voltage value. In either arrangement the highest resistor range is used to calculate the design values. Most manufacturers use a 1.5-V D cell for the lower ranges and a 9-V transistor battery for the higher range. A schematic depicting this design is shown in Figure 12-12.

Complete VOM Meter Layout There are certain requirements for proper VOM meter layout. A simple design is shown in Figure 12-13. The main concern is in proper switch connection. To begin with, understand that the ground side (common) for all of the differing meter sections should be the same. With this in mind, the only other concern is the three necessary levels of switching that must be included. Starting at the positive probe, we must include one level of switches to place the battery in and out of the circuit.

This level of switching may be accomplished in several different manners. One

DESIGN AND CONSTRUCTION OF RESISTANCE METERS

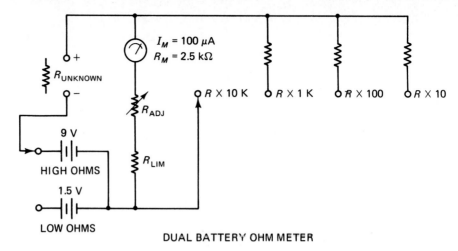

FIGURE 12-12

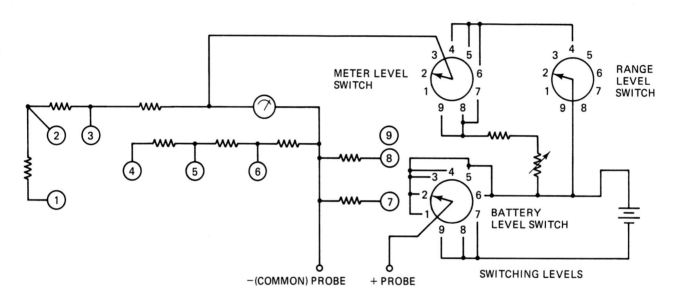

BASIC VOM METER DESIGN

FIGURE 12-13

way would be to use binding posts or female banana jacks. Then have the operator plug into the proper jack. Another way is to use a SPDT switch, where one position allows the circuit to operate without the battery and the other would allow battery insertion. If a three-level rotary switch were used, as in Figure 12-13, one level could be devoted to this task.

A second level would be devoted to connecting the circuit positive probe (from the first-level switch) to the resistor for the desired range (first voltage range, second current range, etc.). This level is almost always accomplished by a rotary switch. In Figure 12-13 this is labeled as the range-level switch. Understand that although the schematic does not show connections leaving this switch, a wire would lead from each numbered connection on this switch to a corresponding number on the meter circuitry.

The third level has the prime function of selecting the desired meter operation (volts, amperes, or resistance). This can again be done in a variety of ways. Anything from three SPST switches to a single three-position switch to a rotary switch can accomplish this task. Again Figure 12-13 uses a three-level rotary switch.

EXTENDING METER RANGES

There are times when a particular meter you are working with does not have a needed or useful range. In most instances a single resistor may extend this meter's capability to include this needed range. A voltmeter, for example, can be extended by connecting a series resistor to one of the probes. This is done by turning the meter to a range setting less than what is desired. If the difference between this range setting and the desired setting is multiplied by the sensitivity, the result is a resistance value that will extend the range to the desired amount. This process is outlined in Example 12-13.

EXAMPLE 12-13

Suppose that we were using a meter whose highest range was 2.5 V. Suppose also that this meter had a sensitivity of 50 kΩ/V. What value of resistance is needed to extend the range of this meter to 50 V?

Solution: Begin by subtracting the meter range from the range desired.

$$50 - 2.5 = 47.5 \text{ V}$$

Then multiply this value by the meter sensitivity.

$$47.5 \times 50 \text{ k}\Omega/\text{V} = 2.375 \text{ M}\Omega$$

Attaching a 2.4-MΩ resistor to one of the probes will provide a 50-V range for this meter. ∎

A current meter, on the other hand, is extended by placing a resistance in parallel with the probes. This is easiest if it is done on the voltmeter, but it can also be done on a current meter. The voltage range indicates V_{meter}, while 1/sensitivity provides the I_M value. The shunt value is generated in a manner similar to that of the individual shunt current meter.

EXAMPLE 12-14

Let's suppose that the 2.5-V meter of Example 12-13 were needed to measure 2 A. The sensitivity is still 50 kΩ/V. Calculate the value of shunt resistance needed to extend this meter's range.

Solution: Begin by determining the I_M of the meter.

$$I_M = \frac{1}{50\text{k}}$$

$$= 20 \text{ }\mu\text{A}$$

Find the shunt current for the range desired.

$$I_{shunt} = 2 - 20\ \mu A$$
$$= 1.99998\ A$$

Next find the shunt resistance.

$$R_{shunt} = \frac{2.5}{1.99998}$$
$$= 1.25\ \Omega$$

Connecting a 1.25-Ω resistor in parallel with both probes will provide a 2-A meter.

One other consideration is the power dissipation required by the resistor at the current level.

$$P = I_{range}^2 \times R_{shunt}$$
$$= 2^2 \times 1.25\ \Omega$$
$$= 5\ W$$

Another concern is the resistance of the meter. Ideally, a current meter should have a low internal resistance. Using a voltmeter for this purpose may create enough resistance to affect the circuit's operation. This should be avoided. ∎

SUMMARY

In summary, the VOM meter is several meters housed in the same case. It is the most used single tool in electronics. A prime consideration when choosing a VOM meter is the sensitivity of the voltmeter section. This specification indicates the meter's accuracy. In general, digital and transistorized meters have a higher sensitivity than other VOM meter types. The higher the sensitivity, the better the meter performance capabilities.

Voltmeters use current-limiting resistors called multipliers. These resistors determine the range application for the voltmeter. A larger multiplier will produce a higher voltage range. The multipliers can be connected as individual resistors or cascaded to form differing voltage ranges. The range of any voltmeter can be extended by adding a series resistance.

Current meters use shunts to control the current flow through the meter movement. The smaller the shunt resistance, the higher the resulting current range. Again, shunt resistors can be used individually or cascaded to form an Ayrton shunt. The range of any current meter can be extended by adding a parallel resistance.

QUESTIONS

12-1. Describe the term *meter loading*.

12-2. How is voltmeter sensitivity related to meter loading?

12-3. Would a typical voltmeter have more or less loading effect when turned to the 5- or 15-V range? Why?

12-4. Why are transistorized and vacuum-tube voltmeters generally said to be better than other types of voltage meters?

12-5. A voltmeter has a sensitivity of 50 kΩ/V. What is its internal resistance when it is turned to the 5-V range? The 25-V range?

12-6. Define the term *multiplier resistor*.

12-7. Describe the difference between an individual multiplier and a cascaded multiplier voltmeter. Include operational and construction differences.

12-8. What is a make-before-brake switch? Where would it be used?

12-9. Why do you think it's not necessary to use exact resistor values when constructing a voltmeter?

12-10. Describe the method used to find an individual multiplier resistor.

12-11. Define the term *shunt resistor*.

12-12. Describe the method used to find the total shunt resistance for an Ayrton shunt current meter.

12-13. When a shunt resistance of 0.11 Ω is needed, what do manufacturers use?

12-14. Describe the difference between an individual shunt and a cascaded shunt current meter. Include operational and construction differences.

12-15. How is the shunt current determined?

12-16. What is meant by the total loop resistance of a current meter?

12-17. Describe how a voltmeter is used.

12-18. Describe how a current meter is used.

12-19. Describe how an ohmmeter is used.

12-20. What is meant by *midscale* on an ohmmeter? Why is midscale so important?

12-21. The text states that $R_{MT} = R_M + R_{\lim} + R_{adj}$ and that $R_{\lim} = 0.95 \times R_{MT}$ and $R_{adj} = 0.1 \times R_{MT}$. How can this be?

12-22. Describe the procedure used to calculate ohmmeter shunts.

12-23. Describe the method used to produce the ohmmeter scale.

12-24. Why do some manufacturers include two batteries in their ohmmeter designs?

12-25. What are the three levels of switching required in a complete VOM meter schematic?

12-26. What information is needed to extend the range of a voltmeter?

12-27. Why is it easier to change a voltmeter into a current meter than it is to extend the range of an existing current meter? What can be a disadvantage of doing this?

PRACTICE PROBLEMS

12-1. A voltmeter has a sensitivity of 25 kΩ/V. What is its I_M?

12-2. A certain meter movement has a current rating of 30 μA. If a voltmeter were constructed using this meter movement, what sensitivity would it have?

12-3. Given that the sensitivity of a voltmeter is 60 kΩ/V, find the current rating for its meter movement.

12-4. If a voltmeter were to be constructed using a 1-mA meter, what sensitivity would it have?

12-5. A voltmeter has a sensitivity of 5 kΩ/V. Find its I_M.

12-6. A voltmeter having a sensitivity of 20 kΩ/V is turned to the 10-V range. What is its internal resistance?

12-7. A 50-kΩ/V meter is connected across a 200-kΩ resistor. The meter is turned to its 5-V range. What is the parallel resistance of the resistor and meter combination?

12-8. A voltmeter is to be used to measure the voltage across a 250-kΩ resistor. What is the minimum internal resistance that this meter should have in order for it to provide an accurate reading? If the voltage to be measured is 10 V, what sensitivity should this meter have?

12-9. Given a meter movement with an $I_M = 20$ μA and an $R_M = 2500$ Ω. Design a voltmeter having 5, 25, 50, and 250 V as ranges. Use either individual or cascaded multiplier design.

12-10. Design a voltmeter having 2.5, 20, and 50 V as ranges. Use a meter movement having an internal resistance of 2000 Ω and a current rating of 500 μA.

12-11. Use a 50-μA meter movement to design a voltmeter. The voltmeter is to have four ranges: 3, 15, 30, and 150 V. The internal resistance of the meter movement is 1500 Ω.

12-12. Given a meter movement with an $I_M = 20$ μA and an $R_M = 2500$ Ω. Design a current meter having 1, 10, and 100 mA as ranges. Use either individual or cascaded shunt design.

12-13. Design a current meter having 2, 20, and 100 mA as ranges. Use a meter movement having an internal resistance of 2000 Ω and a current rating of 500 μA.

12-14. Use a 50-μA meter movement to design a current meter. The current meter is to have four ranges: 1, 10, 100, and 500 mA. The internal meter resistance is 2000 Ω.

12-15. Given a meter movement having an $I_M = 20$ μA and an $R_M = 2500$ Ω. Find R_{MT} if a 9-V battery is used in the construction of an ohmmeter.

12-16. What value would be used as midscale on the ohmmeter of Problem 12-15?

12-17. Use an 18-V battery and a meter movement having a current rating of 1 mA and an internal resistance of 250-Ω design an ohmmeter having: $R \times 1k$, $R \times 100$, $R \times 10$, and $R \times 1$ as ranges.

12-18. The total meter resistance of an ohmmeter is 120 kΩ. The ohmmeter is constructed using a 12-V battery, and the current rating of the meter is 100 μA. Find the current values that will indicate the following resistance points: 1, 5, 12, 20, 50, and 100.

12-19. Given that $R_{MT} = 300$ kΩ and $I_M = 50$ μA. First find the voltage of the battery. Then find the following ohmmeter points: 1, 20, 50, and 100.

12-20. Given $I_M = 50$ μA and midscale = 12. Design an ohmmeter having $R \times 10k$, $R \times 100$, and $R \times 1$ as ranges.

12-21. A voltmeter has 5 V as its maximum range. Its sensitivity is 20 kΩ/V. Calculate the value of the multiplier needed to extend its range to 100 V.

12-22. A voltmeter is constructed using a meter movement that has a current rating of 1 mA. The highest range on this meter is 10 V. Find the value of the multiplier resistor that will extend its range to 25 V.

12-23. Extend the range of a 4-V meter to 50 V. Its sensitivity is 50 kΩ/V.

12-24. A voltmeter has 5 V as its maximum range. Its sensitivity is 20 kΩ/V. Calculate the value of shunt resistance that will allow it to measure 1 A.

12-25. What value of shunt resistance would be needed in order to use a 1-V meter with a sensitivity of 1 kΩ/V to measure 1 A of current? What power rating would this resistor need?

13

Magnetism

Chapter objectives

After reading this chapter and answering the questions and problems, you should be able to:

- Explain the concept of magnetism using magnetic dipoles and magnetic domain.
- Define *flux*, *flux density*, *permeability*, *ferromagnetic*, *paramagnetic*, and *diamagnetic*.
- Identify magnetic poles of an electromagnet using the left-hand rule.
- Define *magnetomotive force*, *field intensity*, and *reluctance*.
- List the SI (MKS) units for mmf, field intensity, flux density, and flux.
- Use Flemming's rule to determine motor action and current generation.
- Describe Lenz's law.
- Describe Faraday's law.
- Discuss the graph of flux density versus field intensity, the B–H curve, and hysteresis.

Magnetism and electricity are very closely related. So much so that it is almost impossible to think of one without thinking of the other. In this chapter we explain the concepts behind magnetism. Also, the terms and units of measure are defined. There are also examples of the physical and mathematic relations that pertain to magnetism.

MAGNETIC CONCEPTS

What is magnetism? Well, magnetism is an electromagnetic phenomenon due to the particular alignment of the electrons in a metal (Figure 13-1). These electrons form small segments called *magnetic dipoles*. When the magnetic dipoles are lined up, the metal is magnetic, which means that it will attract certain other metals. Each magnetic dipole has a north and south pole of its own. This is referred to as a *magnetic domain*.

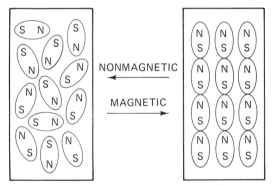

A NONMAGNETIC MATERIAL HAS RANDOM MAGNETIC DOMAINS, WHILE A MAGNETIC MATERIAL HAS UNIFORM MAGNETIC DOMAINS. **FIGURE 13-1**

Flux When a material has the magnetic domains lined up, an invisible force or bond is built up between the groups of north and south poles. These lines of force are called *flux* (Figure 13-2). The symbol for flux is ϕ. By convention flux is said to flow from the north pole to the south pole.

You can think of the flux lines as if they were rubber bands, all having a different length. When certain metals are inserted into the field of flux, the lines (rubber bands) are stretched out to pull the metal closer. As with rubber bands, the weaker lines are broken by the stretch; however, when the metal is close to the magnet, fewer of these lines are broken.

The north pole of a magnet is attracted to the north geographic pole (the south magnetic pole) of the earth; for this reason it is called the *north-seeking pole*. This seems confusing because the north pole of a magnet will repel the north pole of an-

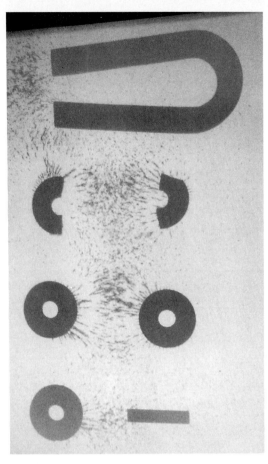

FIGURE 13-2 Flux Lines. Top to bottom single magnet, opposing poles, like poles, single magnet and metal object.

other magnet. The south poles repel each other. Opposite poles attract, while like poles repel.

Flux Density The strength of the magnetic field is referred to as *flux density*, because it is determined by the number of flux lines that are present. The symbol for flux density is B.

$$B = \frac{\phi}{A}$$

The area (A) is a section close to the magnet's pole (usually this is at the diameter or center of the pole). It is important to understand that the magnetic field is stronger at areas closer to the magnet than it is at greater distances. In fact, the strength of the field follows what is known as the *square law principle*. That is, the strength of the field is a direct relation to the square of the distance.

EXAMPLE 13-1

The flux density is known to be 18 units at a distance of 3 in. Find the flux density at 6 and 9 in.

Solution: Six inches is twice (two times) the distance of 3 in. The square law principle means that the flux density will decrease fourfold (2^2).

$$B = \frac{18}{4} = 4.5 \text{ units}$$

Nine inches is three times the distance of 3 in. The square law principle means that the flux density will decrease ninefold (3^2).

$$B = \frac{18}{9} = 2 \text{ units}$$ ∎

Flux density is measured with a magnetometer. A magnetometer measures the field strength (magnetic attraction) at a particular point (Figure 13-3). If this measurement is multiplied by the square of the distance that the magnetometer is located from the magnet, the product will be the flux density.

Permeability Several metals have a quality that allows them to concentrate the flux lines. This quality is referred to as *permeability*. The symbol for permeability is μ. Metals with a high μ can raise the flux density and field strength. One way to rate a metal's magnetic qualities is by its μ. There are three general classes of metals: ferromagnetic, paramagnetic, and diamagnetic.

Ferromagnetic materials have the strongest magnetic capabilities. They generally have a μ of 10 or more. It is not impossible for such a material to have a μ of 10,000 or even 100,000. Several metals fall in this class. Iron is one. The ferromagnetic materials take their name from *ferrous,* which is Latin for iron or fire rock. Other metals in this class are steel, nickel, and cobalt. There are several metal alloys that are also ferromagnetic in nature. Alnico is most used in the manufacture of magnets. Ferrites, nonmetallic man-made magnets, are also included in this class. They can be found, among other places, as the magnet in electric can openers and refrigerator magnets.

Paramagnetic materials are at best only slightly magnetic. Aluminum is the most common of these. When a magnet is held close to aluminum a slight attraction can be felt, but there is not enough attraction to allow the magnet (even a strong

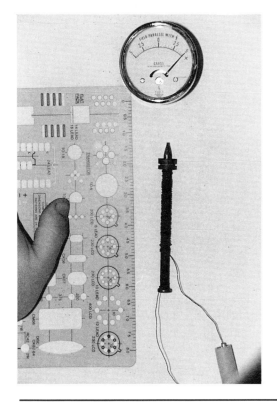

FIGURE 13-3A Magnetometer reading at 1.4 inches.

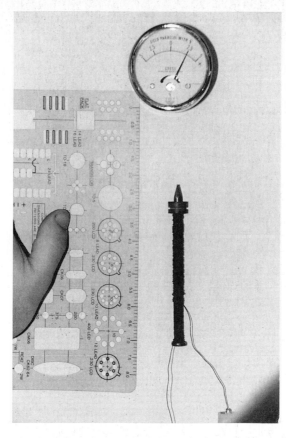

FIGURE 13-3B Magnetometer reading at 2.0 inches.

magnet) to stick to the metal. Other metals falling into this class are platinum, manganese, and chromium. All of these have a μ only slightly more than 1.

Diamagnetic materials are not magnetic. When a magnet is held to them, there is no attraction. In some instances, there is a repelling action that occurs. The most common of these is copper. Other metals in this group are gold, silver, zinc, and mercury. Their permeability is less than 1.

Left-Hand Rule for Pole Identification The majority of terms discussed in the foregoing section have had to do with permanent magnets (ones that continually demonstrated their magnetic qualities). Another type of magnet is an electromagnet (see Figure 13-4). An electromagnet is made by winding a wire around a ferromagnetic material. Current through the winding creates a magnetic field, while the magnetic core concentrates the flux lines. This type of magnet is only strongly magnetic when a current is flowing through the wire; thus it can be turned on or off at will. There are some rules that help us to relate this type of magnet to other types. One of these is called the left-hand rule. It states simply that when the fingers of the left hand are placed around the coil in the direction of current flow, the thumb will point to the north pole of the electromagnet (Figure 13-5). Other rules will be explained later.

Electromagnets and Related Terms Although *flux, flux density,* and *permeability* are used to describe electromagnets, there are other terms with which you should be familiar. The first of which is magnetomotive force (mmf). Mmf refers directly to the number of turns of wire used in the coil construction and the amount of current that flows through the wire. This is the basic description of an electromagnet's magnetic capabilities. The description can be taken a bit further by including the length of the coil's windings (not the total length of the wire but the physical

FIGURE 13-4 Electromagnetic Flux Lines.

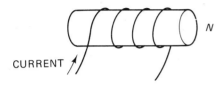

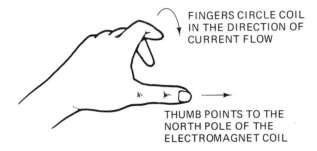

FIGURE 13-5

length of the coil form; see Figure 13-6). For loosely wrapped windings (longer lengths) the strength is weaker. For a strong electromagnet it is necessary for the winding to be as tightly wrapped (short length) as possible. When you consider the coil form length, another term is used. It is *field intensity*. The symbol for field intensity is H. H is equal to the mmf divided by the length of the coil.

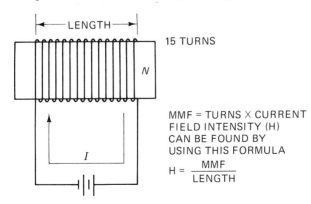

FIGURE 13-6

MAGNETIC CONCEPTS

The flux density is related to the field intensity by a factor equal to the permeability of the coil's core (the material placed in the center of the coil). Naturally, if a material having a high μ value were used as a core, the resultant flux density would be greater than if air were used for a core. In any case, it is necessary for us to have a standard (something to rate other materials by).

Scientists have elected to use air as a standard. The permeability of air is symbolized by μ_0. Other materials use μ_r to symbolize a quality known as relative permeability. To find or calculate the permeability of a material, it is necessary to multiply the permeability of air by the relative permeability of the material. The product is the total permeability of the core. This is the factor that relates field intensity to flux density. In other words,

$$\mu = \mu_0 \times \mu_r$$

and

$$B = \mu \times H$$

Reluctance In a lot of ways, flux is like current. As such, there is an opposition factor called *reluctance*. Reluctance is best explained as the distance between the poles of a magnet. The greater the distance, the more opposition there is to the flow of flux. Ohm's law also works for magnetism. It states that the flux is equal to the mmf divided by the reluctance:

$$\phi = \frac{\text{mmf}}{R}$$

The lines of flux flow from the north pole to the south pole of the magnet. They also serve the function of holding the dipoles in alignment. Any sudden jar or magnetic interference can break this bonding action. As a result, the material will lose a bit of its magnetic qualities. If a magnet is left in an open state for a long period of time, the reluctance that is introduced by the separation of the poles will also lessen the magnetic qualities of the material. For this reason it is recommended that a keeper be placed on or across the poles of a permanent magnet (Figure 13-7).

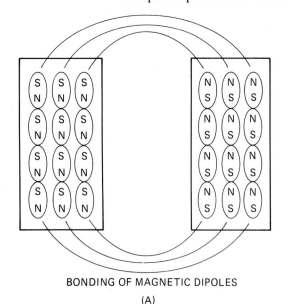

BONDING OF MAGNETIC DIPOLES
(A)

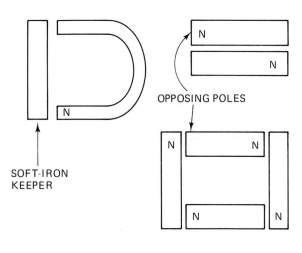

IRON KEEPERS REDUCE RELUCTANCE
(B)

FIGURE 13-7

MAGNETIC UNITS AND SYSTEMS

The most difficult factor to magnetism is not the ideas or concepts of the phenomenon, but instead, the units used to measure them. To begin with, two separate systems that are in use today (in earlier times there were many separate systems). One is the MKS system, generally referred to as the SI system (developed from 1954 through 1960). This is used as an international standard. It consists of larger quantities, in much the same way that a foot length compares to an inch length.

The second system is called the cgs system and is comprised of smaller quantities. Table 13-1 lists these different units, their symbols, and conversion factors. Understand that one line of magnetic flux is equivalent to 1 maxwell.

TABLE 13-1

System	Flux	Flux Density	Permeability of Air	Magnetomotive	Field Intensity
MKS	weber	tesla	1.26×10^{-6}	ampere-turns	ampere-turns/meter
cgs	maxwell	gauss	1	gilbert	oersted
Conversion	100 million	10,000		1.26	0.0126
Symbol	ϕ	B	μ_0	mmf	H

The mathematics that are involved depend largely on converting from one system of units to another. The following set of problems will demonstrate the use of the different systems and provide guidelines for system calculations. Pay particular attention to conversions. As a rule, the calculations are done entirely in one system and conversions, are made to the other as needed. In Example 13-2, the calculations are done in both systems to give you a comparison of the two.

EXAMPLE 13-2

Suppose that an electromagnet was constructed by winding 1500 turns of wire on a coil form. Calculate the amount of magnetomotive force produced when 650 mA of current flows through the coil windings.

Solution: In the MKS system the calculations are simple. You only need to use the equation for mmf. The unit for mmf is ampere-turns in the MKS system. See Table 13-1 for the correct unit.

$$\text{mmf} = \text{turns} \times \text{current}$$
$$= 1500 \times 0.65$$
$$= 975 \text{ ampere-turns}$$

In the cgs system the conversion factor 1.26 must be included in the formula and the unit name is gilberts in the cgs system (see Table 13-1).

$$\text{mmf} = 1.26 \times \text{turns} \times \text{current}$$
$$= 1.26 \times 1500 \times 0.65$$
$$= 1228.5 \text{ gilbert}$$

In Example 13-2 notice that the only difference between calculations for the MKS and cgs systems is the 1.26 conversion factor and the unit name.

EXAMPLE 13-3

Referring to Example 13-2, find the field intensity generated by the coil if the coil length is 15 cm.

Solution: In the MKS system the length must be in meters, not centimeters. So begin by converting the centimeter length into the equivalent number of meters.

$$\text{length} = 15 \text{ cm} = 0.15 \text{ m}$$

Then use the field intensity formula. The MKS unit for field intensity is ampere-turns/meter (see Table 13-1).

$$H = \frac{\text{mmf}}{\text{length}}$$

$$= \frac{975}{0.15}$$

$$= 6500 \text{ ampere-turns/meter}$$

In the cgs system the length must be in centimeters. In this case, no conversion is needed. Simply use the field intensity formula and the known cgs values.

$$H = \frac{\text{mmf}}{\text{length}}$$

$$= \frac{1228.5}{15}$$

$$= 81.9 \text{ oersted}$$

From Table 13-1, the correct cgs unit, oersted, can be found. ∎

In Example 13-3 you start seeing the need for conversions. Most of these conversions will be simple in nature. Be careful to ensure that you work in units which are in the same system. That is, when in the MKS (meter–kilogram–seconds) system, you must work in meters. In the cgs (centimeter–gram–second) system you must use centimeters as the length unit of measure.

EXAMPLE 13-4

Suppose that an iron core was used to make the coil described in Examples 13-2 and 13-3. (The wire is wound around a piece of iron.) Calculate the flux density if the core has a relative permeability of 200.

Solution: In the MKS system you must begin by finding the total permeability of the core.

$$\mu = \mu_0 \times \mu_r$$

$$= 1.26 \times 10^{-6} \times 200$$

$$= 2.52 \times 10^{-4}$$

Then the flux density can be found by using the flux density formula.

$$B = \mu \times H$$

$$= 2.52 \times 10^{-4} \times 6500$$

$$= 1.638 \text{ tesla}$$

See Table 13-1 for the appropriate unit.

In the cgs system, the permeability of air μ_0 is 1, so the relative and the total permeability are equal.

$$\mu = \mu_0 \times \mu_r$$
$$= 1 \times 200$$
$$= 200$$

Use the flux density formula and the cgs units to complete the calculations.

$$B = \mu \times H$$
$$= 200 \times 81.9$$
$$= 16{,}380 \text{ gauss}$$

Again the unit can be found by referring to Table 13-1. ∎

Notice that in these calculations it is necessary to use the proper value for the permeability of air for the particular system in use. Also, you can convert the flux density of the MKS system to the flux density of the cgs system simply by multiplying by 10,000, as Table 13-1 indicates.

Understand that the iron core produces a stronger magnetic field than does air by itself. In Example 13-4 the iron will produce a field 200 times that of air.

Reluctance is an important factor in magnetism. It indicates the resistance to the magnetic flow. The units for reluctance are ampere-turns/weber in the MKS system and gilberts/maxwell the cgs system.

EXAMPLE 13-5

Suppose that the flux in the iron core of Example 13-4 were 25 microweber. Calculate the reluctance of the magnet.

Solution: The formula for finding reluctance is

$$R = \frac{\text{mmf}}{\phi}$$

In the MKS you have

$$\text{mmf} = 975 \text{ ampere-turns (from Example 13-2)}$$

and

$$\phi = 25 \text{ microweber}$$

So

$$R = \frac{975}{0.000025}$$
$$= 39{,}000{,}000 \text{ ampere-turns/weber}$$

In the cgs system you have

$$\text{mmf} = 1228.5 \text{ gilbert (from Example 13-2)}$$

and

$$\phi = 25 \text{ microweber}$$

The flux must be converted from weber units to equivalent maxwell units.

Multiplying by 100,000,000 will do that.

$$\phi = 0.000025 \times 100,000,000$$
$$= 2500 \text{ maxwell}$$

Then

$$R = \frac{1228.5}{2500}$$
$$= 0.4914 \text{ gilbert/maxwell} \quad \blacksquare$$

ELECTROMAGNETIC CONCEPTS

The properties of magnetism are varied and complex; however, they parallel those of electricity, and more important, there are several areas where they cannot be separated. These are the areas that are discussed in the remaining portion of this chapter.

Motor Action and Fleming's Rules Motor action and its inverse operation (current generation) occur when a magnetic field and electromagnetic field interfere. It is possible to cause movement by using magnetic fields. (Remember: Like poles repel.) When you use an electromagnet to provide one pole and a permanent magnet to provide the other, you have the benefit of controlled action.

Suppose that you had a wire laying in a magnetic field. Understand that when a current is made to flow through the wire, a magnetic field will be generated around the wire. This generated field and the permanent field will interfere, causing the wire to move out of the field. This movement is called *motor action*. Flemming's rule for motor action states that when the fingers of the right hand are bent at three right angles (see Figure 13-8): the forefinger pointing in one direction, the middle finger pointing in another, and the thumb pointing in still another direction. When the forefinger points in the direction of mmf (north to south) and the middle finger

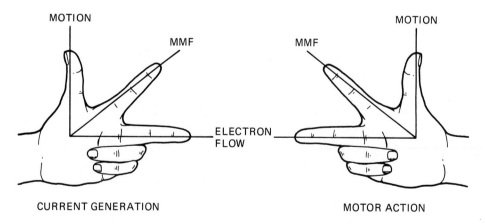

FLEMING'S LEFT-HAND RULE FOR CURRENT GENERATION

FIGURE 13-8 Left-hand Rule. The left hand is used to predict the direction of current produced by a generator. The right hand is used to predict the direction of motor action.

points in the direction of current flow (negative to positive), the thumb will indicate the direction of motor action.

It is almost impossible to draw a diagram that accurately depicts motor action. Most people use a dot to indicate a current that is flowing out of the page (see Figures 13-9 and 13-10). An X indicates a current that is flowing into the page (see Figure 13-10). An easy way to remember which is which is to think of an arrow. When the arrow is coming toward you, you would see the point first (dot). When it is going away from you, you would see the feathers (X) first.

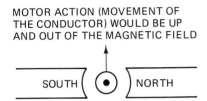

● IN THE CONDUCTOR INDICATES THAT
THE CURRENT IS FLOWING OUT OF THE PAGE   **FIGURE 13-9**

AN X INDICATES THAT THE CURRENT
IS FLOWING INTO THE PAGE WHILE A ●
INDICATES THAT THE CURRENT IS
FLOWING OUT OF THE PAGE   **FIGURE 13-10**

It is also possible to cause a current to flow by moving a wire while it is inside a magnetic field. This is called *current generation*. Flemming's rule also applies here with the exception that it is the left hand that predicts current generation. Again the forefinger points in the direction of mmf (north to south). The thumb points in the direction of movement. The middle finger indicates the direction of current generation (negative to positive).

Lenz's Law *Lenz's law* concerns the direction of current when a magnet is moved into a coil of wire. This is similar to Flemming's rule for current generation. It states that when a magnet is moved into a coil, the current produced will be such that the magnetic field of the coil will oppose that of the magnet. In other words, when the north pole of a magnet is moved into a coil. The coil will generate a current that will produce a magnetic field. The produced magnetic field will have its north pole opposing that of the magnet (see Figure 13-11). Again Lenz's law relates to the polarity of the induced current.

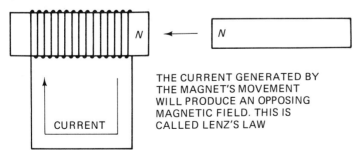

FIGURE 13-11

Faraday's Law An equally important law is Faraday's. Faraday took the conclusions of Lenz a bit further. His in-depth study of electromagnetic generation concluded that there were three determining factors in the production of voltage in a

ELECTROMAGNETIC CONCEPTS

generator. They were the number of turns, the speed with which the magnet is introduced into the coil, and the strength of the magnetic field itself. Mathematically, his formula appears as

$$V_{\text{induced}} = N \times \frac{d\phi}{dt}$$

The terms $d\phi$ and dt are calculus values. They mean roughly the same as $\Delta\phi$ and Δt. In mathematics, delta refers to a change. In this case the changes are in time (speed) and in flux (the amount of flux lines that cross the windings of the coil). Simply stated, changing any of these quantities will directly affect the amount of voltage that is produced. If you double the number of coil windings, the voltage output will also double provided that there are no changes in the other two quantities.

You can increase the voltage output in either of two other ways. Using a stronger magnetic field will increase the voltage output of a generator. The magnetic field is usually increased by running more current through an electromagnet that provides the field. Here again there is a direct ratio between the strength of the magnetic field (field intensity) and the induced voltage. Also, if the magnetic field has a faster movement, the induced voltage will also be increased.

B–H CURVE: FLUX DENSITY VERSUS FIELD INTENSITY

There is normally a direct ratio between the amount of field intensity and flux density. Although magnetic flux is invisible, it does take up space. As such, there is a limited amount of flux that can be placed in a given area. Relate the flux lines to pencils. You can only place a certain number of pencils in a box without breaking the box. The same thing is true of flux. Remember that flux density is the amount of flux in a given area (pencils in a box). We are stating that there is a limitation to that value. In terms of an electromagnet the limiting factors are the permeability of the coil and the diameter of the coil itself (all of the flux lines must go through the center of the coil). These factors also diminish the direct proportionality between flux density and field intensity.

First consider ways in which you can easily vary field intensity.

$$H = \frac{\text{mmf}}{l}$$

Magnetomotive force is measured in ampere-turns. Length is measured in meters. It would not be easy to build a coil that has a controllable length (possibly of a spring construction). Also, it would not be easy to construct a coil where the number of turns could be changed at will (a switch connected to every winding). The easiest of the three would be to change the amount of current through the coil. This would directly affect the amount of field intensity.

In plotting (graphing) the field intensity versus flux density, the simplest manner seems to be to vary the current through the coil and measure the flux density with a magnetometer. Varying the current will change the field intensity. The resultant change in flux density can be measured using the magnetometer.

As the current through the coil is increased (field intensity) from zero to a large amount, we see almost a linear change in flux density until we reach the point where our core can no longer hold any more flux (the pencil box is full). That point is called *saturation* (see Figure 13-12). Saturation relates to the strongest magnetic force that can be produced by our coil. You can increase the current further, but there will be no (or little) resulting increase in flux density. The core is saturated (filled).

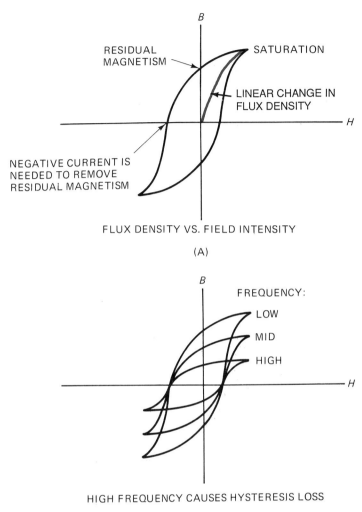

FIGURE 13-12

A peculiar thing happens when the current is later reduced to zero. The core retains some residual magnetism. This is called *retentivity*. Plotted on the graph, this causes a value of *B*, flux density, to exist without the need for field intensity. Also, the difference between the energy (saturated magnetic force) provided the coil and the residual value (the amount remaining after the current was turned off) shows the energy that was lost. This loss is referred to as *hysteresis*.

When the current was present in the coil, all the magnetic dipoles in the core where lined up (all pointing in the same direction). When the current was removed, the dipoles began to drift (some changed direction). This action caused a loss of flux and flux density. Hysteresis refers directly to those dipoles, which lagged behind (remained lined up).

In an alternating current (which flows one direction and then the other) through the coil, a similar loss is experienced. Understand that when the current was removed, it left a residual amount of magnetism. To remove this residual amount, it is necessary to reverse the current. This is shown on the graph (see Figure 13-12) at the point where flux density is zero and the current is negative. In an alternating current this action occurs over and over. If the peak amount of current is maintained and the frequency is increased, notice that the resultant flux density is reduced (see Figure 13-12). This is due to the lagging behind of the magnetic dipoles. At higher and higher frequencies (changing the direction of the current faster), more and more of this hysteresis loss occurs.

SUMMARY

Electricity and magnetism are very closely related. Laws that are used to describe the action of one can be used to describe the action of the other. In magnetism, like poles repel and unlike poles attract. The flow of magnetism is called flux and is symbolized by ϕ. Flux density rates the strength of a magnetic field. The strength of a magnetic field increases by the square law principle as the distance from the magnet decreases.

Some materials are more magnetic than others. Permeability is used to describe this quality. A higher permeability indicates a stronger magnetic capability. The permeability of air is used to rate the permeability of other materials. Permeability of this type is referred to as relative permeability. This factor must be included in most calculations.

Calculations are dependent on the system that is used. The MKS system is the most often used system. In any of these systems, simple conversions, such as centimeter to meters, will cause the greatest difficulty.

Several rules and laws are used to predict the direction of current flow, the polarity of the voltage, and the direction of motor action. Among these are Flemming's rule (motor action), Lenz's law (direction of current flow), and Faraday's law (voltage production). Also, once magnetic dipoles are bonded together, they try to remain bonded. In electromagnets this is seen as hysteresis loss. Reluctance can break these magnetic bonds. Magnetic keepers reduce the reluctance of the magnetic circuit.

QUESTIONS

13-1. Define *magnetic flux*.

13-2. To which direction will the north pole of a magnet point?

13-3. Describe what will happen when the south poles of two different magnets are placed together.

13-4. What is the direction of flux flow?

13-5. Describe flux density. How can it be calculated?

13-6. Describe the square law principle.

13-7. What is a magnetometer? What is it used for?

13-8. If the flux density were measured at a distance of 0.5 m and found to be 0.35 weber, what would the flux density be at a distance of 1 m?

13-9. Define *permeability*.

13-10. Describe how the permeability of a material affects its magnetic qualities.

13-11. Define *ferromagnetic*. List several metals that have this quality.

13-12. Define *paramagnetic*. List several metals that exhibit this quality.

13-13. To what does *diamagnetic* refer? List several metals that show this quality.

13-14. Describe how the left-hand rule can be used to determine the pole of an electromagnet.

13-15. What is meant by the term *magnetomotive force*?

13-16. Define *field intensity*. Describe its relation to magnetomotive force.

13-17. Describe the effect of placing a metal of high permeability in the center of an electromagnet's coil.

13-18. How does reluctance relate to magnetic keepers? Explain the theory behind the use of magnetic keepers.

13-19. Describe the difference between the MKS system of measure and the cgs system of measure.

13-20. When working in the MKS system, in what unit should the length be measured?

13-21. What is the permeability of air in the MKS system? In the cgs system?

13-22. How many square centimeters are in 0.75 m^2?

13-23. Describe motor action. How does it relate to Flemming's rule?

13-24. The right-hand is used to determine motor action. What is the left-hand used for?

13-25. When using Flemming's rule, what does the middle finger point toward?

13-26. What does the X in a conductor represent?

13-27. Describe Lenz's law.

13-28. Briefly state Faraday's law.

13-29. Explain what is meant by the term *B–H* curve?

13-30. Define *hysteresis*. How does it relate to the *B–H* curve?

13-31. How does the term *saturation* relate to magnetic fields?

13-32. Describe the effect of high frequencies on hysteresis and the *B–H* curve.

13-33. Briefly describe the difference between a permanent magnet and an electromagnet.

13-34. Demonstrate the similarities between Ohm's law for electricity and Ohm's law for magnetism.

13-35. In your own words, describe the manner you would use to demagnetize a metal object.

PRACTICE PROBLEMS

Each of these problems is designed to give you maximum practice in unit conversion. Unit conversion is the most difficult and trying portion of studying magnetism. It is suggested that you begin each problem by converting all of the variables into their MKS equivalent units. Perform the needed calculations and convert your answers into the requested system of units (MKS or cgs).

13-1. In a given circular area with a radius of 10 cm, 2000 lines of flux flow. Calculate the flux density of this area.

13-2. In Problem 13-1, calculate the amount of flux. Give your answer in webers.

13-3. A total of 400 microweber passes through 20 cm^2. Find the flux density of this magnetic field.

13-4. A magnetometer measures a field strength of 2.5 gauss/square cm at a distance of 15 cm. Calculate its flux density. Predict its field strength at a distance of 30 cm.

13-5. A total of 525 mA of current flows through 2500 turns of a coil. How much magnetomotive force is produced?

13-6. A coil constructed of 400 turns of wire has 0.225 A of current flowing through it. Calculate the magnetomotive force produced by this coil.

13-7. A coil 25 cm in length is constructed with 5000 turns. Find the field intensity of such a coil when there is 750 mA of current flowing through the coil.

13-8. A coil 0.35 m in length has 0.625 A of current flowing through it. The coil also contains some 2550 turns of wire. What is the field intensity of this coil?

13-9. Exactly 3500 turns of wire are wrapped around a core. The core has a relative permeability of 500 and a length of 15 cm. Calculate the magnetomotive force, field intensity, and flux density of this coil when its current flow is 0.25 A.

13-10. It is desired to use a field intensity of 150 oersted to produce a flux density of 4 tesla. What is the permeability of the needed core?

13-11. A coil is constructed using 2500 turns of wire. The coil is 12.5 cm in length and has a constant 350 mA of current flowing through it. A magnetometer measures approximately 88 gauss. When a soft-iron core is introduced, the magnetometer reading increases to 250 gauss. Find the permeability of the core.

13-12. The flux density of an electromagnet's coil is measured first without a core and then with a core. The magnetometer indicates that the flux density is 20 gauss without the core and 1850 gauss with the core. This core is later used to construct a second electromagnet. The second magnet is to have 2500 turns, a length of 20 cm, and a current flow of 1.25 A. Predict the flux density of this second electromagnet.

13-13. An electromagnet is constructed using 5000 turns and 250 mA of current. This coil uses an air core and generates 3 microweber of flux. Calculate the reluctance of this magnetic circuit.

13-14. The air gap of a certain magnet is known to have a reluctance of 3000 ampere-turns/weber. The flux is known to be 1800 maxwells. Find magnetomotive force generated by this magnet.

13-15. As a generator turns, the flux changes at a constant rate of 80 microweber/second. Calculate the number of turns needed to generate an output voltage of 50 V.

13-16. What rate of change in flux is needed when 2500 turns of wire are used to generate 15 V?

13-17. Given a core having a permeability of 500, and a coil having 3000 turns and a length of 25 cm, calculate the current needed to produce a flux density of 2.5 tesla.

13-18. Given a magnetic coil containing 400 turns that produces a magnetomotive force of 750 gilbert, find the current following through this coil.

13-19. A material has a reluctance of 1000 ampere-turns/weber. How much current must flow through a 100-turn coil to produce 5000 maxwells of flux?

13-20. What is the flux density in a coil 5 cm in diameter if the flux is 0.005 weber? (*Hint:* Use the area of the coil.)

13-21. What is the total flux in a coil having a diameter of 3 cm if the flux density is known to be 250 tesla?

13-22. A 500-turn coil is 25 mm in length and has 10 mA flowing through it. What field strength is produced by the coil?

13-23. What is the length of a coil with 450 turns having 75 mA flowing through it if it is producing 150 oersted?

13-24. What is the relative permeability of a core that allows a field strength of 150 oersted to produce a flux density of 50000 tesla?

13-25. A 750-turn coil is 60 mm in length and has 25 mA flowing through it. What field strength is produced by the coil?

14

AC Voltage

Chapter objectives

After reading this chapter and answering the questions and problems, you should be able to:

- Define *period*, *amplitude*, *quadrant*, *sine*, and *arcsine* and relate each to the sine wave.
- State the effect that changing the amplitude or period has on the sine wave.
- Use a 0 to 90° sine table to find the sine of angles over 90°.
- Define *degree*, *radian*, and *gradient*, and state how each relates to the period of the sine wave.
- Use your calculator to perform the sine and arcsine operations.
- Given any two of quadrant, amplitude, instantaneous voltage, and angle of rotation, find the remaining values.
- Use terms such as *flux*, *rotation*, and *positive* and *negative current* to describe the way a generator produces ac voltage.
- List and describe five ways of measuring ac voltage and state what instruments can be used to measure each.
- Given any one of the following voltage measurements—instantaneous, peak to peak, peak, rms, or average—find the equivalent for each.
- Given any two, use proportionality to find angle of rotation, instantaneous time, and period.
- Define *wavelength*.
- Relate period, frequency, and wavelength to each other using descriptions and conversion methods.
- Define *phase*, *nonsinusiodal*, and *harmonics*.
- Predict the resultant wave when sine waves of different amplitude, frequencies, and phases are combined.

In this chapter we discuss alternating current and how it is produced by a generator. The action of a sine wave and its related mathematics are explained. The different ways in which alternating voltage is measured are discussed and other terms related to ac voltages and frequencies are explained.

GETTING STARTED: SINE WAVES

The theory behind alternating current and its resulting voltage is based on trigonometry, specifically the sine wave and its properties. The sine wave and its major portions are shown in Figure 14-1. It represents the change in voltage (or current) associated with alternating currents. Particular measurements of the sine wave and its amplitude and period will be explained in the section that follow.

Amplitude The amplitude of the sine wave is its peak value. This value is listed as V_P or I_P in an electrical equation. The basic pattern formed by the sine wave remains the same no matter what value the amplitude assumes. The amplitude affects only the height of the wave. The amplitude is the peak value of the alternating voltage (see Figure 14-1).

$$e_{in} = V_P \times \sin \theta$$

This is a typical ac electricity equation. The V_P is the amplitude.

Period The period of the sine wave is the length of time between cycles. The sine wave is a recurring function, meaning that it will repeat itself after every cycle. The period of the cycle can be measured in several units (degrees, radians, and gradients). They all perform the same function and can be used to generate the sine wave. Degrees are the most common form of measurement for the period. There are 360° in one cycle. The radian measure of period relates best to electronics. Most of the electronic formulas have a basis built around the radian mode of measurement. There are 2π radians in one cycle. The gradient mode of measure is not used in electronics. It specifies the percent of grade (slope) in highway construction. A 50% grade is equal to 45°. There are 400 gradients in one cycle.

In years gone by, engineers found it easiest to do the majority of their calculations in the radian mode. They could directly convert radian measure into reactant values (a form of ac resistance covered in Chapter 15). At the end of their calculations, they would change from the radian mode to the degree mode for better interpretation of the results. With the calculators manufactured today, it is no longer necessary to do all the calculations in the radian mode. For that reason, the majority of calculations demonstrated in this book are shown in the degree mode.

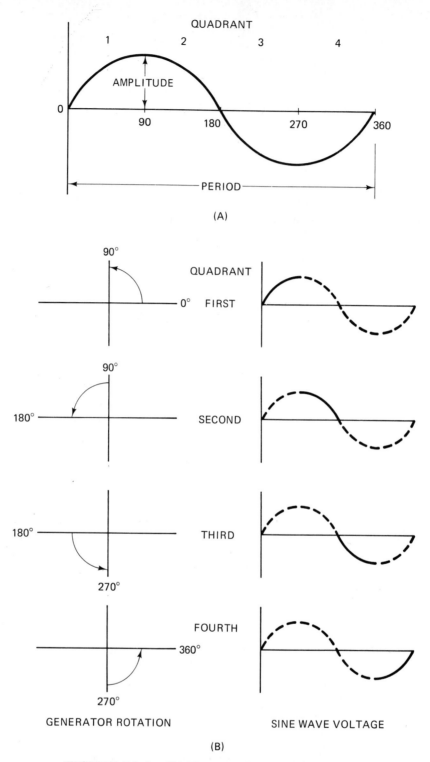

FIGURE 14-1 (A) The terminology for one cycle of a sine wave; (B) Generator rotation and the produced sine wave voltage for each quadrant.

Sine Wave The sine wave itself is divided up into four separate sections called *quadrants*. Each quadrant is then equal to 90° (360°/4 = 90°). The first quadrant begins at 0° and ends at 90°. The second quadrant begins at 90° and ends at 180°. The third is from 180 to 270°. The fourth quadrant begins at 270° and ends at 360°. Understand that the sine wave is a recurring function, which means that when the wave ends, it immediately starts to repeat the same pattern. In other words, the 360° point (the end) has the same sine value as the 0° point (the beginning).

It is only necessary to know the sine values for the first 90°. With the exception of sign (positive or negative), the sine values repeat every 90° (see Table 14-1). The sign wave is a positive value during the first 180° of generator operation (this is called the positive alternation in electronics). It reaches a peak at 90°. During the first 90° the sine value can be found by using the angles as they appear. For the remaining section of the positive alternation (90 to 180°), the sine value can be found by subtracting the angle (α) from 180° (see Figure 14-2) and using the resulting angle (θ) for a reference,

$$\theta = 180° - \alpha$$

for the corresponding sine value.

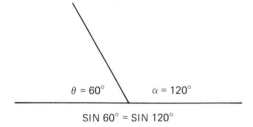

SIN 60° = SIN 120°

FIGURE 14-2

EXAMPLE 14-1

Suppose that you wanted to know the sine value for 110°. What first-quadrant angle has the same sine value?

Solution: You can find the equivalent angle by the following process:

$$\theta = 180° - \alpha$$
$$= 180° - 110°$$
$$= 70°$$

The sine of 110° is equal to the sine of 70°. ■

Similarly, it is possible to find the sine values for angles in the other quadrants using this method. The sine values are negative in both the third and fourth quadrants. In the third quadrant the formula for finding θ is

$$\theta = 180° - \alpha \quad \text{(note that the angle will be negative)}$$

The negative sign is left on the θ value only to indicate that the sine value must also be negative. In the fourth quadrant θ is found by

$$\theta = \alpha - 360° \quad \text{(note that the angle will be negative)}$$

TABLE 14-1 Sine Values for 360°

Angle	sine	Angle	sine	Angle	sine	Angle	sine	Angle	sine	Angle	sine
1	.01745	46	.71933	91	.99984	136	.69465	181	−.01746	226	−.71934
2	.03489	47	.73135	92	.99939	137	.68199	182	−.03490	227	−.73136
3	.05233	48	.74314	93	.99862	138	.66913	183	−.05234	228	−.74315
4	.06975	49	.75470	94	.99756	139	.65605	184	−.06976	229	−.75471
5	.08715	50	.76604	95	.99619	140	.64278	185	−.08716	230	−.76605
6	.10452	51	.77714	96	.99452	141	.62932	186	−.10453	231	−.77715
7	.12186	52	.78801	97	.99254	142	.61566	187	−.12187	232	−.78802
8	.13917	53	.79863	98	.99026	143	.60181	188	−.13918	233	−.79864
9	.15643	54	.80901	99	.98768	144	.58778	189	−.15644	234	−.80902
10	.17364	55	.81915	100	.98480	145	.57357	190	−.17365	235	−.81916
11	.19080	56	.82903	101	.98162	146	.55919	191	−.19081	236	−.82904
12	.20791	57	.83867	102	.97814	147	.54463	192	−.20792	237	−.83868
13	.22495	58	.84804	103	.97437	148	.52991	193	−.22496	238	−.84805
14	.24192	59	.85716	104	.97029	149	.51503	194	−.24193	239	−.85717
15	.25881	60	.86602	105	.96592	150	.50000	195	−.25882	240	−.86603
16	.27563	61	.87461	106	.96126	151	.48480	196	−.27564	241	−.87462
17	.29237	62	.88294	107	.95630	152	.46947	197	−.29238	242	−.88295
18	.30901	63	.89100	108	.95105	153	.45399	198	−.30902	243	−.89101
19	.32556	64	.89879	109	.94551	154	.43837	199	−.32557	244	−.89880
20	.34202	65	.90630	110	.93969	155	.42261	200	−.34203	245	−.90631
21	.35836	66	.91354	111	.93358	156	.40673	201	−.35837	246	−.91355
22	.37460	67	.92050	112	.92718	157	.39173	202	−.37461	247	−.92051
23	.39073	68	.92718	113	.92050	158	.37460	203	−.39074	248	−.92719
24	.40673	69	.93358	114	.91354	159	.35836	204	−.40674	249	−.93359
25	.42261	70	.93969	115	.90630	160	.34202	205	−.42262	250	−.93970
26	.43837	71	.94551	116	.89879	161	.32556	206	−.43838	251	−.94552
27	.45399	72	.95105	117	.89100	162	.30901	207	−.45400	252	−.95106
28	.46947	73	.95630	118	.88294	163	.29237	208	−.46948	253	−.95631
29	.48480	74	.96126	119	.87461	164	.27563	209	−.48481	254	−.96127
30	.49999	75	.96592	120	.86602	165	.25881	210	−.50000	255	−.96593
31	.51503	76	.97029	121	.85715	166	.24192	211	−.51504	256	−.97030
32	.52991	77	.97437	122	.84804	167	.22495	212	−.52992	257	−.97438
33	.54463	78	.97814	123	.83867	168	.20791	213	−.54464	258	−.97815
34	.55919	79	.98162	124	.82903	169	.19080	214	−.55920	259	−.98163
35	.57357	80	.98480	125	.81915	170	.17364	215	−.57358	260	−.98481
36	.58778	81	.98768	126	.80901	171	.15643	216	−.58779	261	−.98769
37	.60181	82	.99026	127	.79863	172	.13917	217	−.60182	262	−.99027
38	.61566	83	.99254	128	.78801	173	.12186	218	−.61567	263	−.99255
39	.62932	84	.99452	129	.77714	174	.10452	219	−.62933	264	−.99453
40	.64278	85	.99619	130	.76604	175	.08715	220	−.64279	265	−.99620
41	.65605	86	.99756	131	.75470	176	.06975	221	−.65606	266	−.99757
42	.66913	87	.99862	132	.74314	177	.05233	222	−.66914	267	−.99863
43	.68199	88	.99939	133	.73135	178	.03489	223	−.68200	268	−.99940
44	.69465	89	.99984	134	.71933	179	.01745	224	−.69466	269	−.99985
45	.70710	90	.99999	135	.70710	180	0	225	−.70711	270	−1

Angle	sine	Angle	sine
271	−.99985	316	−.69466
272	−.99940	317	−.68200
273	−.99863	318	−.66914
274	−.99757	319	−.65606
275	−.99620	320	−.64279
276	−.99453	321	−.62933
277	−.99255	322	−.61567
278	−.99027	323	−.60182
279	−.98769	324	−.58779
280	−.98481	325	−.57358
281	−.98163	326	−.55920
282	−.97815	327	−.54464
283	−.97438	328	−.52992
284	−.97030	329	−.51504
285	−.96593	330	−.50001
286	−.96127	331	−.48481
287	−.95631	332	−.46948
288	−.95106	333	−.45400
289	−.94552	334	−.43838
290	−.93970	335	−.42262
291	−.93359	336	−.40674
292	−.92719	337	−.39074
293	−.92051	338	−.37461
294	−.91355	339	−.35837
295	−.90631	340	−.34203
296	−.89880	341	−.32557
297	−.89101	342	−.30902
298	−.88295	343	−.29238
299	−.87462	344	−.27564
300	−.86603	345	−.25883
301	−.85717	346	−.24193
302	−.84805	347	−.22496
303	−.83868	348	−.20792
304	−.82904	349	−.19081
305	−.81916	350	−.17365
306	−.80902	351	−.15644
307	−.79864	352	−.13918
308	−.78802	353	−.12187
309	−.77715	354	−.10453
310	−.76605	355	−.08716
311	−.75471	356	−.06976
312	−.74315	357	−.05234
313	−.73136	358	−.03499
314	−.71934	359	−.01746
315	−.70711	360	0

When using a calculator this process is not as difficult as it may seem. In fact, the operations are affected only when they are done in reverse order—when you know the sine of the angle and are trying to find the angle.

CALCULATOR USE

The first step to using a calculator to solve these trigonometric problems is to make sure that it is turned to the proper mode (degrees, radians, or gradients). It is best to read the operator's manual for your particular calculator. Most calculators have a button labeled DRG. When this key is pushed, it will cause a different set of small letters (deg, rad, grad) to appear on the display. These letters indicate the present mode.

Again it is best to read the operator's manual for specific information about your calculator. I will outline the most common way to access the sine and other trigonometric functions. The sine function is used as follows:

1. Enter an angle.
2. Press the sin key (or other trig function—cos or tan).

It is important that you wait until the sine value is displayed on the calculator before other entries are made. Otherwise, an incorrect answer will be produced.

EXAMPLE 14-2

Use your calculator to find the sine of 40°.

Solution: The sine of 40° can be found by entering the following sequence of keystrokes:

$$[4]\ [0]\ [\sin]$$

The display should be

$$.6428 \quad \blacksquare$$

It is important to understand that the calculator is a dumb machine. It can only do what you tell it to do. When you enter the number 40, the calculator understands the number (operand) but does not know what you wish it to do to 40 until you press the sine button (operation). Example 14.3 is more complicated.

EXAMPLE 14-3

Use your calculator to solve

$$80 \times \sin 120° =$$

Solution:

$$80 \times \sin 120° =$$
$$[8]\ [0]\ [\times]\ [1]\ [2]\ [0]\ [\sin]\ [=]$$

The display should be

$$69.28$$

Remember to wait after pressing the sine key. $\blacksquare$

The sine function, like every other function, has an inverse function. Remember that the inverse of the multiplication function is divide. The inverse of the addition operation is subtract, and so on. The inverse of the sine function is the arcsine (the angle whose sine is). In textbooks it is written in many different ways:

1. arcsin 0.707 = 45°
2. $\sin^{-1}$ 0.707 = 45°
3. $\widehat{\sin}$ 0.707 = 45°

All of these different manners of writing arcsine may be encountered from time to time. The most common is the $\sin^{-1}$ version. That is the version that is stressed in this book.

Again, it is important that you realize that a calculator is no smarter than its operator. When using the arcsine function it is necessary for you (the operator) to know which particular quadrant the angle is in. In most cases this will be a factor that is given or is evident by the information in the question. When an $\sin^{-1}$ function is done, the calculator will provide an answer that will be meaningful only if the angle in question is located in the first quadrant (also the fourth, if you understand that the negative sign indicates a reverse direction). To place the angle in the proper quadrant, the following calculator procedures can be used.

First quadrant. Calculate θ, then do nothing. The angle is correct.

Second and third quadrants. Calculate θ, then enter this sequence of keystrokes:

$$[+/-] \; [+] \; [1] \; [8] \; [0] \; [=]$$

Fourth quadrant. Calculate θ then enter this sequence of keystrokes:

$$[+] \; [3] \; [6] \; [0] \; [=]$$

These types of calculations are shown in the examples that follow.

EXAMPLE 14-4 (First Quadrant)

Calculate θ in the equation

$$25 = 85 \times \sin \theta$$

Solution:

$$25 = 85 \times \sin \theta$$

$$\sin \theta = \frac{25}{85}$$

$$= 0.2941$$

$$\theta = \sin^{-1} 0.2941$$

$$= 17.10°$$

$$[2] \; [5] \; [\div] \; [8] \; [5] \; [=] \; [\sin^{-1}]$$

The display is

17.10 ∎

Note: It may be necessary to press the second function key before pressing the $\sin^{-1}$.

EXAMPLE 14-5 (Second Quadrant)

Calculate θ in the equation

$$52 = 75 \times \sin \theta$$

Solution:

$$52 = 75 \times \sin \theta$$

$$\sin \theta = \frac{52}{75}$$

$$= 0.6933$$

$$\theta = \sin^{-1} 0.6933$$

$$= 43.89° \quad \text{(first quadrant)}$$

$$= 180° - 43.89$$

$$= 136.11° \quad \text{(second quadrant)}$$

[5] [2] [÷] [7] [5] [=] [sin⁻¹] [+/−] [+] [1] [8] [0]

The display is

$$136.11$$

■

EXAMPLE 14-6 (Third Quadrant)

Calculate θ in the equation.

$$-86 = 105 \times \sin \theta$$

Solution:

$$\sin \theta = \frac{-86}{105}$$

$$= -0.8190$$

$$\theta = \sin^{-1}(-0.8190)$$

$$= -54.99° \quad \text{(fourth-quadrant alternate)}$$

$$= 180° - (-54.99°)$$

$$= 234.99° \quad \text{(third quadrant)}$$

[8] [6] [+/−] [÷] [1] [0] [5] [=]
[sin⁻¹] [+/−] [+] [1] [8] [0] [=]

The display is

$$234.98$$

■

EXAMPLE 14-7 (Fourth Quadrant)

Calculate θ in the equation

$$-45 = 50 \sin \theta$$

Solution:

CALCULATOR USE

$$\sin \theta = \frac{-45}{50}$$
$$= -0.9$$
$$\theta = \sin^{-1}(-0.9)$$
$$= -64.16° \quad \text{(fourth-quadrant alternate)}$$
$$= 360° + -64.16°$$
$$= 295.84° \quad \text{(fourth quadrant)}$$

[4] [5] [+/−] [÷] [5] [0] [=]

[sin⁻¹] [+] [3] [6] [0] [=]

The display is

$$295.84 \quad \blacksquare$$

At first glance some of these calculations appear to be different from the ones outlined previously. They are not. Remember that the +/− key will change the sign of the value displayed on the calculator. As such, it performs the function of subtracting the displayed value from the next number entered. This is the same operation as that done in previous examples.

SELF-TEST

1. $\sin 47° = $ _____
2. $\sin 158° = $ _____
3. $\sin^{-1} 0.8574 = $ _____
4. $\sin^{-1} 0.73829 = $ _____
5. $\sin 243° = $ _____
6. $\sin 58° = $ _____
7. $\sin^{-1} -0.324 = $ _____
8. $\sin^{-1} 0.934 = $ _____
9. $\sin 342° = $ _____
10. $\sin 221° = $ _____
11. $\sin^{-1} -0.574 = $ _____
12. $\sin^{-1} (-0.3739) = $ _____
13. $80 \times \sin 35° = $ _____ is in the _____ quadrant
14. $80 \times \sin 335° = $ _____ is in the _____ quadrant
15. $80 \times \sin 185° = $ _____ is in the _____ quadrant
16. $80 \times \sin 235° = $ _____ is in the _____ quadrant
17. $80 \times \sin 122° = $ _____ is in the _____ quadrant
18. Given that the angle is located in the second quadrant and that $180 \times \sin \theta = 150$. Find θ.
19. Given that the angle is located in the third quadrant and that $80 \times \sin \theta = -70$. Find θ.
20. Given that the angle is located in the fourth quadrant and that $190 \times \sin \theta = -50$. Find θ.
21. Given that the angle is located in the first quadrant and that $63 \times \sin \theta = 50$. Find θ.
22. Given that the angle is located in the first quadrant and that $678 \times \sin \theta = 152$. Find θ.
23. Given that the angle is located in the third quadrant and that $56 \times \sin \theta = -50$. Find θ.

24. Given that the angle is located in the fourth quadrant and that $80 \times \sin \theta = -45$. Find θ.
25. Given that the angle is located in the first quadrant and that $80 \times \sin \theta = 5$. Find θ.

ANSWERS TO SELF-TEST

1. 0.7314
2. 0.3746
3. 59.03°
4. 47.59°
5. −0.8910
6. 0.8480
7. −18.91°
8. 69.07°
9. −0.3090
10. −0.6561
11. −35.03°
12. −21.96°
13. 45.87 *in first quadrant*
14. −33.81 in fourth quadrant
15. −6.97 *in third quadrant*
16. −65.53 in third quadrant
17. 67.84 *in second quadrant*
18. 123.56°
19. 241.04°
20. 344.74°
21. 52.53°
22. 12.96°
23. 243.23°
24. 325.77°
25. 3.58°

AC THEORY

The alternating-current generator is a device that uses a magnetic field to produce voltage. It is important to realize that the generator rotates counterclockwise. Also, the starting point is at the right horizontal position (see Figure 14-3). At this point the windings of the generator are not crossing any of the magnetic flux lines and there is no resulting voltage produced. As the generator is rotated to the vertical position (in a counterclockwise direction), the induced voltage increases (more of the flux lines are crossed by the windings) until a peak voltage is produced. This occurs at the vertical position after the generator has turned through 90° of rotation (at this point the windings are moving perpendicular to the flux lines). The voltage produced as the generator is rotated another 90° in a counterclockwise direction de-

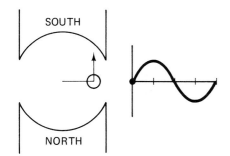

CONDUCTOR MOVEMENT THAT
COINCIDES WITH FLUX DIRECTION
PRODUCES ZERO VOLTAGE

(A)

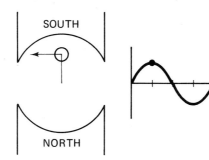

CONDUCTOR MOVEMENT
PERPENDICULAR TO FLUX DIRECTION
PRODUCES MAXIMUM VOLTAGE

(B)

FIGURE 14-3

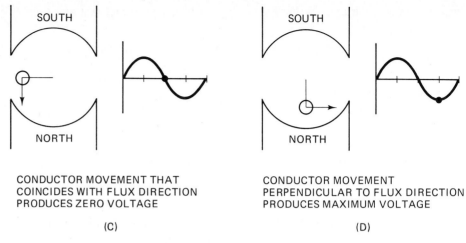

FIGURE 14-3 (*continued*)

creases continually until the generator reaches the 180° position. Here the voltage is zero, because the windings are moving with the lines of flux and do not cross them. As the generator is rotated from the 180° to the 270° position, a voltage is again induced. However, at this time, the polarity of the voltage is reversed (the resulting voltage is negative). The voltage value is at a negative peak at the 270° position. The negative voltage is reduced continually as the generator is rotated back to its starting position, where again the voltage level is zero.

VOLTAGE MEASUREMENT

There are several methods of measuring ac voltage and many instruments that can be used. A VOM meter can measure V_{rms} or V_{ave}. A VTVM can measure V_{rms}, V_P, or V_{ave}. An oscilloscope can measure V_P, V_{PP}, or e_{in}. In the following sections we discuss each of these voltage measurements.

Instantaneous Voltage The pattern followed by the ac generator voltage is a trigonometric sine wave. The formula for calculating the instantaneous voltage is

$$e_{in} = V_P \times \sin \theta$$

An oscilloscope displays the instantaneous voltage. Since the instantaneous voltage is a continually changing value, we do not generally use it as a method of measurement; however, it can be useful to know the shape of the wave that is being produced (irregularities can cause difficulties). There are several other methods for measuring ac voltage.

Peak-to-Peak Voltage One of the simplest voltage measurements to make on an oscilloscope is peak-to-peak voltage. It is the potential difference between the voltage level at positive peak and the voltage level at the negative peak. This value is referred to as peak-to-peak voltage (see Figure 14-4).

Peak Voltage Another method of measuring ac voltage is to use the peak voltage, the amount of voltage between the zero base line voltage and the voltage peak. The peak voltage is half of the peak-to-peak voltage (see Figure 14-4).

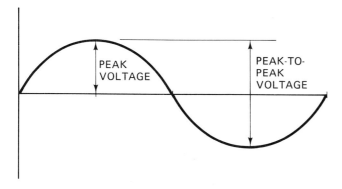

FIGURE 14-4

RMS Voltage Although peak-to-peak and peak voltage are simple measurements. They do not accurately indicate the amount of power actually being supplied to a component. Engineers have developed another method for measuring an ac voltage which does provide an accurate indication of the power supplied and dissipated in the circuit. It is called (root-mean-square) (rms) voltage. It is calculated using statistical methods that provide a standard deviation of the sine wave itself. An exercise at the end of the chapter will demonstrate its calculation. The rms value is measured directly using a voltmeter (VOM, VTVM, etc.). The rms voltage is also referred to as the effective voltage. It is equivalent to the amount of dc voltage it would take to do the same amount of work. Also, the rms voltage is 70.7% of the peak voltage (on a sine wave).

Average or DC Voltage If we add all of the instantaneous voltage values together and divide the total by the number of entries, we get an average value. It is important to understand that if we use the average for a complete sine wave (360°), the average value is zero. The sine wave extends as far positive as it does negative, so the two peaks, as well as the in-between values, cancel. However, if we consider only the positive section, we get a value equal to 63.6% of the peak. When we refer to a voltage that has no negative portion, we use a term called *rectified voltage* (see Figure 14-5). An electronic component called a diode is used to provide this rectified voltage.

There are two basic forms or types of voltage rectification. One is half-wave rectification, the other is full-wave rectification. The half-wave rectification method removes the negative section of the sine wave (the positive section can also be removed). The result is a wave that is flat (at zero volts) for one-half of the cycle. This type of rectification produces a voltage which is (31.8% of peak voltage) half that of full rectification (63.6% of peak voltage). Full-wave rectification uses an electronic circuit called a diode bridge which effectively turns the negative section of the sine wave into a positive voltage. This produces a wave that has no flat spots (consists of a series of humps or peaks). During the remainder of this book only full-wave rectifiers will be considered.

Conversion Factors There is a constant need to convert from one method of voltage measurement to another. For this reason you need to memorize the following equations. They make these conversions much easier.

$$V_{PP} = 2 \times V_P$$

$$V_{rms} = \frac{V_P}{\sqrt{2}}$$

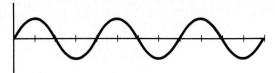

ALTERNATING VOLTAGE

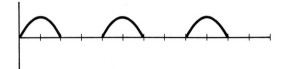

HALF-WAVE RECTIFIED VOLTAGE

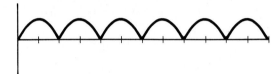

FULL-WAVE RECTIFIED VOLTAGE   **FIGURE 14-5**

Half-wave rectifiers:

$$V_{\text{ave}} = \frac{V_P}{\pi}$$

Full-wave rectifiers:

$$V_{\text{ave}} = 2 \times \frac{V_P}{\pi}$$

These formulas are given in terms of peak voltage. The peak voltage and the rms voltage levels are the two most often used. The numerical factor of 0.707 can be used to convert peak voltage to rms voltage (this has the same effect as dividing by $\sqrt{2}$). Also, the numerical value 1.414 ($\sqrt{2}$) can be used to convert rms voltage to peak voltage. These two values are used for several different areas of electronics. They should be recognized as important constants. The following examples show how to make these voltage conversions.

EXAMPLE 14-8

Suppose that the peak voltage were 7.5 V. Find the peak-to-peak, rms, and average values of voltage.

Solution:

Peak-to-Peak: Begin by inserting the known values into the conversion equation.

$$V_{PP} = 2 \times V_P$$
$$= 2 \times 7.5$$
$$= 15 \text{ V}$$

Rms: Again insert the known values into the conversion equation.

$$V_{rms} = \frac{V_P}{\sqrt{2}}$$

$$= \frac{7.5}{\sqrt{2}}$$

$$= 5.3033 \text{ V}$$

Ave (full wave): Insert the known values into the conversion equation.

$$V_{ave} = 2 \times \frac{V_P}{\pi}$$

$$= 2 \times \frac{7.5}{\pi}$$

$$= 4.7746 \text{ V}$$

Note: Most calculators have a key that is labeled π. Sometimes it is necessary to press a second function key to get this function. If your calculator does not have this function, you can use 3.14 ($\pi = 3.14159$) instead. ∎

EXAMPLE 14-9

Suppose that the rms voltage is 2.56 V. Find the peak-to-peak, peak, and average voltage values.

Solution: Begin by calculating the peak voltage. It can be used to find the other values. Use the rms voltage formula since the rms voltage was given.

$$V_{rms} = \frac{V_P}{\sqrt{2}}$$

$$2.56 = \frac{V_P}{\sqrt{2}}$$

$$V_P = \sqrt{2} \times 2.56$$

$$= 3.62 \text{ V}$$

Use the peak voltage to calculate the peak-to-peak value. As indicated, $2 \times V_P$ is the peak-to-peak voltage.

$$V_{PP} = 2 \times V_P$$

$$= 2 \times 3.62$$

$$= 7.2407 \text{ V}$$

Use the peak voltage to calculate the average value. As indicated, the peak-to-peak value divided by 3.14 is the average value.

$$V_{ave} = 2 \times \frac{V_P}{\pi}$$

$$= 2 \times \frac{3.62}{\pi}$$

$$= 2.3048 \text{ V} \qquad ∎$$

EXAMPLE 14-10

Suppose that the full-wave average voltage is 89.5 V. Find the the peak-to-peak, peak, and rms voltage values.

Solution: Begin by calculating the peak voltage. It can be used to find the other values. Use the average formula since average voltage was given.

$$V_{ave} = 2 \times \frac{V_P}{\pi}$$

$$89.5 = 2 \times \frac{V_P}{\pi}$$

$$89.5 \times \pi = 2 \times V_P$$

$$281.2 = 2 \times V_P$$

$$V_P = \frac{281.2}{2}$$

$$= 140.6$$

Use the peak voltage to calculate the peak-to-peak value. As indicated, $2 \times V_P$ is the peak-to-peak voltage.

$$V_{PP} = 2 \times V_P$$

$$= 2 \times 140.6$$

$$= 281.2 \text{ V}$$

Use the peak voltage to calculate the rms voltage. As indicated, the peak voltage divided by 1.414 is the rms value.

$$V_{rms} = \frac{V_P}{\sqrt{2}}$$

$$= \frac{140.6}{\sqrt{2}}$$

$$= 99.41 \text{ V}$$ ∎

SELF-TEST

Assume full-wave rectification.

1. $V_{PP} = 120$ V
   $V_P = $ _____
   $V_{rms} = $ _____
   $V_{ave} = $ _____

2. $V_{PP} = $ _____
   $V_P = 45$ V
   $V_{rms} = $ _____
   $V_{ave} = $ _____

3. $V_{PP} = $ _____
   $V_P = $ _____
   $V_{rms} = 10.2$ V
   $V_{ave} = $ _____

4. $V_{PP} = $ _____
   $V_P = $ _____
   $V_{rms} = $ _____
   $V_{ave} = 45$ V

5. $V_{PP} = $ _____
   $V_P = 84$ V
   $V_{rms} = $ _____
   $V_{ave} = $ _____

6. $V_{PP} = $ _____
   $V_P = $ _____
   $V_{rms} = 68$ V
   $V_{ave} = $ _____

7. V_{PP} = _____
   V_P = _____
   V_{rms} = _____
   V_{ave} = 47.3 V

8. V_{PP} = 58.75 V
   V_P = _____
   V_{rms} = _____
   V_{ave} = _____

9. V_{PP} = _____
   V_P = 135 V
   V_{rms} = _____
   V_{ave} = _____

10. V_{PP} = _____
    V_P = _____
    V_{rms} = 220 V
    V_{ave} = _____

11. V_{PP} = 450 V
    V_P = _____
    V_{rms} = _____
    V_{ave} = _____

12. V_{PP} = _____
    V_P = 50 mV
    V_{rms} = _____
    V_{ave} = _____

ANSWERS TO SELF TEST

1. V_P = 60 V, V_{rms} = 42.43 V, V_{ave} = 38.20 V
2. V_{PP} = 90 V, V_{rms} = 31.82 V, V_{ave} = 28.65 V
3. V_{PP} = 28.85 V, V_P = 14.42 V, V_{ave} = 9.18 V
4. V_{PP} = 141.4 V, V_P = 70.69 V, V_{rms} = 49.98 V
5. V_{PP} = 168 V, V_{rms} = 59.40 V, V_{ave} = 54.48 V
6. V_{PP} = 192.3 V, V_P = 96.17 V, V_{ave} = 61.22 V
7. V_{PP} = 148.6 V, V_P = 74.30 V, V_{rms} = 52.54 V
8. V_P = 29.38 V, V_{rms} = 20.77 V, V_{ave} = 18.70 V
9. V_{PP} = 270 V, V_{rms} = 95.46 V, V_{ave} = 85.94 V
10. V_{PP} = 622.3 V, V_P = 311.1 V, V_{ave} = 198.1 V
11. V_P = 225 V, V_{rms} = 159.1 V, V_{ave} = 143.2 V
12. V_{PP} = 100 mV, V_{rms} = 35.36 mV, V_{ave} = 31.83 mV

PERIOD AND FREQUENCY

Although mathematically the period is usually measured in degrees, in actuality the period is a measure of time. The normal unit for time is seconds. There is a direct relation between the speed at which the generator is rotating, the length of time the generator has been turning, and the number of degrees which the generator has rotated. The time period is symbolized as T. The instantaneous time (the length of time that the generator has turned) is symbolized as t. A proportionary relation can be used to convert the number of degrees to time, and vice versa. Examine the relations shown in Figure 14-6. The following equation can be used for conversion purposes.

$$\frac{\theta}{360} = \frac{t}{T}$$

The time period is also related to the speed of the generator. The number of times that the generator rotates per second is referred to as the frequency of the sine wave. The time period is the reciprocal of the frequency (and vice versa). The following formulas show this relationship.

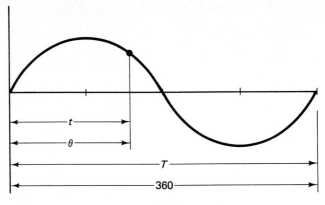

FIGURE 14-6

$$f = \frac{1}{T}$$

$$T = \frac{1}{f}$$

The standard unit for frequency is hertz (Hz), although depending on the particular application, cycles per second (cps) and pulses per second (pps) are sometimes used. Examples of normal conversions and calculations follow.

EXAMPLE 14-11

Given that the frequency is 12.5 kHz. Find the period and the instantaneous time after 128° of rotation of the generator.

Solution: Begin by finding the period of the wave.

$$T = \frac{1}{f}$$

$$= \frac{1}{12.5 \text{ kHz}}$$

$$= 80 \text{ }\mu\text{s}$$

Use the time period to set up a proportionary relation between degrees and time. This will provide the instantaneous time.

$$\frac{\theta}{360} = \frac{t}{T}$$

$$\frac{128}{360} = \frac{t}{0.000080}$$

$$0.3556 = \frac{t}{0.00008}$$

$$t = 0.3556 \times 0.00008$$

$$= 28.44 \text{ }\mu\text{s}$$

EXAMPLE 14-12

Given that the time period is 100 ms and the generator has been rotating for 79 ms. Find the frequency and angle of rotation for the generator.

Solution: Begin by calculating the frequency.

$$f = \frac{1}{T}$$

$$= \frac{1}{0.100}$$

$$= 10 \text{ Hz}$$

Set up a ratio to find the angle of rotation.

$$\frac{\theta}{360} = \frac{0.079}{0.100}$$

$$= 360 \times \frac{0.079}{0.100}$$

$$= 284.4°$$

SELF-TEST

1. $\theta = 60°$
   $t = $ _____
   $T = 10 \ \mu s$
   $f = $ _____

2. $\theta = $ _____
   $t = 4.444 \ \mu s$
   $T = $ _____
   $f = 120 \text{ kHz}$

3. $\theta = $ _____
   $t = 12 \text{ ms}$
   $T = 32 \text{ ms}$
   $f = $ _____

4. $\theta = 232°$
   $t = 24 \ \mu s$
   $T = $ _____
   $f = $ _____

5. $\theta = 167°$
   $t = $ _____
   $T = $ _____
   $f = 225 \text{ kHz}$

6. $\theta = $ _____
   $t = 1 \ \mu s$
   $T = $ _____
   $f = 455 \text{ kHz}$

7. $\theta = $ _____
   $t = 234 \ \mu s$
   $T = 330 \ \mu s$
   $f = $ _____

8. $\theta = 320°$
   $t = 16.7 \text{ ms}$
   $T = $ _____
   $f = $ _____

9. $\theta = $ _____
   $t = 3.333 \text{ ns}$
   $T = $ _____
   $f = 103 \text{ MHz}$

10. $\theta = 225°$
    $t = $ _____
    $T = 10 \ \mu s$
    $f = $ _____

11. $\theta = 280°$
    $t = $ _____
    $T = $ _____
    $f = 1020 \text{ kHz}$

12. $\theta = $ _____
    $t = 0.55 \ \mu s$
    $T = $ _____
    $f = 1380 \text{ kHz}$

ANSWERS TO SELF TEST

1. $t = 1.667\ \mu s, f = 100$ kHz
2. $T = 8.333\ \mu s, \theta = 192.0°$
3. $\theta = 135°, f = 31.25$ Hz
4. $T = 37.24\ \mu s, f = 26.85$ Hz
5. $t = 2.062\ \mu s, T = 4.444\ \mu s$
6. $\theta = 163.8°, T = 2.198\ \mu s$
7. $\theta = 255.3°, f = 3.030$ kHz
8. $T = 18.79$ ms, $f = 53.23$ Hz
9. $\theta = 123°, T = 9.709$ ns
10. $t = 6.25\ \mu s, f = 100$ kHz
11. $T = 980.4$ ns $t = 762.5$ ns
12. $T = 724.6$ ns, $\theta = 273.2°$

Wavelength The frequency plays an important role in the design of antennas for radios, television, and other communication equipment. It is a ratio between the frequency of the signal and the velocity of light that determines which lengths will be best in the reception of a particular channel (radio station). This ratio is called the *wavelength* of the frequency. It is the actual distance between the peaks of the wave. The speed of light is 299.97 million meters per second. For practical purposes we use 300 million meters per second and 984 million feet per second as the rounded values for this constant. An antenna will perform better if it has a length that is an exact multiple or submultiple of this value. For example, if the wavelength were 100 m, an antenna that is 10, 1, or 200 m long will have better reception than one 45 m long.

It is hard to understand the exact reasons for this phenomenon, but the basic idea behind it comes from the notion that the antenna itself is a vibrating object. When an electromagnetic wave strikes the antenna, it sets it into a vibrating mode. Objects of a particular length will vibrate at a particular frequency. If the frequency of the antenna (the length of the antenna) matches the wavelength of the signal, there is very little interference between the vibrations. If the antenna is improperly cut (has a length that will not support the frequency of the signal), a portion of the incoming signal is lost due to an interference between the vibrating waves in the antenna and the electromagnetic wave of the signal. The symbol for wavelength is the Greek lowercase letter lamda (λ). A formula for finding the wavelength is shown below. Examples follow.

$$\lambda = \frac{300\text{ million meters/second}}{f}$$

or

$$\lambda = \frac{984\text{ million feet/second}}{f}$$

EXAMPLE 14-13

Suppose that the frequency of a particular wave is 120 kHz. Find the wavelength for the signal.

Solution: Use the formula to calculate λ.

$$\lambda = \frac{300\text{ million meters/second}}{f}$$

$$= \frac{300\text{ million meters/second}}{120\text{ kHz}}$$

$$= 2500\text{ m}$$

EXAMPLE 14-14

Find the frequency of a signal whose wavelength is 2 m.

Solution:

$$\lambda = \frac{300 \text{ million meters/second}}{f}$$

$$2 = \frac{300{,}000{,}000}{f}$$

$$f = \frac{300{,}000{,}000}{2}$$

$$= 150 \text{ MHz}$$

SELF-TEST

1. $\lambda_{meters} = 200$ m
   $\lambda_{feet} = $ _____
   $f = $ _____

2. $\lambda_{meters} = $ _____
   $\lambda_{feet} = 1000$ ft
   $f = $ _____

3. $\lambda_{meters} = $ _____
   $\lambda_{feet} = $ _____
   $f = 1380$ kHz

4. $\lambda_{meters} = $ _____
   $\lambda_{feet} = $ _____
   $f = 92$ MHz

5. $\lambda_{meters} = 50$ m
   $\lambda_{feet} = $ _____
   $f = $ _____

6. $\lambda_{meters} = $ _____
   $\lambda_{feet} = 100$ ft
   $f = $ _____

ANSWERS TO SELF TEST

1. $f = 1.5$ MHz, $\lambda_{feet} = 656$ ft
2. $f = 984$ kHz, $\lambda_{meters} = 304.9$ m
3. $\lambda_{meters} = 217.4$ m, $\lambda_{feet} = 713$ ft
4. $\lambda_{meters} = 3.26$ m, $\lambda_{feet} = 10.7$ ft
5. $f = 6$ MHz, $\lambda_{feet} = 164$ ft
6. $f = 9.84$ MHz, λ_{meters} 30.49 m

NONSINUSOIDAL WAVEFORMS

One of the more interesting facets of electronics has to do with the combination of sine waves. At first it is hard to picture that every wave, no matter what its shape, can be produced simply by adding sine waves of different frequencies and amplitudes. The combinations of these waves will produce differing signals.

Frequency The basic requirement of a sinusoidal wave is that it follow an exact reproduction of the sine function. When waves of the same frequency are combined, the result is a sine (sinusoidal) wave. When waves of differing frequencies are combined, the resultant is a nonsinusoidal waveform. Remember when sine waves of differing frequencies are combined, the result will not follow the sine function pattern.

Phase If the waves begin at the same time (are in phase), the resultant wave is of the same frequency but will have a larger amplitude. If the waves begin at opposite times (are 180° out of phase), the combined waveform will have a decreased amplitude. Any phase shift will affect the amplitude, but a phase shift of 180° will reduce it the most and a phase shift of 0° will increase it the most.

SELF-TEST

Combine the waveforms and predict which combination will produce a sinusoidal or a nonsinusoidal waveform.

1.

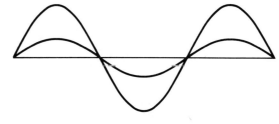

SAME FREQUENCY IN PHASE

2.

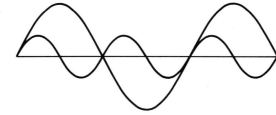

DOUBLE FREQUENCY

3.

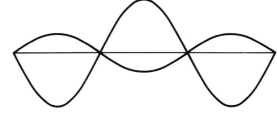

SAME FREQUENCY 180° OUT OF PHASE

4.

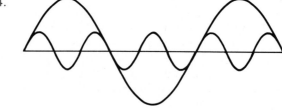

TRIPLE FREQUENCY

ANSWERS TO SELF-TEST

1.
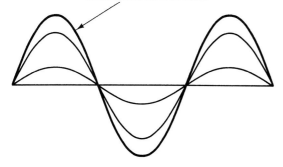
SAME FREQUENCY IN PHASE SINUSOID

2.
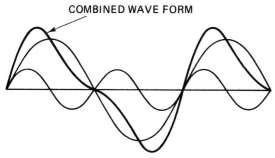
DOUBLE FREQUENCY NONSINUSOID

3.
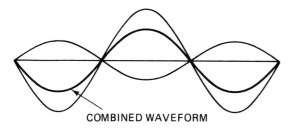
SAME FREQUENCY 180° OUT OF PHASE SINUSOID

4.
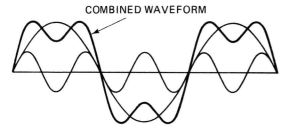
TRIPLE FREQUENCY NONSINUSOID

Harmonics Multiples and submultiples of a basic frequency are called *harmonic frequencies*. For example, 10 Hz and 1000 Hz are harmonic frequencies of the basic frequency 100 Hz because $100 \times (1/10) = 10$ (submultiple) and $100 \times 10 = 1000$ (multiple).

In electronics when the fundamental frequency is doubled, we call this an *octave*. When a frequency is doubled, the tonal quality is very similar (same tone but higher in pitch). In music the fundamental or basic frequency is 440 Hz (concert A.) The 880-Hz note is also called A. It is one octave higher than the A of 440 Hz. Any time that a frequency is doubled, the resultant frequency is called one octave. The frequency 1760 Hz is two octaves higher than the frequency 440 Hz.

Another useful multiple of fundamental frequencies is the *decade*. A decade has a factor of 10 between the fundamental and the harmonic frequency. The frequency 100 kHz is two decades higher than the frequency 1 kHz. The measure of octaves and decades is very useful in the study of the filters used in components such as graphic equalizers.

Square Wave The square wave is the basic waveform used in digital circuits. The square wave is a nonsinusoidal waveform comprised of the sum of the fundamental frequency and all the odd harmonics. For example, a 1000-Hz square wave can be constructed by adding a 1000-Hz sine wave, the odd multiples of 1000 Hz ($3 \times 1000 = 3000$ Hz, $5 \times 1000 = 5000$ Hz, $7 \times 1000 = 7000$ Hz, etc.), and the submultiples ($\frac{1}{3} \times 1000 = 333.3$ Hz, $\frac{1}{5} \times 1000 = 200$ Hz, $\frac{1}{7} \times 1000 = 142.8$ Hz, etc.).

To produce a good square wave is necessary to use at least five of the odd harmonics. This is especially important when selecting an oscilloscope. For instance, if you were working with a digital system that operated at a maximum frequency of 12 MHz (typical high speed), it would seem logical to use a 12-MHz scope; however, a 12-MHz scope will not reproduce a good square wave. The square wave would have rounded corners. To avoid this type of distortion it is necessary to have

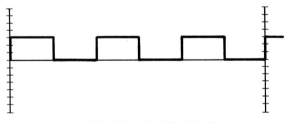

UNDISTORTED SQUARE WAVE

(A)

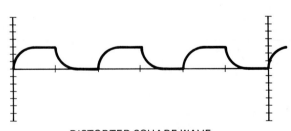

DISTORTED SQUARE WAVE

(B)

FIGURE 14-7

an oscilloscope with a bandwidth of at least 5 × 12 MHz = 60 MHz (see Figure 14-7).

Triangle Waveforms (Sawtooth Waveforms) In video circuits it is necessary to have a triangle wave signal. This signal is commonly called a *voltage ramp*. It is a composite of the fundamental and all of the even harmonics. Again when choosing an oscilloscope it is necessary to pick one that has five times the required frequency (see Figure 14-8).

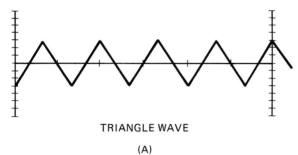

TRIANGLE WAVE

(A)

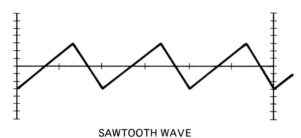

SAWTOOTH WAVE

(B)

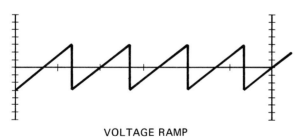

VOLTAGE RAMP

(C)

FIGURE 14-8

SUMMARY

Alternating current is produced by the movement of a coil in a magnetic field. When the movement is perpendicular to the magnetic field, a large voltage is produced; however, when the movement is in the same plane as the magnetic field, no voltage is produced. This causes the voltage to peak and decrease to zero every 90°. The voltage produced follows a pattern that resembles a trigonometric sine wave. The general form of equation for such a wave is

$$e_{in} = V_P \times \sin \theta$$

where e_{in} is the instantaneous voltage.

The amplitude or maximum peak is called the peak voltage. Peak or peak-to-peak voltage is usually measured using an oscilloscope. Rms values are also used. They are measured with an ac voltmeter. Rms values indicate the amount of power received by the component.

The period, the wavelength, and the frequency are used to indicate the amount of time between repetitions of the sine wave. The period is generally measured in degrees of rotation. The frequency is measured in hertz (cycles per second). Wavelength can be indicated in terms of meters or feet.

Combining sine waves of the same frequency but different phases will produce another sine wave; however, if waves of different frequencies are combined, the result will not be a sine wave. Multiples and submultiples of a frequency are referred to as harmonics. When the fundamental and the odd harmonics are combined, the result is a square wave. When the fundamental and the even harmonics are combined, the result is a triangle wave.

QUESTIONS

14-1. Sketch a sine wave and label its period, amplitude, and each quadrant.

14-2. Compare radians, degrees, and gradients. Which do you prefer? Why?

14-3. Relate each of the four quadrants of a sine wave to the four quadrants of a rectangular graph.

14-4. Describe how the angle θ is found in each of the four quadrants.

14-5. What is the inverse of the sine function?

14-6. Describe the method your calculator uses to change from the radian, degree, and gradient modes.

14-7. List each of the four quadrants and indicate whether the sine is positive or negative in that quadrant.

14-8. When referring to the rotation of a generator, which direction is positive?

14-9. What is needed for the movement of the generator windings through a magnetic field if the windings are to produce voltage? Describe the relation between the magnetic field and the generator movement.

14-10. Describe the difference between peak, peak-to-peak, rms, and average voltages. Include instruments used to measure each of these and methods used to convert from one to the other.

14-11. Why is the average value of one complete sine wave, zero volts? Can you think of any ways to prove that this is true?

14-12. What does the term *rectification* mean?

14-13. What is the difference between half-wave and full-wave rectification?

14-14. What electronic component is used to rectify a sine-wave voltage?

14-15. Explain the relation between the frequency and the period of a sine wave.

14-16. Explain the relation between instantaneous time and the degree of rotation.

14-17. Define *octave*. Include several numeric examples.

14-18. Define *decade*. Include several numeric examples.
14-19. Sketch a square wave. Label its amplitude and period.
14-20. Sketch a triangle wave. Label its amplitude and period.
14-21. What is meant by the term *nonsinusoidal?*
14-22. What is needed of two sine waves if their combination is to be nonsinusoidal?
14-23. What is the requirement of two sine waves if their combination is to be sinusoidal?
14-24. What is meant by the term *harmonic?*
14-25. What is a fundamental frequency? A multiple of a fundamental frequency? A submultiple of a fundamental frequency?

PRACTICE PROBLEMS

Find the missing values for each of the following problems.

14-1.
V_{PP} = _____
V_P = 50 μV
V_{rms} = _____
θ = _____
t = 85.64 μs
T = _____
f = _____
λ_{meters} = 25 cm
λ_{feet} = _____
quadrant: _____
e_{in} = _____

14-2.
V_{PP} = 100 μV
V_P = _____
V_{rms} = _____
θ = 220°
t = 1.111 μs
T = _____
f = _____
λ_{meters} = _____
λ_{feet} = _____
quadrant: _____
e_{in} = _____

14-3.
V_{PP} = _____
V_P = _____
V_{rms} = 353.5 mV
θ = _____
t = _____
T = _____
f = _____
λ_{meters} = _____
λ_{feet} = 656 ft
quadrant: fourth
e_{in} = −492.4 mV

14-4.
V_{PP} = _____
V_P = _____
V_{rms} = _____
θ = 105°
t = _____
T = _____
f = 9 GHz
λ_{meters} = _____
λ_{feet} = _____
quadrant: _____
e_{in} = 265.6 mV

14-5.
V_{PP} = _____
V_P = _____
V_{rms} = 91.93 μV
θ = _____
t = _____
T = 2.857 μs
f = _____
λ_{meters} = _____
λ_{feet} = _____
quadrant: third
e_{in} = −88.66 μV

14-6.
V_{PP} = _____
V_P = 2.445 V
V_{rms} = _____
θ = 310°
t = 8.36 μs
T = _____
f = _____
λ_{meters} = _____
λ_{feet} = _____
quadrant: _____
e_{in} = _____

14-7. V_{PP} = 520 mv
V_P = _____
V_{rms} = _____
θ = _____
t = 2.719 μs
T = 5.263 μs
f = _____
λ_{meters} = _____
λ_{feet} = _____
quadrant: _____
e_{in} = _____

14-8. V_{PP} = _____
V_P = 60 V
V_{rms} = _____
θ = _____
t = 1.673 μs
T = _____
f = _____
λ_{meters} = _____
λ_{feet} = 2315 ft
quadrant: _____
e_{in} = _____

14-9. V_{PP} = _____
V_P = _____
V_{rms} = _____
θ = _____
t = 3.017 ns
T = _____
f = _____
λ_{meters} = 1.034 m
λ_{feet} = _____
quadrant: _____
e_{in} = −5.303 V

14-10. V_{PP} = _____
V_P = _____
V_{rms} = 2.227 V
θ = _____
t = _____
T = _____
f = _____
λ_{meters} = _____
λ_{feet} = 2695 ft
quadrant: second
e_{in} = 815.3 mV

14-11. V_{PP} = _____
V_P = _____
V_{rms} = _____
θ = 230°
t = _____
T = 8 μs
f = _____
λ_{meters} = _____
λ_{feet} = _____
quadrant: _____
e_{in} = −14.17 V

14-12. V_{PP} = _____
V_P = _____
V_{rms} = 4.419 V
θ = 65°
t = _____
T = _____
f = 635 kHz
λ_{meters} = _____
λ_{feet} = _____
quadrant: _____
e_{in} = _____

14-13. V_{PP} = _____
V_P = _____
V_{rms} = 5.3 mV
θ = 25.7°
t = _____
T = _____
f = _____
λ_{meters} = 5000 km
λ_{feet} = _____
quadrant: _____
e_{in} = _____

14-14. V_{PP} = _____
V_P = 150 mV
V_{rms} = _____
θ = _____
t = 798.6 μs
T = _____
f = _____
λ_{meters} = 375 m
λ_{feet} = _____
quadrant: _____
e_{in} = _____

14-15. V_{PP} = _____
V_P = _____
V_{rms} = _____
θ = _____
t = 850 ns
T = _____
f = _____
λ_{meters} = _____
λ_{feet} = 1230 ft
quadrant: third
e_{in} = −3.942 V

14-16. V_{PP} = _____
V_P = _____
V_{rms} = 830 mV
θ = _____
t = _____
T = 666.7 ns
f = _____
λ_{meters} = _____
λ_{feet} = _____
quadrant: second
e_{in} = 304.1 mV

15

Inductors

Chapter objectives

After reading this chapter and answering the questions and problems, you should be able to:

- Given the lengths of any two sides of a right triangle. Use the Pythagorean theorem to find the missing side.
- Given one side and one angle of a right triangle, use trigonometry to find the length of the other two sides.
- Define eddy current and skin effect and describe methods of overcoming each when working with high frequencies.
- Describe the construction of typical inductors.
- Use the formula

$$L = \mu \times \frac{N^2 \times A}{l} \times 1.26 \times 10^{-6}$$

- Calculate total inductance in series and parallel circuits.
- Describe mutual inductance, how it effects total inductance, and state how repositioning the coils can control it.
- Describe the construction and operation of a transformer.
- Use the turns ratio to find the primary and secondary voltage and current of a transformer and its reflected impedance.
- Calculate the efficiency of a transformer and describe the phase relation between primary and secondary windings.
- Define *induced voltage, inductive reactance,* and *counter electromotive force*, and state how each is related to the other.
- Calculate X_L and X_{LT} for series and parallel circuits.
- Use the Pythagorean theorem and trigonometry to solve *LR* series and parallel circuits.
- Describe various methods of troubleshooting inductors.

Any conductor that carries current exhibits a quality referred to as *inductance*. Some conductors are designed especially for their inductance properties. They are called *inductors*. In this chapter we discuss the construction and design of inductors and how their properties are determined and tested.

Right triangles are used in analyzing the effects on voltage and current in circuits containing both inductors and resistors. For that reason, this chapter also contains a description and many examples of right-triangle solution.

GETTING STARTED: RIGHT TRIANGLES

The analysis of any ac electrical circuit is simple, provided that you have a good understanding of right triangles and their solution. To solve a right triangle, two mathematical theorems are needed: the Pythagorean theorem and the trigonometric functions (sine, cosine, and tangent). The mathematic section of this chapter is devoted to an explanation of these theorems.

Pythagorean Theorem The ancient Greeks held the triangle in awe. They felt that it held a mystical quality and spent years studying it. They felt that to solve the secrets of the triangle was to solve the mysteries of life itself. They formed many ideas about triangles, one of which was the Pythagorean theorem. The Pythagorean theorem states that the square of the length of the hypotenuse (longest side) of a right triangle is equal to the sum of the squares of the lengths of the other two sides (refer to Figure 15-1). In equation form this is

$$C^2 = A^2 + B^2$$

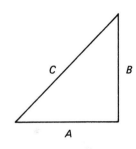

FIGURE 15-1

Examples of the use of the Pythagorean theorem follow.

EXAMPLE 15-1

Given that the length of the two sides of a right triangle are 20 and 15, find the length of the hypotenuse.

Solution: Use the Pythagorean theorem to find the hypotenuse.

$$\begin{aligned} C^2 &= A^2 + B^2 \\ &= 20^2 + 15^2 \\ C &= \sqrt{20^2 + 15^2} \\ &= \sqrt{400 + 225} \\ &= \sqrt{625} \\ &= 25 \end{aligned}$$ ∎

EXAMPLE 15-2

Given that the length of the hypotenuse of a right triangle is 58 and the length of one of the sides is 42, find the length of the remaining side.

Solution: Begin by using the Pythagorean theorem.

$$C^2 = A^2 + B^2$$
$$58^2 = 42^2 + B^2$$

Rearrange the equation so that B^2 is on one side by itself.

$$B^2 = 58^2 - 42^2$$
$$B = \sqrt{58^2 - 42^2}$$
$$= \sqrt{3364 - 1764}$$
$$= \sqrt{1600}$$
$$= 40$$

■

The Pythagorean theorem can be rearranged to give the following three formulas.

Hypotenuse:

$$C = \sqrt{A^2 + B^2}$$

Missing side:

$$A = \sqrt{C^2 - B^2}$$
$$B = \sqrt{C^2 - A^2}$$

SELF-TEST

1. Given that $A = 14.2$ and $B = 25.3$, find C.
2. Given that $A = 24.6$ and $C = 55$, find B.
3. Given that $C = 53.7$ and $B = 37.8$, find A.
4. Given that $A = 23$ and $B = 89$, find C.
5. Given that $C = 42.7$ and $A = 22.1$, find B.
6. Given that $B = 120$ and $C = 300$, find A.

ANSWERS TO SELF-TEST

1. 29.01
2. 49.19
3. 38.14
4. 91.92
5. 36.54
6. 274.9

Trigonometric Relations: Sine, Cosine, and Tangent The first step to understanding the trig functions and how they pertain to a right triangle is to be able to identify the sides as they relate to a particular angle. As stated earlier, the longest side of a right triangle is called the *hypotenuse*. The hypotenuse and the adjacent side join to form the angle. The third side is opposite (across from) the angle. Examine the triangle shown in Figure 15-2 and take particular notice of these sides and how they pertain to the different angles of the right triangle.

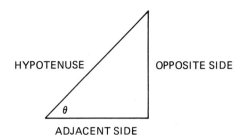

ADJACENT SIDE   FIGURE 15-2

The phrase Soh-Cah-Toa can be used to remember how the trigonometric functions relate to the sides. It is important to realize that the trig functions are only ratios of one side to another. The formulas for calculating these functions are listed below.

Sine   Soh

$$\sin \theta = \frac{\text{opposite}}{\text{hypotenuse}}$$

Cosine   Cah

$$\cos \theta = \frac{\text{adjacent}}{\text{hypotenuse}}$$

Tangent   Toa

$$\tan \theta = \frac{\text{opposite}}{\text{adjacent}}$$

The trig functions can be used to solve many types of triangle problems. The following examples show the power of these functions.

EXAMPLE 15-3

Solve the triangle shown in Figure 15-3 for the missing side and θ.

Solution: Begin by examining the triangle. You are given the adjacent side and the hypotenuse. Relate this to Cah and use the cosine formula.

$$\cos \theta = \frac{\text{adjacent}}{\text{hypotenuse}}$$
$$= \frac{18}{25}$$
$$= 0.72$$

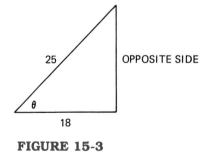

FIGURE 15-3

The cosine of θ is 0.72. Use the $\cos^{-1}$ function to find θ.

$$\theta = \cos^{-1} 0.72 = 43.94°$$

[1] [8] [÷] [2] [5] [=] [2nd] [$\cos^{-1}$]

The display is

4.3945   01

GETTING STARTED: RIGHT TRIANGLES

Note: Most calculators use the same key for both the cos and cos^{-1} functions. It is necessary to press the second function key first. This is also true for the sine and tangent functions.

The angle θ can be used in the sine formula to find the opposite side: Soh.

$$\sin \theta = \frac{\text{opposite}}{\text{hypotenuse}}$$

$$\sin 43.91° = \frac{\text{opposite}}{25}$$

$$\text{opposite} = 25 \times \sin 43.91°$$

$$= 25 \times 0.6940$$

$$= 17.3476 \qquad ■$$

EXAMPLE 15-4

Solve the triangle shown in Figure 15-4 for the hypotenuse and the adjacent sides.

Solution: Begin by examining the triangle. You are given the opposite side and the angle. So you can use Soh and the sine formula to find the hypotenuse.

$$\sin \theta = \frac{\text{opposite}}{\text{hypotenuse}}$$

$$\sin 65° = \frac{49}{\text{hypotenuse}}$$

$$\text{hypotenuse} = \frac{49}{\sin 65°}$$

$$= 54.07$$

[4] [9] [÷] [6] [5] [sin] [=]   display 5.4065   01

Toa and the tangent formula can be used to find the adjacent side.

$$\tan \theta = \frac{\text{opposite}}{\text{adjacent}}$$

$$\tan 65° = \frac{49}{\text{adjacent}}$$

$$\text{adjacent} = \frac{49}{\tan 65°}$$

$$= 22.85$$

[4] [9] [÷] [6] [5] [tan] [=]   display 2.2849   01   ■

Understand that it is not necessary to revert to the trigonometric functions in order to solve every triangle. It is possible to use these functions, however; and you should realize their worth. You should try to reach a point where the trig functions are a comfortable process. They are necessary when analyzing circuits that contain inductors and resistors.

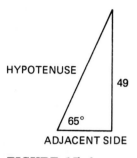

FIGURE 15-4

SELF-TEST

1. Hypotenuse = 100
   adjacent side = 70.7
   opposite side = _____
   θ = _____

2. Hypotenuse = _____
   adjacent side = 70.7
   opposite side = 50.3
   θ = _____

3. Hypotenuse = _____
   adjacent side = _____
   opposite side = 120
   θ = 45°

4. Hypotenuse = 100
   adjacent side = _____
   opposite side = 55.6
   θ = _____

5. Hypotenuse = _____
   adjacent side = 70.7
   opposite side = _____
   θ = 25°

6. Hypotenuse = 100
   adjacent side = _____
   opposite side = _____
   θ = 25°

7. Hypotenuse = 150
   adjacent side = 45.7
   opposite side = _____
   θ = _____

8. Hypotenuse = _____
   adjacent side = 200
   opposite side = 135
   θ = _____

9. Hypotenuse = _____
   adjacent side = _____
   opposite side = 450
   θ = 75°

10. Hypotenuse = 1200
    adjacent side = _____
    opposite side = 850
    θ = _____

11. Hypotenuse = _____
    adjacent side = 2500
    opposite side = _____
    θ = 63°

12. Hypotenuse = 1800
    adjacent side = _____
    opposite side = _____
    θ = 34°

13. Hypotenuse = 2700
    adjacent side = 1170
    opposite side = _____
    θ = _____

14. Hypotenuse = _____
    adjacent side = 8200
    opposite side = 10000
    θ = _____

ANSWERS TO SELF TEST

1. Opposite side = 70.7, θ = 45°
2. Hypotenuse = 86.77, θ = 35.43°
3. Hypotenuse = 169.7, adjacent side = 120
4. Adjacent side = 83.12, θ = 33.78°
5. Hypotenuse = 78.01, opposite side = 32.97
6. Adjacent side = 90.63, opposite side = 42.26
7. Opposite side = 142.9, θ = 72.26°
8. Hypotenuse = 241.3, θ = 34.02°
9. Hypotenuse = 465.9, adjacent side = 120.6
10. Adjacent side = 847.1, θ = 45.1°
11. Hypotenuse = 5507, opposite side = 4907
12. Adjacent side = 1492, opposite side = 1007
13. Opposite side = 2433, θ = 64.32°
14. Hypotenuse = 12932, θ = 50.65°

GETTING STARTED: RIGHT TRIANGLES

CONSTRUCTION AND DESIGN

There are numerous methods for constructing inductors. So many, in fact, that it is not feasible to discuss all of them. We will, however, examine the similarities between these methods of construction. Most of the inductors that are manufactured today are constructed in a tubular coil form (see Figure 15-5). That is, they are nothing more than a long piece of wire that has been wrapped around a supporting frame. They are tubular in shape and have a hollow center. At times a cylindrical metal rod is inserted in their centers to increase the magnetic properties of the inductor.

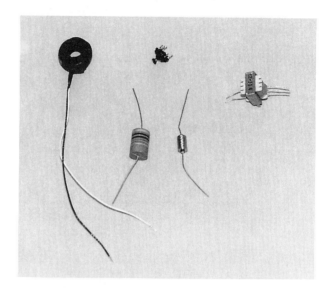

FIGURE 15-5 Typical Inductors. Transformer—right, Adjustable Ferrite Core—top center, Air Core—right center, Molded—right center, Torroid—left.

At first, the construction of an inductor sounds like that of an electromagnet. Well, they are constructed in a similar fashion. The one thing that separates the two is the type of current that powers them. An electromagnet requires a dc voltage. An inductor functions with an ac voltage.

There are two factors (the windings of the coil and the metal core) that limit the efficiency of an inductor. That is, high-frequency operation allows there to be a loss of power in both the windings of the coil and in the magnetic field of the core. With ac voltages operating at high frequencies, the currents begin to flow in an unusual manner. In fact, they tend to flow in a circular motion, like what you would see in a stream when there is a rapid change in direction of water flow (whirlpool). This is called an *eddy current*. Eddy currents in the wire cause most of the current movement to flow only at the surface of the wire (see Figure 15-6). This is known as the *skin effect*. To minimize these effects, a stranded wire is used at frequencies higher than 10 MHz.

By introducing a metal core in the center of the inductor, the magnetic strength of the inductor is increased. This allows a smaller inductor to exhibit the qualities that a larger inductor would normally have (saves space). To combat the weakening effects of higher frequencies, a ferromagnetic powder (dust) is com-

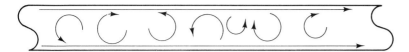

FIGURE 15-6 At high frequencies current tends to flow only at the surface of the conductor. This is referred to as skin effect. Eddie currents are produced at the center of the conductor.

pressed to form the metal core. The use of ferromagnetic powder (dust) in effect minimizes the core losses due to hysteresis.

When designing an inductor you should pay particular attention to the physical dimensions of the coil. It is these dimensions that determine the inductance (magnetic strength) of the coil. The formula shown below will allow you to predict the inductance of an inductor.

$$L = \mu \times \frac{N^2 \times A}{l} \times 1.26 \times 10^{-6}$$

The unit of measure for an inductor is the henry. If the area is measured in square meters and the length is measured in meters, the outcome will be in henrys. A henry corresponds to a volt/(ampere/second) or a volt-second/ampere. The inductance is a composite of the permeability of the core, the number of turns, the area, and the length of the coil. Remember that the permeability is a magnetic property. Also notice the correction factor (1.26×10^{-6}) at the end of the formula. This is the permeability of air. Understand that the permeability of air is 1.26×10^{-6} (in the MKS system), while the relative permeability of air is 1 (in any system). It may be difficult for you to differentiate between the permeability of air and relative permeability. Example 15.5 demonstrates the use of this formula.

EXAMPLE 15-5

A coil of wire consists of 250 turns of wire wrapped around a soft-iron core, which has a relative permeability of 100. The wire is formed around a bobbin that has an internal diameter of 20 mm. The coil has a length of 15 cm. Calculate the inductance of the coil.

Solution: Begin by calculating the area of the center of the bobbin. The diameter is 20 mm and must be converted to meters before the area is calculated, because the area must be in units of square meters.

$$r = 10 \text{ mm} = 0.01 \text{ m}$$

Then the area can be calculated using

$$A = \pi \times r^2$$
$$= \pi \times (0.01)^2$$
$$= 3.14159 \times 10^{-4} \text{ m}^2$$

The length must also be converted to units of meters.

$$l = 15 \text{ cm} = 0.15 \text{ m}$$

CONSTRUCTION AND DESIGN

Then the inductance can be found using the inductance formula.

$$L = \mu \times \frac{N^2 \times A}{l} \times 1.26 \times 10^{-6}$$

$$= 100 \times \frac{(250)^2 \times 3.14159 \times 10^{-4}}{0.15} \times 1.26 \times 10^{-6}$$

$$= 16.49 \text{ mH}$$

■

It is also possible to rearrange this formula to find other missing values. Some examples of these manipulations follow.

EXAMPLE 15-6

It is desired to construct an air-core inductor which has an inductance of 50 μH. There are to be 1000 turns of wire, and the length of the coil is 15 cm. What is the diameter of the coil?

Solution: Begin by realizing that the relative permeability (μ) is equal to 1 because the coil has an air core. Then convert the length to units of meters.

$$1 = 15 \text{ cm} = 0.15 \text{ m}$$

Then insert the known values into the inductance formula and solve for the area.

$$L = \mu \times \frac{N^2 \times A}{l} \times 1.26 \times 10^{-6}$$

$$50 \text{ } \mu\text{H} = 1 \times \frac{(1000)^2 \times A}{0.15} \times 1.26 \times 10^{-6}$$

$$0.00005 \times 0.15 = (1000)^2 \times A \times 1.26 \times 10^{-6}$$

$$7.5 \times 10^{-6} = (1000)^2 \times A \times 1.26 \times 10^{-6}$$

$$= 1.26 \times A$$

$$A = \frac{7.5 \times 10^{-6}}{1.26}$$

$$= 5.952 \times 10^{-6} \text{ m}^2$$

Then use the area to calculate the radius.

$$A = \pi \times r^2$$

$$5.952 \times 10^{-6} = \pi \times r^2$$

$$r^2 = \frac{5.952 \times 10^{-6}}{\pi}$$

$$= 1.895 \times 10^{-6}$$

$$r = \sqrt{1.895 \times 10^{-6}}$$

$$= 1.376 \text{ mm}$$

The diameter is twice the radius.

$$d = 2 \times r$$

$$= 2.753 \text{ mm}$$

■

SELF-TEST

1. Calculate the inductance of a coil having a radius and length that measure 10 cm and 20 cm, respectively. The coil is comprised of 250 turns of wire around a soft-iron core having a relative permeability of 350.
2. An air-core coil is made using 2000 turns of wire around a coil form whose radius is 15 cm. The length of the coil is 0.025 m. What is the inductance value of this coil?
3. A 550-mH coil is to be made using a coil form having an area of 35×10^{-4} m². What is the coil length if 250 turns are to be wrapped around a core whose relative permeability is 200?
4. What is the diameter in centimeters of a coil form whose area is 50×10^{-4} m²?
5. A different core is used in the coil described in problem 1. The inductance value is found to be 2 H. What is the relative permeability of the new core?

ANSWERS TO SELF-TEST

1. $L = 4.329$ H
2. $L = 14.25$ H
3. $l = 0.1002$ m
4. Diameter $= 7.979$ cm
5. $\mu = 161.7$

INDUCTANCE

The magnetic properties of an inductor can be thought of as stored energy. That is, as current flows through a wire, a corresponding magnetic field is generated. The magnetic field itself has the potential to produce current. In physics the typical equation describing this stored energy (elastic energy of a spring) is

$$E = \tfrac{1}{2} \times k \times X^2$$

In this equation the constant k is analogous to the inductance of the coil. The X^2 value (normally, the distance the spring is stretched) corresponds to the current flowing through the coil. This is referred to as the square law principle. The effort required to stretch the spring is increased at the same rate as the square of the distance. So although the distance may be doubled, the effort will be four times that of the original (see Figure 15-7).

Inductance is like the spring constant then. It differs from one inductor to another; however, if you know how one inductor will perform in a circuit and its inductance value (spring constant), it is possible to predict the action of another inductor in the same circuit provided that you know its inductance (spring constant).

When an inductor is inserted into an ac circuit, it acts like a spring that is being compressed. That is, it resists any and all changes in current. The faster the change in current, the more the inductor resists the change. The inductance value of an inductor can be used to predict the amount of resistance (called reactance) exhibited by the inductor. Also, inductors can be connected in series and parallel to form circuits.

Series Inductance In series the inductance values are additive, like resistance values. To find the total inductance in any series circuit, all you would have to do is

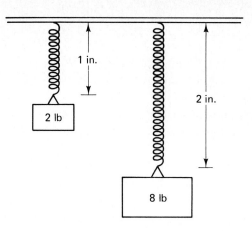

FIGURE 15-7

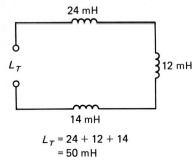

$L_T = 24 + 12 + 14$
$= 50$ mH

INDUCTANCE IS ADDITIVE IN SERIES

FIGURE 15-8

add all the individual inductor values. See the circuit shown in Figure 15-8. The formula for finding total inductance in a series circuit is:

$$L_T = L_1 + L_2 + L_3 + \cdots$$

Parallel Inductance In parallel the inductance is no longer additive. As with resistor circuits, it is the reciprocal of the inductance that is additive. The reciprocal of inductance does not have a name, but it does have a symbol. The Greek capital gamma (Γ) is used to indicate this value. Gamma looks like an upside-down L.

$$\Gamma = \frac{1}{L}$$

In parallel the formula for finding total inductance is

$$\frac{1}{L_T} = \frac{1}{L_1} + \frac{1}{L_2} + \frac{1}{L_3} + \cdots$$

EXAMPLE 15-7

Suppose that three inductors, 20 mH, 40 mH, and 50 mH, were connected in parallel (Figure 15-9). Find the total inductance of the circuit.

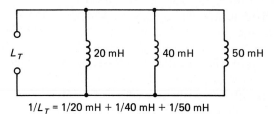

$1/L_T = 1/20$ mH $+ 1/40$ mH $+ 1/50$ mH

FIGURE 15-9

Solution: Use the total inductance formula for parallel circuits.

$$\frac{1}{L_T} = \frac{1}{L_1} + \frac{1}{L_2} + \frac{1}{L_3}$$

$$= \frac{1}{0.02} + \frac{1}{0.04} + \frac{1}{0.05}$$

$$= 50 + 25 + 20 = 95$$

$$L_T = \frac{1}{95}$$

$$= 10.53 \text{ mH}$$

SELF-TEST

1. Find the total inductance when a 27-, 59-, and 88-mH inductor are connected in series.
2. Find the total inductance when a 125-mH, 35-mH, 350-mH, and a 50-μH inductor are connected in series.
3. Find the total inductance when a 27-, 59-, and 88-mH inductor are connected in parallel.
4. Find the total inductance when a 125-mH, 35-mH, 350-mH, and a 50-μH inductor are connected in parallel.

ANSWERS TO SELF-TEST

1. 174 mH
2. 510.1 mH
3. 15.3 mH
4. 49.9 μH

MUTUAL INDUCTANCE

Whenever two (or more) inductors are placed in close proximity (close together), the magnetic field of one will extend into that of the other. Depending on the direction of current flow and the direction of the windings of the coil, the magnetic fields may aid or oppose each other. The left-hand rule can be used to determine whether the coils are aiding or opposing. The sketch in Figure 15-10 shows how to do this.

A number describing how well the magnetic flux lines of separate coils link up is called the *coefficient of coupling*. The coefficient of coupling is positive when the flux lines are aiding. It is negative when they are opposing. The coefficient of coupling is symbolized by k. It is important to realize that k is a percentage. It tells the percentage of total flux lines compared to those that have linked up. Thus k can be no greater than 1. The formula for finding the mutual inductance of two coils (we seldom worry about mutual inductance on more than two coils at a time) is

$$L_M = k \times \sqrt{L_1 \times L_2}$$

An example of how to calculate mutual inductance follows.

EXAMPLE 15-8

Given that the coefficient of coupling is 0.75, L_1 is 25 mH, and L_2 is 47 mH, find the mutual inductance.

Solution: Use the mutual inductance formula as follows:

$$L_M = k \times \sqrt{L_1 \times L_2}$$
$$= 0.75 \times \sqrt{0.025 \times 0.047}$$
$$= 25.70 \text{ mH}$$

The amount of mutual inductance will affect the total inductance of a circuit. It

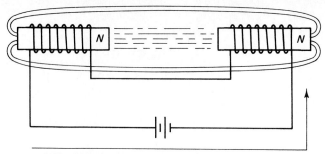

WHEN THE CURRENT AND THE WINDINGS
ALLOW THE MAGNETIC POLES TO LINK,
THEN THE TOTAL INDUCTANCE IS INCREASED
BY THE *MUTUAL INDUCTANCE* OF THE PAIR

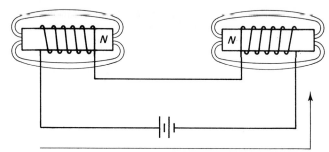

WHEN THE CURRENT AND THE WINDINGS
CAUSE THE MAGNETIC FIELDS TO OPPOSE,
THEN THE TOTAL INDUCTANCE IS DECREASED
BY THE *MUTUAL INDUCTANCE* OF THE PAIR

FIGURE 15-10

will increase the inductance of each inductor if the coupling is aiding and reduce it if the inductance is opposing. Again the type of circuit construction determines exactly how mutual inductance will affect the total inductance. See Figure 15-11.

Mutual Inductance in Series In series the inductance of each inductor will change an equal amount. This amount is what is termed mutual inductance. In other words, suppose that a 35-mH inductor were placed close to a 50-mH inductor, with a resulting mutual inductance of 5 mH. Then the 35-mH inductor would act as if it were a 40-mH inductor, while the 50-mH inductor would act as if it had an inductance of 55 mH (30 mH and 45 mH, respectively, when opposing). In series the total inductance can be calculated using the following formula.

$$L_T = L_1 + L_2 \pm 2 \times L_M$$

The ± sign is placed in the formula to show that both aiding and opposing coupling exist. Naturally, the plus is used when the mutual inductance is aiding, while the minus is used for opposing coupling. When there are two inductors, each affects the other in an equal manner (L_M). That is why two times the mutual inductance is used.

Mutual Inductance in Parallel As in series circuits, the mutual inductance will affect the true value of both inductors. This effect is equal on each inductor. When the mutual inductance is aiding, the value of both inductors is increased by L_M (reduced when the inductors are opposing). In parallel, the value for L_1 becomes $L_1 + L_M$ (when aiding) and L_2 becomes $L_2 + L_M$. Substituting these values into the parallel formula gives the following equation for total inductance:

$$\frac{1}{L_T} = \frac{1}{L_1 \pm L_M} + \frac{1}{L_2 \pm L_M}$$

As before, whether the plus or minus is used is determined by the circuit (aiding uses + and opposing uses −). Examples of the types of mutual inductor problems to be encountered follow.

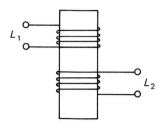

WHEN AIR CORE COILS ARE PLACED IN CLOSE PROXIMITY, THE RESULTANT COEFFICIENT OF COUPLING, k, IS APPROXIMATELY 0.1

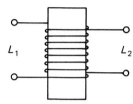

WHEN THE WINDINGS OVERLAP THE COEFFICIENT OF COUPLING, k, INCREASES TO APPROXIMATELY 0.3

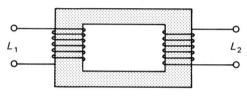

WHEN THE COILS ARE WOUND AROUND THE SAME IRON CORE, THE COEFFICIENT OF COUPLING, k, IS APPROXIMATELY 1.0

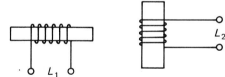

WHEN THE COILS ARE PLACED PERPENDICULAR TO EACH OTHER, THE COEFFICIENT OF COUPLING, k, CAN BE CONSIDERED AS 0

FIGURE 15-11

EXAMPLE 15-9

A 12-mH coil and a 50-mH coil are connected in series. The coefficient of coupling for the two is −0.45. Find the total inductance for the circuit.

Solution: Begin by calculating the mutual inductance of the circuit.

$$L_M = k \times \sqrt{L_1 \times L_2}$$
$$= -0.45 \times \sqrt{0.012 \times 0.05}$$
$$= -11.02 \text{ mH}$$

(Note that the negative sign indicates that the mutual inductance is opposing.) Then use the total inductance formula as follows:

$$L_T = L_1 + L_2 - 2 \times L_M$$
$$= 12 \text{ mH} + 50 \text{ mH} - 2 \times 11.02 \text{ mH}$$
$$= 39.95 \text{ mH}$$

EXAMPLE 15-10

One inductor has an inductance of 75 mH. The other has an inductance of 30 mH. When the two are connected together in series, the total inductance is 145 mH. Find the mutual inductance and the coefficient of coupling.

Solution: Begin by calculating the mutual inductance. Insert the known values in the total series inductance formula. Use the + of the ± in the formula because the total is more than the sum of the individual values.

$$L_T = L_1 + L_2 + 2 \times L_M$$
$$0.145 = 0.075 + 0.03 + 2 \times L_M$$
$$= 0.105 + 2 \times L_M$$
$$2 \times L_M = 0.04$$
$$L_M = \frac{0.04}{2}$$
$$= 0.02$$
$$= 20 \text{ mH}$$

Use the mutual inductance value to calculate the coefficient of coupling for the inductors.

$$L_M = k \times \sqrt{L_1 \times L_2}$$
$$0.02 = k \times \sqrt{0.075 \times 0.03}$$
$$= k \times 0.04743$$
$$k = \frac{0.02}{0.04743}$$
$$= 0.4216$$

Note: The coefficient of coupling is a ratio and as such should not be written using scientific notation, but instead should be written as a decimal fraction.

∎

EXAMPLE 15-11

Find the total inductance when a 27-mH inductor is connected in parallel to a 56-mH inductor to form a coupling with a coefficient of 0.68.

Solution: Begin by calculating the mutual inductance of the pair of inductors.

$$L_M = k \times \sqrt{L_1 \times L_2}$$
$$= 0.68 \times \sqrt{0.027 \times 0.056}$$
$$= 26.44 \text{ mH}$$

Then use the total parallel inductance formula to find the total inductance of the circuit.

$$\frac{1}{L_T} = \frac{1}{L_1 + L_M} + \frac{1}{L_2 + L_M} \quad \text{(plus since } k \text{ is positive)}$$

$$= \frac{1}{27 \text{ mH} + 26.44 \text{ mH}} + \frac{1}{56 \text{ mH} + 26.44 \text{ mH}}$$

$$= \frac{1}{53.44 \text{ mH}} + \frac{1}{82.44 \text{ mH}}$$

$$L_T = 32.42 \text{ mH}$$

SELF-TEST

1. Find the total inductance when a 27-mH and a 88-mH inductor are connected in series, given that the coefficient of coupling is +0.589.
2. Find the total inductance when a 125-mH and a 350-mH inductor are connected in series, given that the coefficient of coupling is −0.487.
3. Find the total inductance when a 27-mH and a 88-mH inductor are connected in parallel, given that the coefficient of coupling is −0.796.
4. Find the total inductance when a 125-mH and a 350-mH inductor are connected in parallel, given that the coefficient of coupling is +0.569.

ANSWERS TO SELF-TEST

1. 172.4 mH
3. 15.5 mH
2. 271.3 mH
4. 160 mH

TRANSFORMERS

The best example of mutual inductance is a transformer. A transformer consists of a pair of inductors which are so close to each other that at times one set of windings is wrapped around the other. This allows a full linkage between the magnetic fields of each. In other words, practically all of the flux lines link up and the coefficient of coupling is approximately 1. For the purposes of our discussion we assume that $k = 1$ (although most transformers are not this efficient).

Transformers are used to change the voltage level of alternating-current circuits. The input coil of a transformer is called the *primary winding*. The *secondary winding* is made up of the output coil. When a transformer has a higher secondary (output) voltage than its primary (input) voltage, it is called a *step-up transformer*. If the secondary voltage is lower than the primary, the transformer is referred to as a *step-down transformer*.

A step-up transformer will increase the secondary voltage, but the current in the secondary will be less than that in the primary. In essence, the input power ($V_1 \times I_1$) must equal the output power ($V_2 \times I_2$). This is the basis for all of the transformer calculations.

Turns Ratio The ratio of the number of turns in the primary to the number of turns in the secondary is called the *turns ratio*.

$$\text{TR} = \frac{N_2}{N_1}$$

The turns ratio is a very important number. It can be used to convert any of the primary quantities to secondary quantities. The turns ratio can also be found in several other ways. They are all listed in the formulas that follow.

$$\frac{N_2}{N_1} = \frac{V_2}{V_1} = \frac{I_1}{I_2}$$

Notice that current is the only quantity that does not use secondary over primary as a ratio. Instead, it uses primary over secondary. Different textbooks list these formulas in reciprocal form. Most manufacturers label their transformers in the manner shown here.

Table 15-1 contains conversion factors that make work with transformers less complicated. The TR is shown as a conversion factor for all of the quantities. Examples of these calculations follow.

TABLE 15-1 Transformer Conversion Factors

Primary	Conversion	= Secondary
Voltage	× turns ratio	= voltage
Current	÷ turns ratio	= current
Power		= power
Turns	× turns ratio	= turns
Impedance	× (turns ratio)^2	= impedance

EXAMPLE 15-12

The primary voltage of a transformer is 120 V, and the secondary voltage is 12.6 V. The current in the secondary windings is 2.5 A. There are 2500 turns in the primary coil windings. Find the following values: primary current, primary and secondary power, turns ratio, number of turns in the primary and secondary windings, and both primary and secondary impedance.

Solution: Begin by calculating the turns ratio. To find the turns ratio, you must know the number of primary and secondary turns, the primary and secondary voltage values, or the primary and secondary current levels. Notice that both the primary and secondary voltage values are known.

$$\text{TR} = \frac{V_2}{V_1}$$

$$= \frac{12.6}{120}$$

$$= 0.105$$

The turns ratio and Table 15-1 can be used to find the primary current. Insert the known values in the formula.

$$\frac{I_1}{\text{TR}} = I_2$$

$$\frac{I_1}{0.105} = 2.5$$

$$I_1 = 2.5 \times 0.105$$

$$= 262.5 \text{ mA}$$

The turns ratio and Table 15-1 can also be used to find the secondary turns. Again insert the known values into the formula.

$$N_1 \times TR = N_2$$

$$2500 \times 0.105 = 262.5 \text{ turns}$$

The primary and secondary powers are equal and can be calculated using the power formula and either the primary or secondary voltages and currents.

$$12.6 \text{ V} \times 2.5 \text{ A} = 31.5 \text{ W}$$

Ohm's law can be used to find the primary and secondary impedance.

$$Z_2 = \frac{V_2}{I_2}$$

$$= \frac{12.6 \text{ V}}{2.5 \text{ A}}$$

$$= 5.04 \text{ }\Omega$$

(Impedance Z is the ac equivalent to resistance R.)

$$Z_1 = \frac{V_1}{I_1}$$

$$= \frac{120 \text{ V}}{0.2625 \text{ A}}$$

$$= 457.1 \text{ }\Omega \qquad \blacksquare$$

SELF-TEST

1. Primary turns = 150   secondary turns = 10,000
   primary voltage = 12 V   secondary voltage = _____
   primary current = 3 A   secondary current = _____
   primary power = _____   secondary power = _____
   primary impedance = _____ secondary impedance = _____
   turns ratio = _____

2. Primary turns = 10,000   secondary turns = 250
   primary voltage = 120 V   secondary voltage = _____
   primary current = 0.3 A   secondary current = _____
   primary power = _____   secondary power = _____
   primary impedance = _____ secondary impedance = _____
   turns ratio = _____

3. Primary turns = 500   secondary turns = _____
   primary voltage = _____   secondary voltage = 50 V
   primary current = _____   secondary current = _____
   primary power = _____   secondary power = _____
   primary impedance = _____ secondary impedance = 8 Ω
   turns ratio = 0.25

4. Primary turns = 800   secondary turns = _____
   primary voltage = _____   secondary voltage = _____
   primary current = _____   secondary current = _____
   primary power = 5 W   secondary power = _____
   primary impedance = 12 Ω secondary impedance = _____
   turns ratio = 8

TRANSFORMERS

5. Primary turns = 250   secondary turns = 1000
    primary voltage = 60 V   secondary voltage = _____
    primary current = 0.1 A   secondary current = _____
    primary power = _____   secondary power = _____
    primary impedance = _____   secondary impedance = _____
    turns ratio = _____

6. Primary turns = 1250   secondary turns = _____
    primary voltage = 120 V   secondary voltage = 6.3 V
    primary current = _____   secondary current = 2.5 A
    primary power = _____   secondary power = _____
    primary impedance = _____   secondary impedance = _____
    turns ratio = _____

7. Primary turns = _____   secondary turns = 1000
    primary voltage = 15 V   secondary voltage = _____
    primary current = 0.250 A   secondary current = 0.01 A
    primary power = _____   secondary power = _____
    primary impedance = _____   secondary impedance = _____
    turns ratio = _____

8. Primary turns = 12,500   secondary turns = _____
    primary voltage = _____   secondary voltage = _____
    primary current = _____   secondary current = _____
    primary power = 0.5 W   secondary power = _____
    primary impedance = 2000 Ω   secondary impedance = 8000 Ω
    turns ratio = _____

9. Primary turns = 10,000   secondary turns = 200
    primary voltage = 120 V   secondary voltage = _____
    primary current = 0.125 A   secondary current = _____
    primary power = _____   secondary power = _____
    primary impedance = _____   secondary impedance = _____
    turns ratio = _____

10. Primary turns = _____   secondary turns = 1200
    primary voltage = 20 V   secondary voltage = _____
    primary current = _____   secondary current = _____
    primary power = 10 W   secondary power = _____
    primary impedance = _____   secondary impedance = _____
    turns ratio = 2.5

ANSWERS TO SELF-TEST

1. Secondary voltage = 800 V   secondary current = 45 mA
    primary power = 36 W   secondary power = 36 W
    primary impedance = 4 Ω   secondary impedance = 17.8 kΩ
    turns ratio = 67

2. Secondary voltage = 3 V   secondary current = 12 A
    primary power = 36 W   secondary power = 36 W
    primary impedance = 400 Ω   secondary impedance = 0.25 Ω
    turns ratio = 0.025

3. Secondary turns = 125   primary voltage = 200 V
    primary current = 1.563 A   secondary current = 6.25 A
    primary power = 312.5 W   primary power = 312.5 W
    primary impedance = 128 Ω

4. Secondary turns = 6400   primary voltage = 7.746 V
    secondary voltage = 61.97 V   primary current = 645.5 mA
    secondary current = 80.69 mA   secondary power = 5 W
    secondary impedance = 768 Ω
5. Secondary voltage = 240 V   secondary current = 25 mA
    primary power = 6 W   secondary power = 6 W
    primary impedance = 600 Ω   secondary impedance = 9.6 kΩ
    turns ratio = 4
6. Secondary turns = 65.6   primary current = 131.3 mA
    primary power = 15.75 W   secondary power = 15.75 W
    primary impedance = 914.3 Ω   secondary impedance = 2.52 Ω
    turns ratio = 0.0525
7. Primary turns = 40   secondary voltage = 375 V
    primary power = 3.75 W   secondary power = 3.75 W
    primary impedance = 60 Ω   secondary impedance = 37.5 kΩ
    turns ratio = 25
8. Secondary turns = 25000   primary voltage = 31.62 V
    secondary voltage = 63.24 V   primary current = 15.81 mA
    secondary current = 7.9 mA   secondary power = 0.5 W
    turns ratio = 2
9. Secondary voltage = 2.4 V   secondary current = 6.25 A
    primary power = 15 W   secondary power = 15 W
    primary impedance = 960 Ω   secondary impedance = 0.384 Ω
    turns ratio = 0.02
10. Primary turns = 480   secondary voltage = 50 V
    primary current = 0.5 A   secondary current = 0.2 A
    secondary power = 10 W   primary impedance = 40 Ω
    secondary impedance = 250 Ω

Efficiency The main thing that makes one transformer better than another is the amount of power that is transmitted from the primary to the secondary windings. This rating is called the *efficiency*. The Greek lowercase letter eta (η) is used to symbolize efficiency. The basic formula for calculating the efficiency is

$$\eta = \frac{P_{out}}{P_{in}} = \frac{P_2}{P_1}$$

Understand that the efficiency rating is a percentage. Your calculator will provide the decimal equivalent to the percentage figure. It is a simple matter to convert a decimal fraction into a percentage. All that is needed is to move the decimal point two places to the right. That is, $0.15 = 15\%$.

SELF-TEST

1. The input of a transformer receives 250 mW of power. The output of this transformer is only 200 mWatts. What is the efficiency rating for the transformer?
2. The efficiency rating of a transformer is 85%. What is the power output when the power input is 2.5 W?
3. The efficiency rating of a transformer is 95%. What power input is required for an output power of 250 mW?

4. What is the efficiency of a transformer whose input is 500 mW and whose output is 480 mW?
5. Calculate the efficiency of a transformer given that the input power is 1.5 W and the output power is 1.2 W.

ANSWERS TO SELF-TEST

1. 80%
2. 2.125 W
3. 263.2 mW
4. 96%
5. 80%

Phase Relations The schematic symbol for a transformer is shown in Figure 15-12. The presence of lines indicates that a metallic core is used. The absence of lines means that the transformer uses an air core (refer to Figure 15-12).

TRANSFORMER SYMBOLS   **FIGURE 15-12**

The phase relationship between the input and output of a transformer may not be zero. Most transformers have a 180° phase shift from primary to secondary (see Figure 15-13). A phase shift is usually indicated by dots over the primary and secondary windings. Dots directly across form each other indicate that there is no phase shift. Dots placed diagonally across from each other indicate a 180° phase shift. The absence of dots will usually indicate a 180° phase shift.

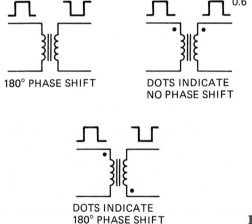

FIGURE 15-13

INDUCTIVE REACTANCE

There are two things that you must remember when trying to understand how inductive reactance is created and what it is.

1. Current flowing through a coil of wire will produce a magnetic field.
2. As a magnetic field enters a coil of wire it produces a current.

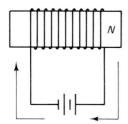

THE CURRENT FLOWING THROUGH THIS COIL PRODUCES A MAGNETIC FIELD WHICH HAS ITS NORTH POLE LOCATED ON THE RIGHT SIDE

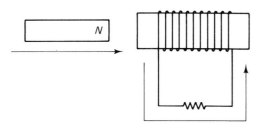

WHEN THE NORTH POLE OF THIS MAGNET IS PUSHED INTO THIS COIL, CURRENT IS GENERATED. NOTICE THAT THE CURRENT PRODUCED FLOWS IN THE OPPOSITE DIRECTION, WHILE THE MAGNETIC FIELD STILL HAS THE NORTH POLE ON THE RIGHT SIDE

FIGURE 15-14

See Figure 15-14.

When an ac current flows through a coil the amount of current is continually changing. As it changes there is a corresponding change in the magnetic field being produced, similar to a magnet (like the one in Figure 15-14) being moved into the coil. The changing magnetic field generates a current that opposes the original current (the one that produced the magnetic field in the first place). This begins a never-ending cycle. Current produces magnetic field. Magnetic field produces current. This phenomenon is called counterelectromotive force (cemf). Also, in some ways, it is like pushing against a wall. The harder you push against the wall, the harder, in turn, the wall pushes back. Voltage (emf) provides the push against the wall. Cemf works like the wall pushing back against emf. When cemf and emf are equal there is no current flow and both are at a peak (see Figure 15-15).

It is important for you to understand that it is not the current flowing through the coil that produces cemf; instead, it is the change in the current that produces this force. The rate of change in the current fluctuates also. When the current is at a peak, there is a moment when the corresponding rate of change is zero, like the moment of weightlessness as a roller coaster pops over a hill. The maximum rate of change occurs as the current changes direction. This corresponds to the point where current is zero. Examine the sine wave shown in Figure 15-15.

There are several ways to look at cemf. One way is to predict the voltage that will be produced. Calculus is usually used in an analysis of this voltage. The formula for calculating the induced voltage is

$$v_{\text{ind}} = L \times \frac{di}{dt}$$

The di/dt in this formula indicates the rate of change in the current. Remember that the d stands for change. Here we have the change in current divided by the change in time. The induced voltage is symbolized as v_{ind}. An example of this type of problem follows.

INDUCTIVE REACTANCE

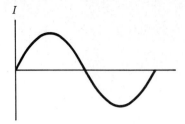

THE CURRENT AND MAGNETOMOTIVE FORCE ARE NORMALLY IN PHASE WITH EACH OTHER

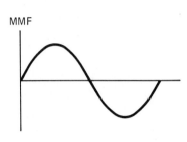

THE COUNTER ELECTROMOTIVE FORCE (INDUCED VOLTAGE) IS OUT OF PHASE WITH CURRENT AND MAGNETOMOTIVE FORCE. IT LEADS THE CURRENT BY 90°

FIGURE 15-15

EXAMPLE 15-13

Calculate the voltage produced when the current in a 25-mH inductor changes from 40 mA to 75 mA in 30 μs.

Solution: Begin by calculating the rate of change (di/dt) of current flow. This can be accomplished by the following process.

di:

$$75 \text{ mA} - 40 \text{ mA} = 30 \text{ mA}$$

di/dt:

$$\frac{30 \text{ mA}}{30 \text{ }\mu\text{s}} = 1000 \text{ A/s}$$

This rate can be used in the induced voltage formula to predict the cemf of the coil.

$$v_{\text{ind}} = L \times \frac{di}{dt}$$
$$= 25 \text{ mH} \times 1000 \text{ A/s}$$
$$= 25 \text{ V}$$

SELF-TEST

1. Given that the rate of change in current though a 2.5-mH choke is 2500 A/s, find the induced voltage produced by this inductor.

2. The current through a 50-mH coil changes from 0.25 A to 1 A every 5 ms. Calculate the amount of voltage induced by this circuit.

3. Two hundred and fifty volts are produced by a coil when the current flow is decreased from 500 mA to 10 mA durring a 30-ns span. What is the inductance value of the coil?

4. Find the rate of change in current needed to produce a 50-V induced voltage across a 35-mH coil.

ANSWERS TO SELF-TEST

1. 6.25 V
3. 15.3 μH
2. 7.5 V
4. 1429 A/s

Counterelectromotive force is also an opposition to current flow. In an ac voltage circuit we call this opposition reactance. Reactance, one might guess, is measured in ohms, like any other opposition to current flow. The formula for calculating the amount of reactance a particular inductor might have is

$$X_L = 2 \times \pi \times f \times L$$

The symbol for reactance is X. The amount of reactance that any inductor will show depends greatly on the frequency of the signal generator. An inductor will show more reactance for a high frequency than for a low frequency. The quantity $2 \times \pi \times f$ is called the *angular velocity*. Some textbooks symbolize angular velocity by the Greek lowercase letter omega (ω). They write the reactance formula as

$$X_L = \omega \times L$$

A typical problem involving X_L is demonstrated in the following example.

EXAMPLE 15-14

Find the reactance provided by a 35-mH choke (another name for an inductor) for a 25-kHz signal.

Solution: Insert the known values into the reactance formula.

$$X_L = 2 \times \pi \times f \times L$$
$$= 2 \times \pi \times 25,000 \times 0.035$$
$$= 5497.78 \ \Omega$$

SELF-TEST

1. Calculate the inductive reactance of a 50-mH choke at the following frequencies: 1000 Hz, 2500 Hz, 12,500 Hz, and 100 kHz.
2. Calculate the inductive reactance of a 200-mH coil at the following frequencies: 1500 Hz, 3500 Hz, 25 kHz, and 250 kHz.
3. At what frequency will a 33-mH choke have an inductive reactance of 2500 Ω?
4. At what frequency will a 500-μH coil show an inductive reactance of 5000 Ω?

ANSWERS TO SELF-TEST

1. 314 Ω, 785 Ω, 3927 Ω, 31.4 kΩ
2. 1885 Ω, 4398 Ω, 31,416 Ω, 314.2 kΩ
3. 12.06 kHz
4. 1.592 MHz

Reactance in Series Reactance in series, like any other ohm quantity, is additive in series. The formula for finding total series reactance is

$$X_{LT} = X_{L1} + X_{L2} + X_{L3} + \cdots$$

Reactance in Parallel Again, ohms in parallel are combined in the same manner no matter what their source may be. The formula for finding total parallel reactance is

$$\frac{1}{X_{LT}} = \frac{1}{X_{L1}} + \frac{1}{X_{L2}} + \frac{1}{X_{L3}} + \cdots$$

The reciprocal of reactance is called *susceptance* (Figure 15-16).

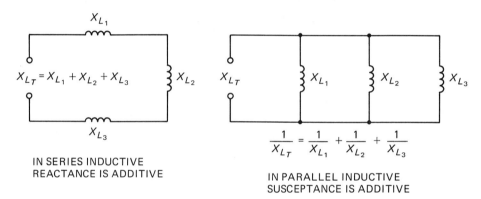

FIGURE 15-16

Reactance and Ohm's Law Ohm's law will still apply to circuits that contain inductors. The basic idea is that one must first find the inductive reactance (the ohm value) for the inductor. Once this value has been found, it can be treated like a resistance, provided that the circuit contains only inductors. There are times when resistors and inductors are combined in the same circuit.

SELF-TEST

1. A 30-, 150-, and 75-mH coil are connected in series. Find the current and voltage drop across each when they are powered by a 10-V signal operating at 2.5 kH.
2. A 250-, 180-, and 130-μH coil are connected in series. Find the current and voltage across each when they are powered by a 5.6-V signal operating at 50 kH.
3. A 30-, a 150-, and 75-mH coil are connected in parallel. Find the current through each when they are powered by a 10-V signal operating at 2.5 kH.
4. A 250-, 180-, and 130-μH coil are connected in parallel. Find the current through each when they are powered by a 5.6-V signal operating at 50 kHz.

ANSWERS TO SELF-TEST

1. $I = 2.497$ mA, $V_{30\,mH} = 1.176$ V, $V_{150\,mH} = 5.882$ V, $V_{75\,mH} = 2.941$ V
2. $I = 31.83$ mA, $V_{250\,mH} = 2.5$ V, $V_{180\,mH} = 1.8$ V, $V_{130\,mH} = 1.3$ V
3. $I_{30\,mH} = 21.22$ mA, $I_{75\,mH} = 8.488$ mA, $I_{150\,mH} = 4.244$ mA
4. $I_{250\,\mu H} = 71.30$ mA, $I_{180\,\mu H} = 99.03$ mA, $I_{130\,\mu H} = 137.1$ mA

SOLUTION OF *LR* CIRCUITS

The solution of *LR* circuits can be simple provided that you know how to solve a right triangle and can follow a few simple rules. In the next sections we describe and explain those rules.

ELI the ICEman The phrase ELI the ICEman has been around for a long time and has helped many an electronics student through hard times. This memory device can be used for at least a dozen clues about reactive circuits (ac and dc included). The phrase is usually taught to help one remember that the voltage leads the current in an inductive circuit (ELI) by 90° and that current leads voltage in a capacitive circuit (ICE) by 90°. Students experience several misconceptions about these letters. Understand that the *L* in ELI stands for inductance, and that ELI is used only for inductive circuits. The *C* in ICE stands for capacitance, and similarly, ICE is used only for capacitive circuits. Capacitive reactance is explained in Chapter 16.

Triangles All of the triangles used in this book have a horizontal side, a vertical side, and a hypotenuse. Some triangles will have their points upward, others downward. How to determine which way a particular triangle points is explained later in the book under the topic of phase angles. For now, we concentrate on what values are placed on which sides.

The resistive component is always placed on the horizontal side. The resistive component may be resistance, voltage across the resistor, current through the resistor, or conductance ($1/R$), depending on the triangle being used.

The vertical side is reserved for the reactive component. Again the reactive component may be reactance, voltage across the inductor, current through the inductor, or susceptance ($1/X$). This depends on the individual triangle.

The hypotenuse is used to represent the total quantities. These may be impedance (Z) (combined resistance and reactance), total applied voltage, total circuit current, or admittance ($1/Z$). Again these values depend on the triangle being implemented (see Figure 15-17).

Series-Circuit Triangles An early step in circuit analysis is to determine whether the circuit is series or parallel. Any series circuit can have two triangles. These are a voltage triangle and an impedance triangle. The phase angle is the same (equal) for both of these triangles. This is sometimes referred to as lying on the same plane.

ELI can be used to remind you that a voltage triangle is possible. Look at ELI; if you eliminate the *I* (current is the same in a series circuit) and *L* (the circuit contains an inductor), then *E* (the voltage) is the remaining quantity. Understand that this indicates that a voltage triangle does exist for a series circuit. Remember that *impedance and voltage lie on the same plane*. So an impedance triangle can also be used when analyzing a series circuit.

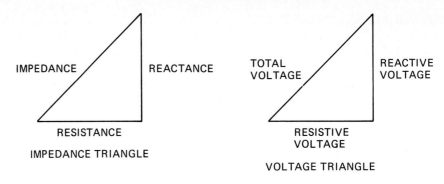

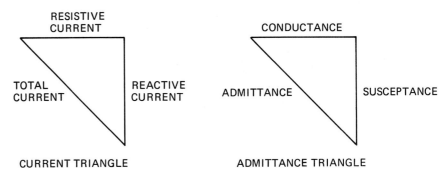

FIGURE 15-17

Parallel-Circuit Triangles The current and admittance triangles are devoted to parallel circuit analysis. The admittance triangle does exist; however, it is best for beginning students to promptly forget this. Trying to solve an admittance triangle requires the use of many, many reciprocal functions. Most beginning students experience too many difficulties in these calculations. For that reason it is best to use only the current triangle when working with parallel circuits.

ELI can be used on a parallel circuit to remind you that a current triangle is possible. Again, look at ELI; if you eliminate the E (voltage is the same in a parallel circuit) and L (the circuit contains an inductor), then I (the current) is the remaining quantity. This indicates that a current triangle exists for a parallel circuit. Remember: *try to avoid using an admittance triangle*.

Phase Angle of Triangles The angle between the hypotenuse and the horizontal side is called the *phase angle*. It represents the number of degrees between the peak values of these components. This angle may be positive (the hypotenuse quantity peaks first) or negative (the horizontal quantity peaks first). ELI can be used to determine if this angle is positive or negative. Draw ELI in the manner shown in Figure 15-18.

FIGURE 15-18

In a series circuit the current is the same everywhere. When you use ELI as a memory device, a line is drawn out from the constant quantity (I) and then upward to the remaining component (E). This shows that the triangle should be drawn with the point upward (indicating that the phase is positive). In a parallel circuit the process in the same except that the voltage (E) is constant and has the line drawn outward, while the current (I) is the remaining quantity. The resulting triangle would point downward.

The resistive voltage is usually used as a reference point. This causes no difficulty when you are working a series circuit; however, when you are working a parallel circuit, your current triangle uses the resistive current as a reference. All that is necessary to change the reference from resistive current to resistive voltage is to change the sign of the phase angle, because the phase angle of the current on a parallel circuit is equal and opposite the phase angle of the impedance.

RL Circuits Following are several examples of series and parallel circuit analysis.

EXAMPLE 15-15

Analyze the circuit shown in Figure 15-19.

FIGURE 15-19

Solution: Begin by examining the circuit. This is a series circuit. ELI indicates the there is a voltage triangle. (I is the same in series; this leaves E.) Voltage and impedance lie on the same plane, so there is also an impedance triangle. ELI also indicates that the phase angle is positive. Draw the triangles as shown in Figure 15-20. Use the Pythagorean theorem to solve the impedance triangle for impedance.

$$Z = \sqrt{2500^2 + 4000^2} = 4717 \, \Omega$$

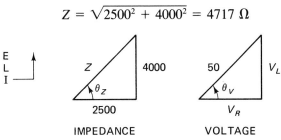

FIGURE 15-20

The tangent formula can be used to find the phase angle of the impedance.

$$\theta = \tan^{-1} \frac{4000}{2500}$$
$$= \tan^{-1} 1.6$$
$$= 57.99°$$

[4] [EE] [3] [÷] [2] [.] [5] [EE] [3] [=] [2nd] [tan⁻¹] [mem]   display   5.7999   01

SOLUTION OF *LR* CIRCUITS

Remember that voltage and impedance lie on the same plane, so

(θ_z is equal to θ_v)

Then Cah can be used to find the resistor voltage.

$$V_R = V_T \times \cos \theta$$
$$= 50 \times \cos 57.99°$$
$$= 26.503 \text{ V}$$

[5] [0] [×] [rcl] [cos] [=]   display   2.6503   01

Soh can be used to find the inductor voltage.

$$V_L = V_T \times \sin \theta$$
$$= 50 \times \sin 57.99°$$
$$= 42.398 \text{ V}$$

[5] [0] [×] [rcl] [sin] [=]   display   4.2398   01

Ohm's law can be used to find the current.

$$I = \frac{26.503}{2500}$$
$$= 10.601 \text{ mA}$$ ∎

Note the current can be found in any of several ways. Current can be thought of as a common factor between the voltage and current triangles. That is, as in other circuits, Ohm's law will apply but only at individual levels. In this case, the individual levels refer to the corresponding sides of the triangle (voltage horizontal/impedance horizontal or voltage vertical/impedance vertical or voltage hypotenuse/impedance hypotenuse).

EXAMPLE 15-16

Analyze the circuit shown in Figure 15-21.

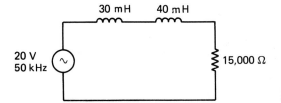

FIGURE 15-21

Solution: Begin by examining the circuit. This is a series circuit. ELI indicates the there is a voltage triangle. (I is the same in series; this leaves E.) Voltage and impedance lie on the same plane, so there is also an impedance triangle. ELI also indicates that the phase angle is positive. Draw the triangles as shown in Figure 15-22. Begin by calculating the total inductance of the two inductors in the circuit.

$$L_T = L_1 + L_2$$
$$= 30 \text{ mH} + 40 \text{ mH}$$
$$= 70 \text{ mH}$$

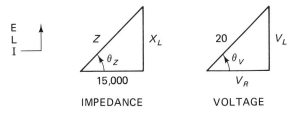

FIGURE 15-22

Then use this value to find the inductive reactance of this circuit.

$$X_L = 2 \times \pi \times 50 \text{ kHz} \times 70 \text{ mH}$$
$$= 21.99 \text{ k}\Omega$$

Now use the impedance triangle and the Pythagorean theorem to find the impedance for this circuit.

$$Z = \sqrt{15{,}000^2 + 21{,}990^2} = 26.61 \text{ k}\Omega$$

Use the tangent function to find the θ_z.

$$\theta = \tan^{-1} \frac{15{,}000}{21{,}990}$$
$$= \tan^{-1} 0.6821$$
$$= 34.3°$$

[1] [5] [EE] [3] [÷] [rcl] [=] [2nd] [$\tan^{-1}$] [mem]
$$\qquad\qquad\qquad\qquad\qquad\text{display}\quad 3.4279 \qquad 01$$

Remember that voltage and impedance lie on the same plane, so

$$(\theta_z \text{ is equal } \theta_v)$$

Then use Cah to find the resistor voltage.

$$V_R = V_T \times \cos \theta$$
$$= 20 \times \cos 34.3°$$
$$= 16.52 \text{ V}$$

[2] [0] [×] [rcl] [cos] [=]   display   1.6521   01

Use the sine function find the inductor voltage.

$$V_L = V_T \times \sin \theta$$
$$= 20 \times \sin 34.3$$
$$= 11.27 \text{ V}$$

[2] [0] [×] [rcl] [sin] [=]   display   1.1270   01

Use Ohm's law to find the current for the circuit.

$$I = \frac{V_R}{R} \qquad \text{or} \qquad I = \frac{V_L}{X_L} \qquad \text{or} \qquad I = \frac{V_T}{Z}$$

$$I = \frac{16.52}{15{,}000}$$
$$= 1.101 \text{ mA} \qquad\qquad\qquad\qquad\qquad\qquad ■$$

SOLUTION OF LR CIRCUITS

EXAMPLE 15-17

Analyze the circuit shown in Figure 15-23.

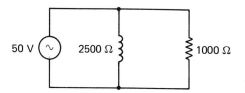

FIGURE 15-23

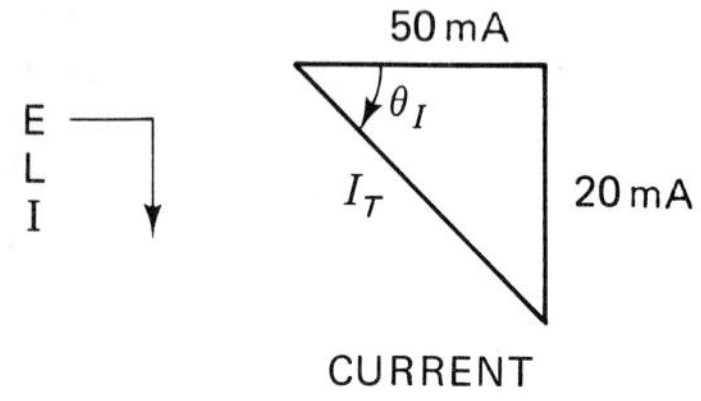

Solution: Begin by examining the circuit. This is a parallel circuit. ELI indicates that there is a current triangle. (E is the same in parallel; this leaves I.) ELI also indicates that the phase angle is negative. Draw the triangle as shown in Figure 15-24. Begin by using Ohm's law to calculate the resistor current.

$$I_R = \frac{V}{R}$$

$$= \frac{50}{1000}$$

$$= 50 \text{ mA}$$

Then calculate the inductor current.

$$I_L = \frac{V}{X_L}$$

$$= \frac{50}{2500}$$

$$= 20 \text{ mA}$$

FIGURE 15-24

Insert these values in the appropriate places on the current triangle and use the Pythagorean theorem to find the total circuit current.

$$I_T = \sqrt{I_R^2 + I_L^2}$$

$$= \sqrt{0.05^2 + 0.02^2}$$

$$= 53.85 \text{ mA}$$

Use the tangent function Toa to find the θ_I.

$$\theta = \tan^{-1}\frac{0.02}{0.05}$$

$$= \tan^{-1}(0.4)$$

$$= -21.8°$$

(Note that the triangle points downward, so this angle is negative.)

[2] [0] [EE] [3] [+/−] [÷]

[5] [0] [EE] [3] [+/−] [=] [2nd] [tan⁻¹]   display   2.1801   01

Use Ohm's law and the total values to find the impedance.

$$Z = \frac{V}{I_T}$$

$$= \frac{50}{0.05385}$$

$$= 928.5 \; \Omega \qquad \blacksquare$$

SELF-TEST

1. A 2000-Ω inductive reactance is connected in series with an 1800-Ω resistor. Calculate the total impedance and phase of the circuit.
2. A 5000-Ω resistor is connected in series with a 2500-Ω inductive reactance. Calculate the total impedance and phase of the circuit.
3. A 20-V source powers a 250-Ω resistor that is connected in parallel with a 500-Ω inductive reactance. Calculate the total current and phase of the circuit.
4. A 600-Ω inductive reactance is connected in parallel with a 600-Ω resistor. Calculate the total current and phase of the circuit when it is powered by a 25-V source.

ANSWERS TO SELF-TEST

1. $Z = 2.69$ kΩ, $\theta = 48.01°$ 2. $Z = 5.59$ kΩ, $\theta = 26.57°$
3. $I_R = 80$ mA, $I_L = 40$ mA, $I_T = 89.44$ mA, $\theta = -26.56°$
4. $I_R = I_L = 41.67$ mA, $I_T = 58.92$ mA, $\theta = -45°$

TROUBLESHOOTING INDUCTORS

The accepted manner in which to test an inductor, transformer, or other coil is by using a resistance test. A large number of faults found in coils are caused by an over-current (overheat) condition. Too much heat will cause the insulation on the coil winds to melt. This will allow the coil windings to make contact in places where they are not meant to touch. A problem of this sort will cause the resistance of the coil to be lower than the specified amount. The resistance value for various coils can be anywhere from less than 1 Ω to several thousands of ohms. The range of acceptable values is so wide that you must check the specification for each coil that you check.

Larger coils are sometimes tested by checking the number of oscillating rings that the coil produces when a voltage is suddenly applied or removed. All inductors have a basic desire to maintain the same amount of current flowing through them. When you suddenly apply voltage, the inductor resists the sudden change in current and produces a large opposing voltage. This induced voltage appears as several diminishing oscillations when viewed on an oscilloscope (Figure 15-25). This is called *ringing*. A good coil will ring five to seven times.

The inductance value of an inductor can be measured by using an *RCL* bridge network (Figure 15-26). The use of an *RCL* bridge is more of an art than a science. For the most part, technicians today are caught up in a high-tech mode. As such, it is important that you make the necessary checks quickly and accurately. The *RCL* bridge seems lacking in both of the categories. Instead, most technicians will use an

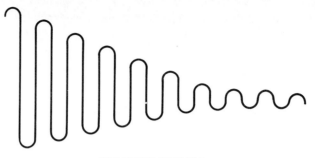

INDUCTOR RINGING

FIGURE 15-25

FIGURE 15-26 Impedance (Capacitance and Inductance) LCR Bridge—left, Z-meter—right.

impedance meter (Z meter; Figure 15-26). This device is used in a manner similar to an ohmmeter. The leads are connected to the coil. A button is pushed and a digital display almost instantly reads out the inductance value. When a second button is pushed, the number of rings is displayed, making the test of the inductor a simple task.

SUMMARY

The solution of right triangles using both the Pythagorean theorem and trigonometric formulas are used in the analysis of inductors and inductor circuits. The physical measurements and number of turns of wire used to make the coil determine its inductance value. This is indicated in the formula

$$L = \mu \times \frac{N^2 \times A}{l} \times 1.26 \times 10^{-6}$$

The circuit inductance is additive in a series circuit, while the reciprocal of the inductance value, Γ, is additive in a parallel circuit.

When two inductors are placed physically close together the magnetic fields of each interact with the other. This linkage can either increase (aid) or decrease (oppose) the total inductance of the pair. The formula is

$$L_M = k \times \sqrt{L_1 \times L_2}$$

One very useful application of mutual inductance is the transformer. A transformer is used to change the ac voltage level. The input of the transformer is called the primary side, while the output is the secondary side. It is possible to both step up and step down voltage from the primary to the secondary. When the voltage is stepped up, the current is stepped down, and vice versa. The turns ratio and efficiency of the transformer determine the size of the step.

Any change in the rate of current flowing through the inductor creates a magnetic field, which in turn produces an induced voltage that will oppose the supply voltage. This induced voltage in an ac circuit is refered to as *inductive reactance*. Reactance is an opposition to current flow and is measured in ohms. Reactance is additive in a series circuit, while susceptance (the reciprocal of reactance) is additive in a parallel circuit. Reactance is frequency dependent and can be calculated using the formula

$$X_L = 2 \times \pi \times f \times L$$

The voltage across an inductor leads the current through that inductor by a phase angle of 90°. This allows the use of right triangles to solve ac circuits which contain both inductors and resistors. ELI is used to determine which triangle (voltage, impedance, or current) should be used for circuit analysis and to determine if the phase angle is positive or negative. The resistive component is placed on the horizontal side of the triangle. The reactive component is placed on the vertical side of the triangle. The total value is placed on the hypotenuse of the triangle.

The ohmmeter is the most common method for testing inductors and transformers. A resistance measurement of the primary and the secondary windings is compared to specified measurements to determine the coil's acceptability. Sophisticated equipment such as *Z* meters and *RCL* bridges may be used to determine the coil's inductance value.

QUESTIONS

15-1. In your own words, state the Pythagorean theorem. Explain how it is used to solve right triangles.

15-2. What does Soh-Cah-Toa refer to?

15-3. Describe how the $\cos^{-1}$ function is accessed on your calculator.

15-4. Why is it important to know how to find the area of a circle when you are designing an inductor?

15-5. Why must the area be calculated in square meters?

15-6. Describe the construction of an inductor.

15-7. What is meant by *eddy current?*

15-8. What is the skin effect?

15-9. What measures are taken to reduce the effect of eddy current and skin effect in high-frequency inductors?

15-10. How does hysteresis affect high-frequency coil operation?

15-11. What can be done to reduce the effects of hysteresis on high-frequency coils?

15-12. If the length of an inductor were doubled and everything else were kept the same, what effect would this have on the inductance of the coil?

15-13. If the number of turns were doubled and everything else were kept the same, what effect would this have on the inductance of the coil?

15-14. If the area were doubled and everything were kept the same, what effect would this have on the inductance of the coil?

15-15. When given a long formula such as the inductance formula, describe how you would solve for an indirect variable such as area.

15-16. Define *inductance*. Use your own words.

15-17. Describe how you would calculate total inductance in a series circuit if the individual values were known.

15-18. Describe how you would calculate total inductance in a parallel circuit if the individual values were known.

15-19. What is mutual inductance? How is it calculated?

15-20. What causes mutual inductance?

15-21. How can mutual inductance be minimized?

15-22. How can mutual inductance be maximized?

15-23. How can you determine whether flux lines are aiding or opposing each other?

15-24. When the coefficient of coupling, k, is a negative value, what is indicated?

15-25. Describe the effect of mutual inductance on a series circuit; on a parallel circuit. Include statements about voltage, current, and total inductance of the circuit.

15-26. What does the $\pm$ sign in the mutual inductance formula mean?

15-27. Describe the construction of a transformer.

15-28. What is the purpose of a transformer?

15-29. Which winding is referred to as the primary? Why do you think this is?

15-30. Define a step-up transformer.

15-31. Define a step-down transformer.

15-32. Do some research. See how many uses you can find for both a step-up and a step-down transformer.

15-33. If input and output power are equal, what is the efficiency rating for a transformer? What is the coefficient of coupling?

15-34. What is meant by *turns ratio?*

15-35. Describe how primary and secondary voltage can be used to determine the turns ratio.

15-36. Describe how primary and secondary current can be used to determine the turns ratio.

15-37. How do you convert a decimal into a percentage?

15-38. Define *inductive reactance*.

15-39. What is cemf? What causes it?

15-40. Explain what is meant by di/dt.

15-41. How does di/dt affect the induced voltage?

15-42. Of what importance is $2 \times \pi \times f$? How does it effect an ac inductive circuit?

15-43. If the frequency is doubled, what happens to the inductance of the circuit?

15-44. If the frequency is doubled, what happens to the inductive reactance of the circuit?

15-45. Describe how total inductive reactance is calculated in a series circuit; in a parallel circuit.

15-46. Define *inductive reactance*.

15-47. Define *inductive susceptance*.

15-48. What does ELI the ICEman mean?

15-49. When analyzing an ac circuit containing a resistor and an inductor, explain how right triangles are used. Include which value is placed on each side of the triangle.

15-50. Describe how ELI is used to determine whether the phase angle is positive or negative in an *RL* circuit.

15-51. What is admittance?

15-52. Why should you avoid using an admittance triangle?

PRACTICE PROBLEMS

For Problems 15-1 through 15-5, refer to Figure 15-27.

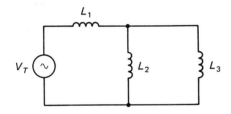

FIGURE 15-27

15-1. $L_1 = 50\ \mu H$ $I_1 = $ _____
$L_2 = 30\ \mu H$ $I_2 = $ _____
$L_3 = 80\ \mu H$ $I_3 = $ _____
$X_{L1} = $ _____ $I_T = $ _____
$X_{L2} = $ _____ $L_T = $ _____
$X_{L3} = $ _____ $X_{LT} = $ _____
$f = 85$ kHz $V_T = 20$ V

15-2. $L_1 = $ _____ $I_1 = $ _____
$L_2 = 50$ mH $I_2 = $ _____
$L_3 = 40$ mH $I_3 = $ _____
$X_{L1} = 1036\ \Omega$ $I_T = 2.88$ mA
$X_{L2} = $ _____ $L_T = $ _____
$X_{L3} = $ _____ $X_{LT} = $ _____
$f = 5$ kHz $V_T = $ _____

15-3. $L_1 = $ _____ $I_1 = $ _____
$L_2 = $ _____ $I_2 = $ _____
$L_3 = $ _____ $I_3 = $ _____
$X_{L1} = 56.5\ \Omega$ $I_T = $ _____
$X_{L2} = 94.2\ \Omega$ $L_T = 0.3166$ mH
$X_{L3} = 188\ \Omega$ $X_{LT} = $ _____
$f = $ _____ $V_T = 100$ mV

15-4. L_1 = _____ I_1 = 30.94 μA
       L_2 = _____ I_2 = _____
       L_3 = _____ I_3 = _____
       X_{L1} = 9424 Ω I_T = _____
       X_{L2} = 11,780 Ω L_T = _____
       X_{L3} = _____ X_{LT} = 16,156 Ω
       f = 2.5 kHz V_T = 0.5 V

15-5. L_1 = _____ I_1 = _____
       L_2 = _____ I_2 = _____
       L_3 = _____ I_3 = _____
       X_{L1} = 78.53 Ω I_T = _____
       X_{L2} = 157 Ω L_T = _____
       X_{L3} = 314 Ω X_{LT} = _____
       f = 2.5 kHz V_T = 100 V

For Problems 15-6 through 15-10, refer to Figure 15-28.

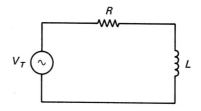

FIGURE 15-28

15-6. V_T = 15 V L = 33 mH
       V_R = _____ f = 2.5 kHz
       V_L = _____ R = 1000 Ω
       X_L = _____ I = _____
       Z = _____ θ = _____

15-7. V_T = _____ L = 25 μH
       V_R = 52.9 V f = 850 kHz
       V_L = _____ R = 250 Ω
       X_L = 133.5 Ω I = _____
       Z = _____ θ = _____

15-8. V_T = _____ L = _____
       V_R = _____ f = 56 kHz
       V_L = _____ R = 15 kΩ
       X_L = _____ I = 141 μA
       Z = 31.895 kΩ θ = _____

15-9. V_T = 30 V L = _____
       V_R = _____ f = 2.5 kHz
       V_L = _____ R = 25 kΩ
       X_L = _____ I = _____
       Z = 29.52 kΩ θ = 32.14°

15-10. V_T = 20 L = _____
       V_R = _____ f = 455 kHz
       V_L = _____ R = _____
       X_L = _____ I = _____
       Z = 933 Ω θ = 50°

For Problems 15-11 through 15-15, refer to Figure 15-29.

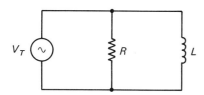

FIGURE 15-29

15-11. $V_T =$ _____ $I_L = 0.4493$ A
$L =$ _____ $I_R = 0.24$ A
$f = 850$ kHz $I_T =$ _____
$R = 250$ Ω $X_L =$ _____
$Z =$ _____ $\theta =$ _____

15-12. $V_T = 450$ mV $I_L =$ _____
$L = 80$ mH $I_R =$ _____
$f = 56$ kHz $I_T =$ _____
$R = 15$ kΩ $X_L =$ _____
$Z =$ _____ $\theta =$ _____

15-13. $V_T =$ _____ $I_L =$ _____
$L = 33$ mH $I_R = 15$ mA
$f =$ _____ $I_т = 32.59$ mA
$R = 1000$ Ω $X_L = 518$ Ω
$Z =$ _____ $\theta =$ _____

15-14. $V_T =$ _____ $I_L =$ _____
$L =$ _____ $I_R =$ _____
$f = 2.5$ kHz $I_T = 1.618$ mA
$R =$ _____ $X_L =$ _____
$Z = 18.53$ kΩ $\theta = -551.85°$

15-15. $V_T = 20$ mV $I_L =$ _____
$L = 250$ μH $I_R =$ _____
$f = 455$ kHz $I_T =$ _____
$R = 600$ Ω $X_L =$ _____
$Z =$ _____ $\theta =$ _____

PRACTICE PROBLEMS

16

Capacitors

Chapter Objectives

After reading this chapter and answering the questions and problems, you should be able to:

- Describe the effect that changing the plate area, plate distance, and dielectric will have on the capacitance and maximum voltage rating of a capacitor.
- List several types of capacitors and state the applications of each.
- Given any three of C, I, T, and V, find the missing quantity.
- Calculate the total capacitance in series and parallel circuits.
- Describe the effect frequency has on capacitive reactance.
- Calculate total capacitive reactance in series and parallel circuits.
- Calculate angular velocity, ω, and give its unit of measure.
- List the advantages and disadvantages of capacitive voltage dividers.
- Given the capacitor values and one voltage of components in a capacitive voltage divider, find the missing voltages.
- Solve an RC circuit using the Pythagorean theorem and trigonometry.
- List and describe the three checks used to determine the condition of a capacitor.
- List several tools (pieces of equipment) that can be used to test the condition of a capacitor.

In this chapter we discuss the construction and design of capacitors. We also describe the manner in which total capacitance is calculated in both series and parallel circuits. At low frequencies capacitors show a large amount of resistance (reactance). We examine the cause of this phenomenon and how to predict its effect at specific frequencies. Many appliances use capacitive voltage dividers. A portion of this chapter is devoted to calculating the voltages of these devices. Resistor–capacitor combination circuits are also discussed, together with methods of troubleshooting capacitive faults.

CONSTRUCTION AND DESIGN

The construction of capacitors is varied; however, all designs have two plates, which are separated by an insulator (dielectric). The dielectric may be any of a number of insulators. Air, ceramic, mica, and paper are among the most common insulators used in capacitors.

The operation of a capacitor is similar to that of a battery. Like a battery, a capacitor can be charged. The one very important difference is that an internal chemical action of battery can actually produce voltage. A capacitor cannot produce voltage.

When a capacitor is connected to a battery, electrons flow from the negative terminal of the battery through the circuit. They collect on the negative plate of the capacitor (see Figure 16-1). The electron collection on the negative plate produces a

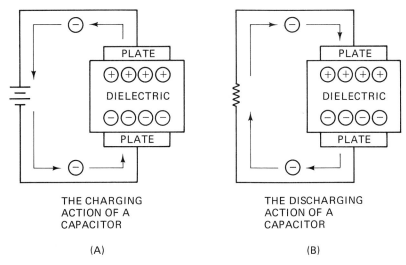

FIGURE 16-1

negative charge that repels (like charges repel) the electrons remaining on the positive plate of the capacitor. This leaves a collection of electron holes on the positive plate.

After the charging voltage has been removed, the holes remain on the positive plate and electrons remain on the negative plate, because the dielectric serves as an insulator preventing their return. The capacitor is charged to a voltage about equal to the charging battery and will remain so until a load is connected. The capacitor will provide current to the load for a time. The length of time is determined in part by the resistance value of the load.

There are four basic considerations involved in capacitor design. The dielectric's ability to hold back (without leaking) voltage is a main concern. This ability is called the *dielectric strength*. If you examine several capacitors you will find a voltage specification attached to each. It is important that this voltage rating not be exceeded. Too high a voltage level can caused a capacitor to break down, and under extreme circumstances, even explode.

Another consideration is the dielectric constant, K, associated with the insulating material separating the plates. This constant can be any value (depending on the insulator) from 1 to several thousands (see Table 16-1).

TABLE 16-1 Dielectric Properties

Material	Dielectric Strength (V/mil)	Dielectric Constant
Plastics		
Mica	—	7.4–7.85
Nylon	100–350	5.0–7.0
Shellac	200–600	—
Vinyl butyral	350–400	3.92
Ceramics		
Alumina	40–160	4.5–8.4
Porcelain dry	40–240	6.0–8.0
Porcelain wet	90–400	6.0–7.0
Titanium dioxide	100–210	14–110
Waxes		
White beeswax	—	2.75–3.0
Yellow beeswax	—	2.90
Paraffin	250.00	

The physical dimensions comprise the other considerations in capacitor design. They are the surface area of the plates and the distance between (separating) the plates. A formula for calculating the capacitance of a capacitor is

$$C = K \times \frac{A}{d} \times 8.85 \times 10^{-12}$$

The unit for capacitance is farad. A farad is equal to 1 ampere-second/volt. In normal electronic terms a farad is enormous, so typical units are the microfarad (μF) and picofarad (pF). Millifarad and nanofarads are not commonly used terms, and capacitance in these ranges should be converted to either microfarads or picofarads. In the formula, K is the dielectric constant, A is the surface area (must be square meters) of the plates, and d is the distance (must be meters) between the plates; the term 8.85×10^{-12} is the permittivity of air. An application of this formula follows.

EXAMPLE 16-1

Suppose that two plates, each having a surface area of 25 cm² is separated by a distance of 1 cm. Find the resulting capacitance if the insulator has a dielectric constant of 5.6.

Solution: Begin by converting the area from square centimeters to square meters.

$$A = 25 \text{ cm}^2$$
$$= \frac{25}{10,000}$$
$$= 0.0025 \text{ m}^2$$

Also convert the distance between the plate from centimeters to meters.

$$d = 1 \text{ cm}$$
$$= 1/100$$
$$= 0.01 \text{ m}$$

Next insert the known values into the capacitance formula.

$$C = K \times \frac{A}{d} \times 8.85 \times 10^{-12}$$
$$= 5.6 \times \frac{0.0025}{0.01} \times 8.85 \times 10^{-12}$$
$$= 12.39 \text{ pF}$$

SELF-TEST

1. Calculate the breakdown voltage of an air-core capacitor if the plates are positioned 0.020 in. apart. The dielectric strength of air is 20 V/mil.
2. Calculate the minimum distance between the plates of an air-core capacitor if it is to have a breakdown voltage of 200 V.
3. Calculate the minimum distance between the plates of a capacitor if it is to have a breakdown voltage of 1200 V. The plates are separated by paraffin.
4. Calculate the breakdown voltage of a capacitor if the plates are positioned 0.040 in. apart and separated by vinyl butyral. Give maximum and minimum values.
5. Given that the dielectric constant of a material is 6.2 and the area of the plates is 0.03 m². The plates are separated by a distance of 0.001 m. Find the capacitance of this capacitor.
6. Given that white beeswax is used as the dielectric material and the area of the plates is 0.03 m². The plates are separated by a distance of 0.001 m. Find the capacitance of this capacitor.
7. Calculate the dielectric constant needed to provide a capacitance of 12 pF when 0.02-m² plates are separated by a distance of 0.015 m.
8. Given that yellow beeswax is used as a dielectric, find the plate area needed to manufacture a 470-μF capacitor if the distance between the plates is 5 mm.

CONSTRUCTION AND DESIGN

ANSWERS TO SELF-TEST

1. 400 V
2. 0.010 in. or 10 mils
3. 0.0048 in. or 4.8 mils
4. 14 to 16 kV
5. 1646 pF
6. 730 to 796.5 pF
6. $K = 1.017$
8. $A = 91{,}564$ m^2

There are several different types of capacitors on the market today (Figure 16-2), too many, in fact, to make it possible to include a description of each. The following are some of the more popular types of capacitors.

Ceramic disk capacitors are cheap and durable. They are the most commonly used capacitor. Disk capacitors usually have a small capacitance value. The larger ones are only 0.1 μF. They are used as coupling (connecting) devices between stages of amplifiers. They are also used in filters (devices that remove unwanted frequencies) and in tank circuits (a circuit used in the tuning of radios). Disk capacitors do not usually have a polarity. That means that they can be connected to the circuit without worrying about positive or negative terminals. These are determined by the circuit, not the component.

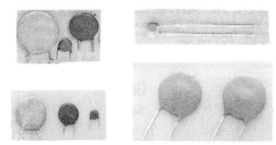

FIGURE 16-2 Assorted Disk Capacitors.

Electrolytic capacitors (Figure 16-3) have a doped (an added or enhanced electrical charge) insulator. This allows the capacitor to start operation with a partial charge. The advantage of such a capacitor is that it has a large capacitance value and

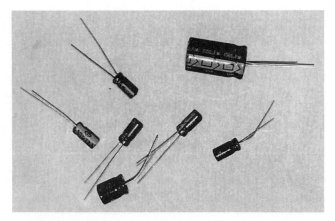

FIGURE 16-3 Assorted Electrolytic Capacitors.

still maintains a small physical size. Electrolytic capacitors are commonly made from aluminum and use a liquid electrolyte to provide a starting potential.

Electrolytic capacitors are polar devices. It is important that they be connected to the circuit with correct polarity (positive terminal to the positive voltage and negative terminal to the negative voltage) or the capacitor will heat up unduly and *possibly explode*. Electrolytic capacitors will generally have a stripe painted above their negative terminal. In any case, the negative terminal is clearly marked.

CAPACITANCE

The capacitance of a capacitor indicates how fast the capacitor will accumulate a charge. There are several different formulas for calculating capacitance. This is one:

$$C = I \times \frac{T}{V}$$

This formula indicates that capacitance is determined by current, time, and voltage. Remember that voltage is the equivalent to pressure in electricity. More voltage can push more current into a capacitor. Like any other container, it takes a particular length of time for a capacitor to fill. It takes longer to fill a larger-farad-value capacitor than it does a smaller one, provided that voltage and current are the same. An example of this type of calculation follows.

EXAMPLE 16-2

Suppose that a current of 6 mA charges a 47-μF capacitor for 8 ms. What is the voltage charge of the capacitor?

Solution: Begin by inserting the known values into the equation.

$$C = I \times \frac{T}{V}$$

$$0.000047 = 0.006 \times \frac{0.008}{V}$$

$$V = \frac{0.000048}{0.000047}$$

$$= 1.021 \ V \qquad \blacksquare$$

Understand that an ampere is a coulomb per second. Multiplying the current by time converts current to charge. The capacitance formula is sometimes written as

$$C = \frac{Q}{V}$$

since

$$I \times T = Q$$

SELF-TEST

1. A 9-V battery charges a 10-μF capacitor at a 12-mA rate. How long will it take for the capacitor to charge to 9 V?

2. A current of 12 mA charges a 4.7-µF capacitor for 2 s. What is the voltage of the capacitor?

ANSWERS TO SELF-TEST

1. $T = 7.5\ mS$
2. $V = 5.106\ kV$

Capacitors in Series The capacitance values of series capacitance are not additive. Instead, you must add the reciprocals of the capacitance. The reciprocal of capacitance is called *elasticity*. The mathematic formula for calculating total series capacitance is

$$\frac{1}{C_T} = \frac{1}{C_1} + \frac{1}{C_2} + \frac{1}{C_3} + \cdots$$

In other words, total series capacitance is calculated in the same manner as parallel resistance.

Capacitors in Parallel In parallel total capacitance can be found by adding individual capacitor values, just as you would find the total resistance in a series circuit. The formula for this calculation is

$$C_T = C_1 + C_2 + C_3 + \cdots$$

SELF-TEST

1. Three capacitors of 47, 20, and 4.7 µF are connected in series. What is the total capacitance of this string?
2. Given capacitors of 2.2, 0.47, 3.3, 0.66, and 4.7 µF connected in series, find the total capacitance of the string.
3. Three capacitors of 47, 20, and 4.7 µF are connected in parallel. What is the total capacitance of this circuit?
4. Given capacitors of 2.2, 0.47, 3.3, 0.66, and 4.7 µF connected in parallel, find the total capacitance of the circuit.

ANSWERS TO SELF-TEST

1. 3.521 µF
2. 0.2167 µF
3. 71.7 µF
4. 11.33 µF

CAPACITIVE REACTANCE

Capacitors have a tendency to reject low frequencies. That is, they appear to be opens, or high impedances, to low frequencies, while they (the capacitors) will appear as a short to high frequencies. This is due to the capacitive reactance of the capacitor.

The capacitor serves as a shock absorber to the circuit. Any sudden voltage change will result in a momentary charge of the capacitor. Hydraulic circuits use a device called an accumulator, which serves the same function as the capacitor in an electric circuit.

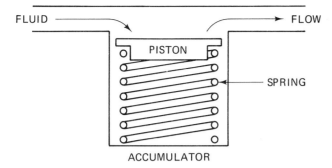

FIGURE 16-4 The accumulator (capacitor) allows a gradual increase in hydraulic pressure (voltage). The accumulator (capacitor) must be filled with fluid (electron current flow) before pressure (voltage) can increase. Hydraulic pressure against the piston slowly compresses the spring. This accumulator action (capacitor action) prevents any sudden change in hydraulic pressure (voltage).

Figure 16-4 is an analogy of a capacitor to an accumulator. When the pressure (voltage) in a hydraulic circuit rises, the fluid (current) will flow into the accumulator (capacitor). Little pressure (voltage) or force can build up until the accumulator (capacitor) is filled. As the pressure (voltage) on the hydraulic system is lowered, the accumulator (capacitor) spring will gradually force fluid (current) back out into the system.

The sine wave in Figure 16-5 shows that as the pressure on the system (emf) increases, the fluid (current) will flow into the accumulator (capacitor). When the pressure (emf) decreases, the accumulator (capacitor) will force fluid (current) out into the circuit.

The resulting phase difference between voltage and current appears as an impedance (reactance) to the circuit. The formula for calculating the capacitive reactance for a circuit is

$$X_C = \frac{1}{2 \times \pi \times f \times C}$$

A careful examination of this equation will indicate that X_C is high for low frequencies and low for high frequencies (f is a term in the denominator and will lower the value if large).

The angular velocity ($2 \times \pi \times f$) plays an important role in determining the capacitive reactance of a particular circuit. This value is the same for the circuit as a whole and will change only when the frequency of a circuit is changed.

Capacitive Reactance in Series Again reactance is an impedance (resistance) value. It is measured in ohms, but unlike capacitance, is additive in series. For that reason it is usually easier to convert the capacitance values to reactance values to find the total capacitive reactance. Then use the reactance value to find the total capacitance. The formula for finding the total capacitive reactance in a series circuit is

$$X_{CT} = X_{C1} + X_{C2} + X_{C3} + \cdots$$

CAPACITIVE REACTANCE

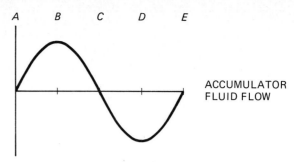

FROM A TO B FLUID ENTERS ACCUMULATOR
FROM B TO D FLUID EXITS ACCUMULATOR
FROM D TO E FLUID ENTERS ACCUMULATOR

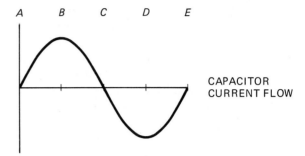

FROM A TO B CURRENT ENTERS CAPACITOR
POINT B IS MAXIMUM CURRENT FLOW
FROM B TO D CURRENT EXITS CAPACITOR
POINT C IS ZERO CURRENT FLOW
FROM D TO E CURRENT ENTERS CAPACITOR

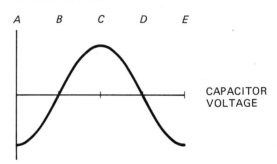

FROM A TO C VOLTAGE INCREASES
POINT B IS ZERO VOLTAGE
POINT C IS MAXIMUM VOLTAGE
FROM C TO E VOLTAGE DECREASES
CURRENT LEADS VOLTAGE (ICE)

FIGURE 16-5

An example of this type of solution follows.

EXAMPLE 16-3

A circuit which is supplied with a 10-kHz signal consists of three series capacitors. The capacitors are 0.01, 0.03, and 0.005 μF, respectively. Find the total capacitive reactance of the circuit.

Solution: Begin by calculating the angular velocity.

$$\omega = 2 \times \pi \times f$$
$$= 2 \times \pi \times 10{,}000$$
$$= 62{,}831 \text{ rad/s}$$

Then calculate the capacitive reactance of each capacitor.

$$X_{C1} = \frac{1}{62{,}831 \text{ rad/s} \times 0.01 \ \mu\text{F}}$$
$$= 1592 \ \Omega$$

$$X_{C2} = \frac{1}{62{,}831 \text{ rad/s} \times 0.03 \ \mu\text{F}}$$
$$= 531 \ \Omega$$

$$X_{C3} = \frac{1}{62{,}831 \text{ rad/s} \times 0.005 \ \mu\text{F}}$$
$$= 3183 \ \Omega$$

Then use the total reactance formula.

$$X_{CT} = X_{C1} + X_{C2} + X_{C3}$$
$$= 1592 \ \Omega + 531 \ \Omega + 3183 \ \Omega$$
$$= 5306 \ \Omega$$

and the total capacitance would be

$$X_{CT} = \frac{1}{2 \times \pi \times f \times C_T}$$
$$5306 \ \Omega = \frac{1}{62{,}831 \text{ rad/s} \times C_T}$$
$$C_T = \frac{1}{62{,}831 \text{ rad/s} \times 5306 \ \Omega}$$
$$= 0.003 \ \mu\text{F}$$ ■

Capacitive Reactance in Parallel Ohms in parallel are not additive. In parallel the reciprocal of the capacitive reactance (susceptance), is additive. The formula shown below will find the total Xc in a parallel circuit.

$$\frac{1}{Xc_T} = \frac{1}{Xc_1} + \frac{1}{Xc_2} + \frac{1}{Xc_3} + \cdots$$

In parallel capacitance, is additive. In general, it is easier to find the total capacitance. Then use this total to find the total reactance of the circuit. This is the reverse order of that used for series circuits. A demonstration of this process follows.

EXAMPLE 16-4

A circuit which is supplied with a 10-kHz signal consists of three parallel capacitors. The capacitors are 0.01, 0.03, and 0.005 μF, respectively. Find the total capacitive reactance for this circuit.

Solution: Begin by calculating the angular velocity.

$$\omega = 2 \times \pi \times f$$
$$= 2 \times \pi \times 10{,}000$$
$$= 62{,}831 \text{ rad/s}$$

Next calculate the total capacitance of the circuit.

$$C_T = C_1 + C_2 + C_3$$
$$= 0.00000001 + 0.00000003 + 0.000000005$$
$$= 0.045 \ \mu F$$

Use this value to find the total capacitive reactance.

$$X_{CT} = \frac{1}{2 \times \pi \times f \times C_T}$$
$$= \frac{1}{62{,}831 \text{ rad/s} \times 0.045 \ \mu F}$$
$$= 354 \ \Omega$$

■

Reactance and Ohm's Law Ohm's law will still apply to circuits that contain capacitors. The basic idea is that you must first find the capacitive reactance (the ohm value) for the capacitor. Once this value has been found, it can be treated like a resistance, provided that the circuit contains only capacitors. That is, when the circuit is purely resistive (having only resistors), purely capacitive (having only capacitors), or purely inductive (having only inductors), Ohm's law and series and parallel rules will apply. However, there are times when resistors and capacitors are combined in the same circuit. You must use other rules to solve these circuits.

SELF-TEST

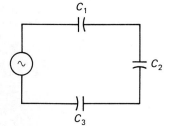

FIGURE 16-6

Refer to Figure 16-6.

1. f = 10 kHz
   C_1 = 2.2 μF
   C_2 = 4.7 μF
   C_3 = 6.6 μF
   X_{C1} = _____
   X_{C2} = _____
   X_{C3} = _____
   X_{CT} = _____

2. f = _____
   C_1 = _____
   C_2 = 0.022 μF
   C_3 = _____
   X_{C1} = 192.9 Ω
   X_{C2} = 99.5 Ω
   X_{C3} = _____
   X_{CT} = 391.9 Ω

3. f = _____
   C_1 = 0.047 μF
   C_2 = 0.022 μF
   C_3 = 0.033 μF
   X_{C1} = 6773 Ω
   X_2 = _____
   X_3 = _____
   X_T = _____

ANSWERS TO SELF-TEST

1. $X_{C1} = 7.234\ \Omega$, $X_{C2} = 3.386\ \Omega$, $X_{C3} = 2.411\ \Omega$, $X_{CT} = 13.031\ \Omega$
2. $f = 72.71$ kHz, $C_1 = 0.0113\ \mu F$, $X_{C3} = 99.5\ \Omega$, $C_3 = 0.002\ \mu F$
3. $f = 500$ Hz, $X_{C2} = 14.468$ kΩ, $X_{C3} = 9.646$ kΩ, $X_{CT} = 30.887$ kΩ

CAPACITIVE VOLTAGE DIVIDER

In dc circuits capacitors are sometimes used as voltage dividers. Since capacitors store electricity, once they are charged, they can provide a fairly constant voltage to a load. However, they have a limited current capacity and require a minimum drain. Also, it can take minutes for these components to charge to the desired voltage. For these reasons, they are not often used in modern technology.

Capacitance and voltage are inversely proportional relationships. (These two quantities react in much the same way as resistance and current do in a parallel circuit.) Remember that a voltage divider is a series circuit. When two capacitors are combined in a series circuit, the smaller one (least capacitance) will charge to the highest voltage level. There are a number of formulas that can be used to indicate this inverse proportion. Some of them follow.

$$\frac{C_2}{C_1} = \frac{V_1}{V_2} \quad \text{and} \quad \frac{C_1}{C_2} = \frac{V_2}{V_1}$$

$$\frac{C_T}{C_1} = \frac{V_1}{V_T} \quad \text{and} \quad \frac{C_1}{C_T} = \frac{V_T}{V_1}$$

EXAMPLE 16-5

Find the voltage across each capacitor in the circuit shown in Figure 16-7.

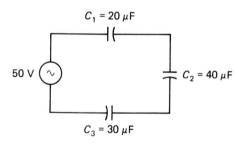

FIGURE 16-7

Solution: Begin by calculating the total capacitance of the circuit.

$$\frac{1}{C_T} = \frac{1}{C_1} + \frac{1}{C_2} + \frac{1}{C_3}$$

$$= \frac{1}{0.00002} + \frac{1}{0.00004} + \frac{1}{0.00003}$$

$$= 50{,}000 + 25{,}000 + 33{,}333 = 108{,}333$$

$$C_T = \frac{1}{108{,}333}$$

$$= 9.231\ \mu F$$

Next set up an inverse relationship between capacitance and voltage for each capacitor.

$$\frac{C_T}{C_1} = \frac{V_1}{V_T}$$

$$\frac{0.000009231}{0.00002} = \frac{V_1}{50}$$

$$V_1 = 50 \times \frac{0.000009231}{0.00002}$$

$$= 23.08 \text{ V}$$

$$\frac{C_T}{C_2} = \frac{V_2}{V_T}$$

$$\frac{0.000009231}{0.00004} = \frac{V_2}{50}$$

$$V_2 = 50 \times \frac{0.000009231}{0.00004}$$

$$= 11.54 \text{ V}$$

$$\frac{C_T}{C_3} = \frac{V_3}{V_T}$$

$$\frac{0.000009231}{0.00003} = \frac{V_3}{50}$$

$$V_3 = 50 \times \frac{0.000009231}{0.00003}$$

$$= 15.38 \text{ V} \quad\blacksquare$$

Although these formulas are usually provided for dc circuits, they apply just as well to ac circuits that consist entirely of capacitances. Remember that the smaller capacitance values will drop the majority of the voltage in any circuit.

SELF-TEST

1. Suppose that you want to use a 1.5-V battery to provide a 0.5-V source for a circuit. How can three 10-μF capacitors be connected to provide this source? Sketch the circuit with the load.
2. Refer to question 1. Suppose that you had a 4.7-μF capacitor. What other two values are needed to provide the 0.5-V source if only two capacitors are to be used?

ANSWERS TO SELF-TEST

1. The three capacitors can be connected in series. The 0.5-V output would be in parallel to one of these capacitors.
2. A 2.35-μF or a 9.4-μF capacitor could provide this circuit.

SOLUTION OF *RC* CIRCUITS

The solution of *RC* circuits can be simple provided that you know how to solve a triangle and can follow a few simple rules. In the next sections we describe and explain those rules. You will note that these rules are almost the same as those of the *LR* circuits of Chapter 15.

ELI the ICEman The phrase ELI the ICEman has been around for a long time and has helped many electronics students through hard times. This memory device can be used for at least a dozen clues about reactive circuits (ac and dc included). The phrase is usually taught to help you remember that the voltage leads the current in an inductive circuit (ELI) by 90° and that current leads voltage in a capacitive circuit (ICE) by 90°. Students experience several misconceptions about these letters. Understand that the *L* in ELI stands for inductance and that ELI is used only for inductive circuits. The *C* in ICE stands for capacitance, and similarly, ICE is used only for capacitive circuits.

Triangles All of the triangles used in this book will have a horizontal side, a vertical side, and a hypotenuse. Some triangles will have their points upward, others downward. How to determine which way a particular triangle point is explained later under the topic of phase angles. For now, we will concentrate on what values are placed on which side (refer to Figure 16-8).

The resistive component is always placed on the horizontal side. The resistive component may be resistance, voltage across the resistor, current through the resistor, or conductance ($1/R$), depending on the triangle being used.

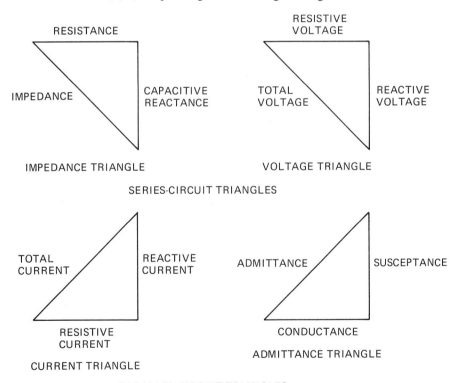

FIGURE 16-8

SOLUTION OF *RC* CIRCUITS

The vertical side is reserved for the reactive component. Again the reactive component may be reactance, voltage across the capacitor, current through the capacitor, or susceptance $(1/X)$. This depends on the individual triangle.

The hypotenuse is used to represent the total quantities. These may be impedance (combined resistance and reactance), total applied voltage, total circuit current, or admittance $(1/Z)$. Again, these values depend on the triangle being implemented.

Series-Circuit Triangles An early step in circuit analysis is to determine if the circuit is series or parallel. Any series circuit can have two triangles: a voltage triangle and an impedance triangle. The phase angle is the same (equal) for both of these triangles. ICE can be used to remind you that the a voltage triangle is possible. Current I is the same everywhere in a series circuit. This lets the voltage (E) form a triangle. Look at ICE; if you eliminate the I (current is the same) and C (the circuit contains an capacitor), E (the voltage) is the remaining quantity. Remember: *Impedance and voltage lie on the same plane*.

Parallel-Circuit Triangles The current and admittance triangles are devoted to parallel-circuit analysis. The admittance triangle does exist; however, it is best for beginning students to forget this promptly. Trying to solve an admittance triangle requires the use of many, many reciprocal functions. Most beginning students experience too many difficulties in these calculations. For that reason it is best to use only the current triangle when working with parallel circuits.

Phase Angle of Triangles The angle between the hypotenuse and the horizontal side is called the phase angle. It represents the number of degrees between the peak values of these components. This angle may be positive (the hypotenuse quantity peaks first) or negative (the horizontal quantity peaks first). ICE can be used to determine if this angle is positive or negative. Draw ICE in the following manner (see Figure 16-9). In a series circuit the current is the same everywhere. When you use ICE as a memory device, a line is drawn out from the constant quantity (I) and then downward to the remaining component (E). This shows that the triangle should be drawn with the point downward (indicating that the phase is negative). In a parallel circuit the process in the same except that the voltage (E) is constant and has the line drawn outward while the current (I) is the remaining quantity. The resulting triangle would point upward.

IN A SERIES CIRCUIT I IS THE SAME   IN A PARALLEL CIRCUIT E IS THE SAME

FIGURE 16-9

***RC* Circuits** The following are several examples of series and parallel circuit analysis.

EXAMPLE 16-6

Analyze the circuit in Figure 16-10. Find its impedance, phase angle, voltage across the resistor and the capacitor, and current.

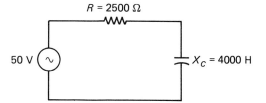

FIGURE 16-10

Solution: Begin by referring to Figure 16-10. This is a series circuit, so there is a voltage and an impedance triangle. Sketch each of these and place the known values on each (Figure 16-11). Next calculate the impedance.

$$Z = \sqrt{2500^2 + 4000^2}$$
$$= 4717 \ \Omega$$

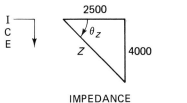

IMPEDANCE   VOLTAGE   **FIGURE 16-11**

Then calculate the phase angle of the impedance.

$$\theta = \tan^{-1} \frac{4000}{2500}$$
$$= \tan^{-1} 1.6$$
$$= -57.99°$$

Note: The triangle in Figure 16-11 indicates that the phase is negative.

Remember that the phase angle of impedance and the phase angle of voltage are equal. Use this to find the resistive voltage

$$V_R = V_T \times \cos \theta$$
$$= 50 \times \cos -57.99$$
$$= 26.50 \ V$$

and the capacitive voltage

$$V_C = V_T \times \sin \theta$$
$$= 50 \times \sin -57.99$$
$$= -42.40 \ V$$

Note: The minus indicates that the capacitive voltage is lagging the resistive voltage.

Current can be found in any one of several manners.

$$I = \frac{V_R}{R} \quad \text{or} \quad I = \frac{V_C}{X_C} \quad \text{or} \quad I = \frac{V_T}{Z}$$

$$I = \frac{26.50}{2500}$$
$$= 10.6 \ mA$$

SOLUTION OF *RC* CIRCUITS

Current can be thought of as a common factor between the voltage and current triangles. That is, as in other circuits, Ohm's law will apply but only at individual levels. In this case, the individual levels refer to the corresponding sides of the triangle (voltage horizontal/impedance horizontal or voltage vertical/impedance vertical or voltage hypotenuse/impedance hypotenuse).

EXAMPLE 16-7

Analyze the circuit in Figure 16-12. Find the total impedance, current, and voltage across each component.

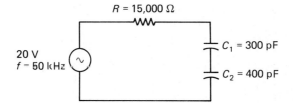

FIGURE 16-12

Solution: Begin by finding the capacitive reactance for each capacitor. Use this to find the total capacitive reactance of the circuit.

$$X_C = \frac{1}{2 \times \pi \times f \times C}$$

$$X_{C1} = \frac{1}{2 \times \pi \times 50 \text{ kHz} \times 300 \text{ pF}}$$

$$= 10{,}610 \ \Omega$$

$$X_{C2} = \frac{1}{2 \times \pi \times 50 \text{ kHz} \times 400 \text{ pF}}$$

$$= 7{,}958 \ \Omega$$

$$X_{CT} = X_{C1} + X_{C2}$$

$$= 10{,}610 \ \Omega + 7{,}958 \ \Omega$$

$$= 18{,}568 \ \Omega$$

Next, since this is a series circuit, draw the voltage and impedance triangles and place circuit values on the correct sides (Figure 16-13).

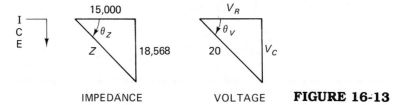

IMPEDANCE　　　VOLTAGE　　**FIGURE 16-13**

Use the Pythagorean theorem to calculate the impedance

$$Z = \sqrt{15{,}000^2 + 18{,}568^2}$$

$$= 23.87 \text{ k}\Omega$$

and trigonometry to calculate the phase angle.

$$\theta = \tan^{-1} \frac{18{,}568}{15{,}000}$$

$$= \tan^{-1} 1.238$$
$$= -51.07° \quad \text{(minus because phase is negative)}$$

The phase angle of the voltage and impedance are equal. Use this to calculate the resistor voltage.

$$V_R = V_T \times \cos\theta$$
$$= 20 \times \cos -51.07$$
$$= 12.57 \text{ V}$$

You cannot use this method to calculate the capacitor voltage because that would be the total voltage of both capacitors. Instead, use Ohm's law to calculate the current and the current to calculate the voltage across each capacitor.

$$I = \frac{V_R}{R}$$
$$= \frac{12.57}{15,000}$$
$$= 837.8 \text{ μA}$$
$$V_{C1} = 837.8 \text{ μA} \times 10,610 \text{ Ω}$$
$$= 8.889 \text{ V}$$
$$V_{C2} = 837.8 \text{ μA} \times 7,958 \text{ Ω}$$
$$= 6.667 \text{ V} \qquad \blacksquare$$

EXAMPLE 16-8

Analyze the circuit in Figure 16-14. Find the current through each component, the total current, and the impedance of the circuit.

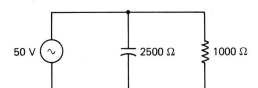

FIGURE 16-14

Solution: Use Ohm's law to find the current through the resistor and the capacitor.

$$I_R = \frac{V}{R}$$
$$= \frac{50}{1000}$$
$$= 50 \text{ mA}$$
$$I_C = \frac{V}{X_C}$$
$$= \frac{50}{2500}$$
$$= 20 \text{ mA}$$

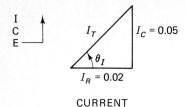

FIGURE 16-15

This is a parallel circuit, so there is a current triangle. Place the calculated current values on the correct sides of the triangle (Figure 16-15) and use the Pythagorean theorem to find the total current.

$$I_T = \sqrt{I_R^2 + I_C^2}$$
$$= \sqrt{0.05^2 + 0.02^2}$$
$$= 53.85 \text{ mA}$$

The phase angle of the current would be

$$\theta = \tan^{-1} \frac{0.02}{0.05}$$
$$= \tan^{-1} 0.4$$
$$= 21.8°$$

Note that the triangle points upward, so this angle is positive. Use Ohm's law to find the impedance.

$$Z = \frac{V}{I_T}$$
$$= \frac{50}{0.05385}$$
$$= 928.5 \text{ }\Omega$$

TROUBLESHOOTING CAPACITORS

There are three tests or checks that need be made in order for one to determine the condition of a capacitor. These checks and instruments for testing them are discussed in this section.

Series Resistance A capacitor has an ac resistance. This is a resistance other than the corresponding capacitive reactance normally associated with an ac voltage and capacitor. This ac resistance is referred to as *series resistance*. Series resistance limits the charge and discharge rate (speed) of a capacitor. Series resistance of a capacitor is not generally a problem in most circuits (unless the capacitor is shorted); however, there are industrial applications where the rate of capacitor charge and discharge is of particular importance. An example of this is when a capacitor is used to remove large surges of voltage, as in circuits containing numerous relays and solenoids. Series resistance can slow the charge and discharge rate sufficiently enough to allow the resulting inductive kickback to damage other circuit components. In some instances this is referred to as the *quality* of the capacitor. The quality is defined as the ratio between the X_C of a capacitor and its series resistance.

An examination of the circuit (signal tracing) and experience seem to be the best tools for identifying this problem, although several manufacturers do provide a means for testing series resistance in capacitors. A comparison ohmmeter test between a known-good capacitor and the capacitor in question can also be used to identify this problem. As a rule, if in doubt, replace the capacitor.

Voltage Leakage In normal operation a capacitor is charged with a voltage. The capacitor is designed and intended to hold that voltage level until the circuit requires that it discharge. Sometimes the dielectric in the capacitor will weaken and allow a portion of this charged voltage to leak across the insulator (dielectric). This voltage leakage is a common problem with capacitors. Voltage leakage is seen primarily on high-voltage capacitors. The operation of these circuits is normal for a period of time and then the circuit will appear to short, when the capacitor is leaky. There are several means for testing for voltage leakage; however, an impedance meter is the most common tool for this test. An impedance meter is used in the same manner as an ohmmeter. That is, the meter leads are connected to the capacitor. The meter is turned to the appropriate voltage level, and a button is pushed. A digital display shows a reading that indicates whether the capacitor is good or bad.

Capacitance   As a capacitor ages it will lose a bit of its capacity. This is especially true of paper capacitors. When a capacitor is used either as a coupling or a bypass capacitor, it is important that the microfarad (picofarad) value be to specifications. An *LCR* bridge can be used to measure the capacitance; however, in today's high-tech world, this test is usually done with a *Z* meter (impedance meter; Figure 16-16). The test with a *Z* meter is a simple one. The leads are connected to the capacitor and a button is pushed. A digital display shows the capacitance value for the capacitor. Several manufacturers include a capacitor tester on their digital voltmeters.

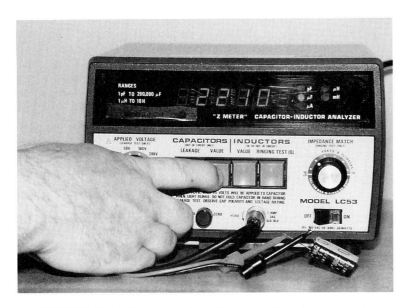

FIGURE 16-16   Capacitor Testing. Z-meter used to measure capacitance.

SUMMARY

There are two ratings associated with any capacitor: its voltage rating and its capacitance. The voltage rating is determined by:

1. The dielectric strength of the material used to separate the plates
2. The distance between the plates

The capacitance is determined by:

1. The area of the plates
2. The distance separating the plates
3. The dielectric constant

The permittivity of air, 8.85×10^{-12}, also determines the capacitance of a capacitor.

There are many types of capacitors. Disk capacitors are used for most applications requiring small values. Electrolytic capacitors are used in most applications where large values of capacitors are needed. Other types are used when charging speed or physical size are of particular importance.

The capacitance value determines how fast a capacitor will except a charge. The formula

$$C = I \times \frac{T}{V}$$

can be used to predict the charge rate, the voltage, and the length of time it will take for a capacitor to charge.

In series the elasticity (reciprocal of capacitance) is additive. In parallel the capacitance is additive.

The capacitor acts like a shock absorber for the circuit. The current and voltage through the circuit are 90° out of phase and allow the capacitor to absorb sudden changes in voltage. This quality appears as a resistance to ac current flow and is called the capacitive reactance. The formula

$$X_C = \frac{1}{2 \times \pi \times f \times C}$$

can be used to predict the capacitive reactance for any given frequency. In general, high frequencies have lower capacitive reactance values, and low frequencies have higher capacitive reactance values.

Capacitors are sometimes used to form voltage dividers. Capacitance and voltage are inversely proportional in such circuits. The smaller value capacitor will drop the most voltage. The capacitive voltage divider works well in situations where the current draw is not excessive.

ELI the ICEman can be used in *RC* circuits to predict the the sign of the phase angle. The resistive quantity is placed on the horizontal side of the triangle, the reactive quantity is placed on the vertical side of the triangle, and the total quantity is placed on the hypotenuse of the triangle. The Pythagorean theorem is used to find any missing sides, and trigonometry is used to find the phase angle of the circuit.

QUESTIONS

16-1. What is an application of a coupling capacitor?

16-2. What does the permittivity of air in the capacitance formula relate to in the inductance formula? How do they compare?

16-3. What is the function of a filter capacitor?

16-4. How does the distance between the plates affect the breakdown voltage of a capacitor?

16-5. How does the distance between the plates affect the capacitance of a capacitor?

16-6. An increase in the distance between the plates of a capacitor will have what effect on breakdown voltage? On capacitance?

16-7. List several dielectrics. Which has the highest dielectric strength? Which has the largest dielectric constant?

16-8. In your own words, define *dielectric*.

16-9. What does the dielectric strength determine?

16-10. What does the dielectric constant determine?

16-11. What is the function of a tank circuit?

16-12. List several types of capacitors. Include typical values.

16-13. Describe the construction of an electrolytic capacitor.

16-14. What effect does reverse polarity have on an electrolytic capacitor?

16-15. What is the equivalent unit for farad?

16-16. The text describes a capacitor as a shock absorber. Explain what is meant.

16-17. What is the lowest possible frequency? Is this ac or dc?

16-18. What is angular velocity? What is used as the symbol for it?

16-19. What is the reciprocal of capacitive reactance?

16-20. What is the reciprocal of capacitance?

16-21. Describe the manner in which total series capacitance is calculated.

16-22. Describe the manner in which total parallel capacitance is calculated.

16-23. Describe the manner in which a capacitor charges and discharge.

16-24. In a capacitor voltage divider, which capacitor will charge higher: the larger capacitance or the smaller capacitance?

16-25. How does ELI the ICEman relate to *RC* circuits?

16-26. List the triangles used in series-circuit analysis. Indicate their phase.

16-27. List the triangles used in parallel-circuit analysis. Indicated their phase.

16-28. How can ICE be used to determine the phase of these triangles?

16-29. Which component is placed on the vertical side of the triangle? The horizontal? The hypotenuse?

16-30. Describe how Ohm's law relates to triangles.

16-31. What differences do you see between the analysis of *LR* circuits and *RC* circuits?

16-32. What three factors determine if a capacitor is faulty?

16-33. List several pieces of equipment that can be used to test a capacitor. Describe the use of each.

16-34. In the event that you cannot determine the condition of a capacitor, what should be done?

16-35. Describe how a capacitor charges as the applied voltage increases and discharges as the applied voltage decreases.

16-36. What is the phase difference between voltage and current in a capacitor?

16-37. In your own words, explain the effect of frequency on capacitance; on capacitive reactance.

16-38. Use your calculator to find the sine, cosine, and tangent of 18°. Divide the sine of 18° by the cos of 18°. What is the result? Explain.

QUESTIONS

PRACTICE PROBLEMS

Problems 16-1 through 16-3 refer to Figure 16-17.

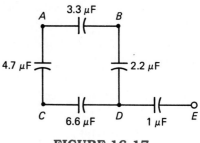

FIGURE 16-17

16-1. Calculate the total capacitance between point A and point E.
16-2. Calculate the total capacitance between point B and point E.
16-3. Calculate the total capacitance between point C and point E.

Problems 16-4 through 16-9 refer to Figure 16-18.

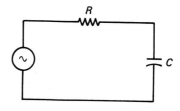

FIGURE 16-18

16-4. $C = 0.002\ \mu F$
 $f = 550$ kHz
 $V_A = 10$ V
 $V_R = $ _____
 $V_C = $ _____
 $R = 120\ \Omega$
 $X_C = $ _____
 $Z = $ _____
 $\theta = $ _____
 $I = $ _____

16-5. $C = $ _____
 $f = 5$ kHz
 $V_A = $ _____
 $V_R = 3.234$ V
 $V_C = $ _____
 $R = 2.7$ kΩ
 $X_C = 3.183$ kΩ
 $Z = $ _____
 $\theta = $ _____
 $I = $ _____

16-6. $C = 100$ pF
 $f = $ _____
 $V_A = 50$ mV
 $V_R = $ _____
 $V_C = 32.57$ mV
 $R = $ _____
 $X_C = 17.17\ \Omega$
 $Z = $ _____
 $\theta = $ _____
 $I = $ _____

16-7. $C = 330$ pF
 $f = $ _____
 $V_A = $ _____
 $V_R = $ _____
 $V_C = $ _____
 $R = 500\ \Omega$
 $X_C = $ _____
 $Z = 660\ \Omega$
 $\theta = $ _____
 $I = 378.9\ \mu A$

16-8. $C = 3330$ pF
$f = $ _____
$V_A = $ _____
$V_R = $ _____
$V_C = 0.7523$ V
$R = $ _____
$X_C = $ _____
$Z = 9.2469$ Ω
$\theta = -30.1°$
$I = $ _____

16-9. $C = $ _____
$f = 330$ kHz
$V_A = $ _____
$V_R = $ _____
$V_C = $ _____
$R = $ _____
$X_C = 731$ Ω
$Z = 1668$ Ω
$\theta = $ _____
$I = 5.394$ mA

Problems 16-10 through 16-15 refer to Figure 16-19.

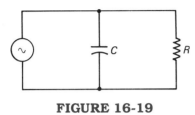

FIGURE 16-19

16-10. $C = $ _____
$f = 60$ Hz
$V_A = 120$ V
$I_C = $ _____
$I_R = $ _____
$I_T = $ _____
$R = 200$ Ω
$X_C = 265.25$ Ω
$\theta = $ _____
$Z = $ _____

16-11. $C = 4.7$ μF
$f = 120$ Hz
$V_A = 20$ V
$I_C = $ _____
$I_R = $ _____
$I_T = $ _____
$R = 180$ Ω
$X_C = $ _____
$\theta = $ _____
$Z = $ _____

16-12. $C = 0.0047$ μF
$f = $ _____
$V_A = 300$ mV
$I_C = $ _____
$I_R = $ _____
$I_T = 43.55$ μA
$R = $ _____
$X_C = $ _____
$\theta = 30.5°$
$Z = $ _____

16-13. $C = $ _____
$f = 480$ kHz
$V_A = $ _____
$I_C = $ _____
$I_R = 166.7$ μA
$I_T = 218.7$ μA
$R = $ _____
$X_C = $ _____
$\theta = $ _____
$Z = 457.1$ Ω

16-14. $C = 470$ pF
$f = $ _____
$V_A = $ _____
$I_C = $ _____
$I_R = 0.2222$ A
$I_T = $ _____
$R = 18$ Ω
$X_C = 32.56$ Ω
$\theta = $ _____
$Z = $ _____

16-15. $C = $ _____
$f = 56$ kHz
$V_A = $ _____
$I_C = $ _____
$I_R = $ _____
$I_T = 85.81$ mA
$R = 270$ Ω
$X_C = $ _____
$\theta = 64.4°$
$Z = $ _____

PRACTICE PROBLEMS

17

Transients in Current and Voltage

Chapter objectives

After reading this chapter and answering the questions and problems, you should be able to:

- Use logarithms to solve exponential equations.
- Define *common* and *natural logarithms*.
- Given a circuit containing a resistor and capacitor or an inductor and a resistor, indicate if the voltage and current are rising or decaying functions of time.
- Define and calculate time constants for *LR* and *RC* circuits.
- Use graphic and mathematic methods for calculating transient voltage and current levels.
- Define and describe the circuitry necessary to build and operate a differentiator and integrator.
- Use *RC* time constants to calculate the time it will take a capacitor to charge (or discharge) to a particular voltage level.
- Calculate the amount of inductive kickback generated as an *LR* circuit is activated (or deactivated).

When an *RC* or *LR* circuit is first turned on (or off), there is a short period of time before current and voltage stabilizes. During that time the current and voltages are constantly changing. These are called *current* or *voltage transients*. The length of time before the steady-state levels are reached is determined by the value of *R*, *C*, or *L* in the circuit. In this chapter we explain some methods of determining the voltage or current at any instant during this warm-up (or shutdown) period.

GETTING STARTED: LOGARITHMS

Exponential relations are of a particular importance to time-constant relationships. In the mathematic section we explain logarithms as they relate to exponents, define the number e, and discuss the use of calculators in finding exponential values.

Power Functions (Logarithms) In the early 1600s John Napier and other mathematicians were trying to find easier ways to calculate complex astronomy relations. The solution they found was logarithms. From that time until the invention of hand calculators, scientists and mathematicians have used logarithm tables faithfully.

A logarithm table contains two number systems. One is an arithmetic system (the next number in the series can be found by adding a value to the preceding one). The other is a geometric system (the next number in the series can be found by multiplying the preceding one by a value).

Arithmetic series:

   0, 1, 2, 3, 4, 5, 6, . . .   add 1 for next member

Geometric series:

   1, 3, 9, 27, 51, 153, 459, . . .   multiply by 3 for next member

Logarithms allow arithmetic to be performed at a simpler level. That is, by using logarithms you can find the product of two numbers by adding.

EXAMPLE 17-1

Use the arithmetic and geometric series to find the product of 8×51.

Solution: Begin by finding 9 on the geometric series. Notice that the 9 corresponds with the 2 on the arithmetic series. Also find 51 on the geometric series. This corresponds with 4 on the arithmetic series. Then add.

$$2 + 4 = 6$$

The 6 is then cross-referenced from the arithmetic series to the geometric series to find the product 459. ∎

Similarly, logarithms turn division into a subtraction problem, finding powers into multiplication, and finding roots into division.

Logarithms are typically defined as an exponential function.

$$\text{base}^{\text{logarithm}} = \text{number}$$

In Example 17-1 the base was 3, the arithmetic series contained the logarithms, and the geometric series contained the numbers. Pure mathematics tells us that the base can be any number, but over the years only logarithms of base 10 and base e (2.71828) have proved to be useful.

Your calculator has in its memory the logarithm tables for both of these bases. Examine your calculator. There should be a key labeled "log." When you enter 100 and press log, your calculator should display 2 (because $100 = 10^2$). The log key is called a common logarithm function. *Common logarithms* are based on powers of 10.

The inverse operation of logarithm is antilogarithm. On most calculators, this function is obtained by entering the number, pressing the second function key and then the log key (others have a key labeled " 10^x ").

Your calculator also has logarithms that are based on the number e (2.71828). The logarithms associated with these power functions are called *natural logarithms*. Most calculators have a key labeled ln (or lnx), which provides this function.

EXAMPLE 17-2

Use logarithms to find the value of e raised to the 3.95 power.

Solution: Use natural logarithms (lnx) and enter the following sequence of keystrokes to acquire the answer.

$$e^{3.95} = \text{antiln } 3.95$$

$$= 51.935$$

[3] [.] [9] [5] [2nd] [ln]   display   5.1935   01   ∎

EXAMPLE 17-3

Use logarithms to find the power of 10 that is equivalent to 7893.

Solution: Use common logarithms (log) and enter the following sequence of keystrokes to acquire the answer.

$$10^x = 7893$$

$$x = \log 7893$$

$$= 3.8972$$

[7] [8] [9] [3] [log]   display   3.8972   00   ∎

Notice that the logarithm in Example 17-3, is 3.8972. Logarithms are expressed in two parts, the characteristic and the mantissa. The whole number is the *characteristic* and serves to indicate the position of the decimal point. The 3 in Example 17.3 shows that the number (7893) is in the range 1000 to 10,000. The decimal fraction is *the mantissa* and relates to the numerical value.

EXAMPLE 17-4

Find the value of 10 raised to the 0.8972 power.

Solution: Use common logarithms (log) and enter the following sequence of keystrokes.

$$10^{0.8972} = \text{antilog } 0.8972$$
$$= 7.8922$$

[.] [8] [9] [7] [2] [2nd] [log]   display   7.8922   00   ■

Notice that when the decimal portion of the logarithm in Example 17.3 was used in Example 17.4, the numerical value was the same in both problems; however, the decimal point is not correctly placed. The characteristic is needed in order to place this point correctly. The characteristic is 3 and if the decimal point is moved three places, the values are approximately equivalent.

SELF-TEST

1. 10 to the _____ power = 12,098.
2. The base 10 logarithm of 2875 is _____ .
3. *e* raised to the 4.867 power is _____ .
4. The natural logarithm of 345 is _____ .
5. The base 10 antilog of 4.6783 is _____ .
6. The base *e* antilog of 2.7894 is _____ .
7. log 3.876 is a base _____ logarithm.
8. ln 3.876 is a base _____ logarithm.
9. antilog 0.9876 is a base _____ antilogarithm.
10. antiln 0.9876 is a base _____ antilogarithm.
11. Find the values of the logarithms in problems 7 and 8.
12. Find the values of the antilogarithms in problems 9 and 10.

ANSWER TO SELF-TEST

1. 4.0827
2. 3.4586
3. 129.9
4. 5.8435
5. 47676
6. 16.27
7. 10
8. *e*
9. 10
10. *e*
11. 0.5884, 1.3548
12. 9.7185, 2.6849

VOLTAGE LEVELS AND CURRENT LEVELS

Use of ICE and ELI   On a circuit that contains a resistor and a capacitor, ICE can be used to indicate whether the voltage level is a rising or decaying function of time. The sketch shown in Figure 17-1 refers to a series circuit containing a capacitor. When the circuit is activated, the voltage across the capacitor rises and the current throughout the circuit decays. Remember that the current is the same at all points on a series circuit. When considering the action of the current in a series cir-

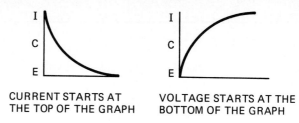

FIGURE 17-1

cuit, understand that the current must be doing the same thing (decaying or rising) through all of the components ($I_R = I_C$). ICE can be used to remember this action. See the sketch shown in Figure 17-1. Since current is the same at all points in the circuit and the voltage across the resistor is the product of the resistance and the current, the voltage across the resistor assumes the same function as the current (decay).

If the circuit contains an inductor instead of a capacitor, the action can be predicted by ELI by following a procedure similar to that outlined above. The diagrams in Figure 17-2 indicate that the inductor voltage is decaying and the current is rising. Therefore, the resistor voltage is also rising.

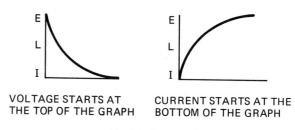

FIGURE 17-2

In short, ICE and ELI (drawn in the manner shown) indicate the starting position of the capacitor and inductor voltages (E) and currents (I). As time progresses these levels move to the opposite extremes. The current though both components is the same ($I_L = I_R$ or $I_C = I_R$). The resistor voltage follows the same function (rise or decay) as the current.

Time Constants Consider the circuit shown in Figure 17-3. After the circuit is activated, the capacitor will slowly charge to a voltage equal to that of the supply. At this point in time, the current will cease to flow. When we discuss current and voltage transients, we are referring to the time period between circuit activation and the steady-state voltage and current conditions, where the current and voltages begin at peak (or zero) and then change continually until the capacitor is charged and the current ceases to flow.

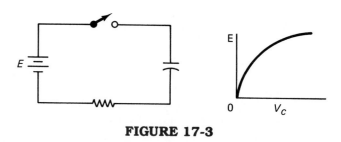

FIGURE 17-3

Don't get a misconception about the importance of this time period. It is not a waste of time to study something that takes only a few milli- or microseconds to occur. Since the invention of IC chips, electronic engineers are relieved of some of the more difficult tasks of design. One IC chip can be used for literally hundreds of different applications. Usually, the design of these applications only entails finding the correct value of an external resistor (one that is connected to an IC chip) and an external capacitor for the particular job (see Figure 17-4). The discussion in this chapter is intended to help you understand the necessary ideas behind the choice of the resistor and capacitor. In most cases the choice is determined by the time constant for the *RC* network being used.

The length of time that a circuit will be transient is a predictable value. That prediction is based on a particular calculation called a *time constant*. It takes roughly five time constants for a circuit to become stable. The formulas for calculating the time constants are

$$TC = \begin{cases} R \times C & \text{for an } RC \text{ circuit} \\ \dfrac{L}{R} & \text{for an } LR \text{ circuit} \end{cases}$$

SELF-TEST

Find the missing value for each set.

1. $R = 2700\ \Omega$, $C = 220$ pF TC = _____
2. $R = 12$ kΩ, $C = 3300$ pF TC = _____
3. $R = 2700\ \Omega$, $L = 35$ mH TC = _____
4. $R = 12$ kΩ, $L = 50\ \mu$H, TC = _____
5. $R =$ _____, $C = 470\ \mu$F, TC = 7.05 s
6. $R =$ _____, $C = 4700$ pF, TC = 117.5 μs
7. $R = 22$ kΩ, $C =$ _____, TC = 145.2 μs
8. $R = 86$ kΩ, $C =$ _____, TC = 0.4042 s

ANSWERS TO SELF TEST

1. TC = 594 ns
2. TC = 39.6 μs
3. TC = 12.96 μs
4. TC = 4.167 ns
5. $R = 15$ kΩ
6. $R = 25$ kΩ
7. $C = 6600$ pF
8. $C = 4.7\ \mu$F

METHODS FOR FINDING VOLTAGE AND CURRENT LEVELS

There are two basic approaches to time constants. One is to calculate the exact voltages at any particular time. The other is to solve for approximate values and construct a graph using these approximations. The graph can be used to provide voltage and current values for other times. It is the graphic approach that we discuss first.

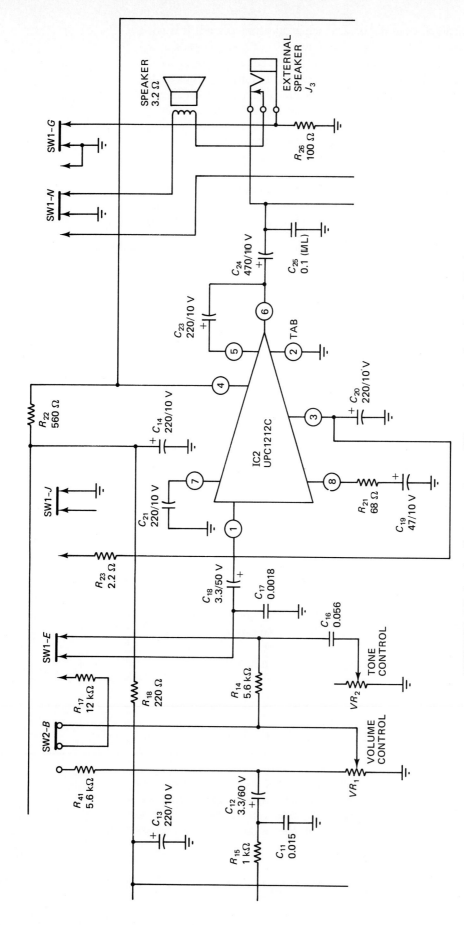

FIGURE 17-4 Audio output amplifier section of a circuit. Notice that the basic circuit consists of 1 IC chip and several resistors and capacitors.

Graphic Method For each time constant the voltage and current are associated with a particular percentage. This is a percentage of the maximum voltage or current level for the circuit. Understand that some of the values are increasing (rising) while others are decreasing (decaying). The first five time constants and their associated percentages are listed in Table 17-1.

TABLE 17-1

Time Constant	Decay (%)	Rise (%)
First	37	63
Second	14	86
Third	5	95
Fourth	2	98
Fifth	1	99

Example 17-5 will demonstrate how to create a graph to be used in making predictions about the transient period of a circuit.

EXAMPLE 17-5

Graph the resulting voltages and currents after the circuit in Figure 17-5 is activated.

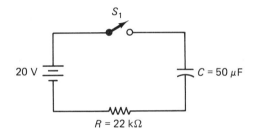

FIGURE 17-5

Solution: Begin by sketching the current and voltage transients (Figure 17-6). Then calculate the value of the time constant.

$$TC = R \times C$$
$$= 20 \text{ k}\Omega \times 50 \text{ }\mu\text{F}$$
$$= 1 \text{ s}$$

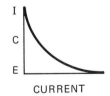
CURRENT

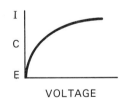
VOLTAGE **FIGURE 17-6**

Next use Table 17-1 to calculate and plot the time and capacitor voltage relationships. (See Figure 17.7(A).)

After 1 s:

$$\text{first TC} = 63\%$$
$$v_C = 20 \text{ V} \times 63\%$$
$$= 12.6 \text{ V}$$

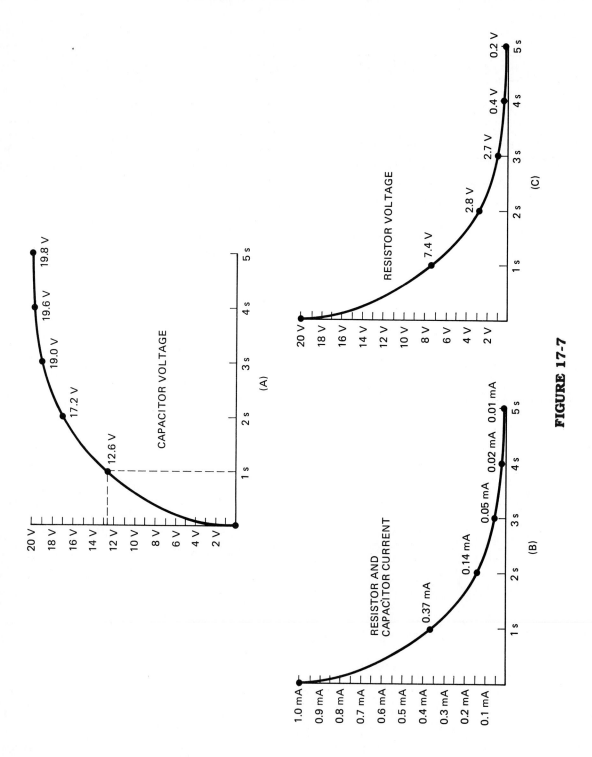

FIGURE 17-7

After 2 s:

$$\text{second TC} = 86\%$$
$$v_C = 20 \text{ V} \times 86\%$$
$$= 17.2 \text{ V}$$

After 3 s:

$$\text{third TC} = 95\%$$
$$v_C = 20 \text{ V} \times 95\%$$
$$= 19 \text{ V}$$

After 4 s:

$$\text{fourth TC} = 98\%$$
$$v_C = 20 \text{ V} \times 98\%$$
$$= 19.6 \text{ V}$$

After 5 s:

$$\text{fifth TC} = 99\%$$
$$v_C = 20 \text{ V} \times 99\%$$
$$= 19.8 \text{ V}$$

Next use Table 17-1 to calculate and plot the time and current relationships. Remember that the maximum current flow occurs at the onset of operation. At this time, the circuit is reduced to the battery and resistor (the capacitor is momentarily shorted). (See Figure 17-7(B).)

$$I_{\max} = \frac{20 \text{ V}}{20 \text{ k}\Omega}$$
$$= 1 \text{ mA}$$

After 1 s:

$$\text{first TC} = 37\%$$
$$i_C \text{ and } i_R = 37\% \times 1 \text{ mA}$$
$$= 0.37 \text{ mA}$$

After 2 s:

$$\text{second TC} = 14\%$$
$$i_C \text{ and } i_R = 14\% \times 1 \text{ mA}$$
$$= 0.14 \text{ mA}$$

After 3 s:

$$\text{third TC} = 5\%$$
$$i_C \text{ and } i_R = 5\% \times 1 \text{ mA}$$
$$= 0.05 \text{ mA}$$

After 4 s:

$$\text{fourth TC} = 2\%$$
$$i_C \text{ and } i_R = 2\% \times 1 \text{ mA}$$
$$= 0.02 \text{ mA}$$

After 5 s:

$$\text{fifth TC} = 1\%$$
$$i_C \text{ and } i_R = 1\% \times 1 \text{ mA}$$
$$= 0.01 \text{ mA}$$

Finally the time and resistor voltage relationships are calculated using Table 17-1. The resistor voltage follows the same decaying pattern as the current.

After 1 s:

$$\text{first TC} = 37\%$$
$$v_R = 20 \text{ V} \times 37\%$$
$$= 7.4 \text{ V}$$

After 2 s:

$$\text{second TC} = 14\%$$
$$v_R = 14\% \times 20 \text{ V}$$
$$= 2.8 \text{ V}$$

After 3 s:

$$\text{third TC} = 5\%$$
$$v_R = 5\% \times 20 \text{ V}$$
$$= 1 \text{ V}$$

After 4 s:

$$\text{fourth TC} = 2\%$$
$$v_R = 2\% \times 20 \text{ V}$$
$$= 0.4 \text{ V}$$

After 5 s:

$$\text{fifth TC} = 1\%$$
$$v_R = 1\% \times 20 \text{ V}$$
$$= 0.2 \text{ V} \qquad \blacksquare$$

These percentages give a quick and fairly accurate approximation of the voltage and current levels during the transient period. Plotting them on a graph will allow an approximation for all of the other times involved. When the graphs in Figure 17-7 were developed, the point in time was found on the horizontal axis. Then the corresponding voltage (or current) level was located on the vertical axis. A vertical line was drawn upward from the horizontal point (this is usually an imaginary line). A horizontal line was drawn from the vertical point across to the vertical line (again an imaginary line). The point where the two lines intersect is plotted (a dot is place at this location). Other points are plotted in a similar fashion. The dots are then connected together to construct the graph.

The graphs shown in Figure 17-7 can be used to approximate the voltage and current levels at various times. This is done in almost the reverse of the plotting procedure that created the graph. The desired point in time is found on the horizontal axis. A vertical line is draw upward from this point until the line crosses the plotted function. A horizontal line is drawn from this intersection to the vertical axis. The value is read from where the horizontal line crosses the vertical axis.

EXAMPLE 17-6

Use the graphs shown in Figure 17-7 to predict the values of the resistor voltage after 2.5 s, the capacitor voltage after 3.75 s, and the current after 1.25 s.

Solution: Find the desired time on the horizontal axis and read the voltage level at this time. That is how the voltage level of 19.5 V for the capacitor after 3.75 s, the voltage level of 1.5 v for the resistor after 2.5 s, and the circuit current of 0.26 mA after 1.25 s were determined. ∎

SELF-TEST

Use the percentages given in Table 17-1 to find the missing values for each of the following sets of data, which refer to Figure 17-8 immediately after the switch is closed.

1. $R = 10\ k\Omega$
   $C = 330\ pF$
   $t = 6.6\ \mu s$
   $TC = \underline{}$
   $V = 12\ V$
   $v_R = \underline{}$
   $v_C = \underline{}$
   $I = \underline{}$

2. $R = 2.7\ k\Omega$
   $C = \underline{}$
   $t = 10.8\ \mu s$
   $TC = 2.7\ \mu s$
   $V = \underline{}$
   $v_R = 0.3\ V$
   $v_C = \underline{}$
   $I = \underline{}$

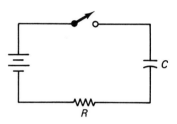

FIGURE 17-8

3. $R = 330\ \Omega$
   $C = 4.7\ \mu F$
   $t = 1.55\ ms$
   $TC = \underline{}$
   $V = \underline{}$
   $v_R = \underline{}$
   $v_C = \underline{}$
   $I = 22.42\ mA$

4. $R = 25\ k\Omega$
   $C = 3000\ pF$
   $t = \underline{}$
   $TC = \underline{}$
   $V = 100\ V$
   $v_R = 5\ V$
   $v_C = \underline{}$
   $I = \underline{}$

The following sets of data refer to Figure 17-9 immediately after the switch is closed.

5. $R = 10\ k\Omega$
   $L = 1\ H$
   $t = 300\ \mu s$
   $TC = \underline{}$
   $V = 12\ V$
   $v_R = \underline{}$
   $v_L = \underline{}$
   $I = \underline{}$

6. $R = 2.7\ k\Omega$
   $L = \underline{}$
   $t = 6.963\ ms$
   $TC = 1.741\ ms$
   $V = \underline{}$
   $v_R = 0.3\ V$
   $v_L = \underline{}$
   $I = \underline{}$

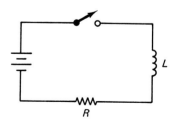

FIGURE 17-9

7. $R = 330\ \Omega$
    $L = 35$ mH
    $t = 106.1\ \mu s$
    TC = _____
    V = _____
    v_R = _____
    v_L = _____
    $I = 22.42$ mA

8. $R = 25$ kΩ
    $L = 300\ \mu H$
    t = _____
    TC = _____
    $V = 100$ V
    $v_R = 5$ V
    v_L = _____
    I = _____

ANSWERS TO SELF-TEST

1. TC = $3.3\ \mu s$, $v_R = 1.68$ V, $v_C = 10.32$ V, $I = 168\ \mu A$
2. $C = 1000$ pF, $V = 15$ V, $v_C = 14.7$ V, $I = 111.1\ \mu A$
3. TC = $1.55\ \mu s$, $V = 20$ V, $v_R = 7.39$ V, $v_C = 12.61$ V
4. $t = 225\ \mu s$, TC = $75\ \mu s$, $v_C = 95$ V, $I = 200\ \mu A$
5. TC = $100\ \mu s$, $v_R = 11.4$ V, $v_L = 0.6$ V, $I = 1.14\ \mu A$
6. $L = 4.7$ H, $V = 306.1$ mV, $v_L = 6.122$ mV, $I = 111.1\ \mu A$
7. TC = $106.1\ \mu s$, $V = 11.74$ V, $v_R = 7.39$ V, $v_L = 4.35$ V
8. $t = 36$ ns, TC = 12 ns, $v_L = 95$ V, $I = 200\ \mu A$

Mathematic Method The second approach to solving time-constant problems is a mathematic approach. It does give an exact answer; however, it can be time consuming to calculate the voltage (or current) level for every particular time. There are instances where only one or two particular points are needed. These warrant a mathematic method rather than a graphic one. The following equations are used to calculate the instantaneous voltages.

Decay:

$$v = V \times e^{-t/\text{TC}}$$

Rise:

$$v = V \times (1 - e^{-t/\text{TC}})$$

The TC in the equations refers to L/R in an inductive circuit and $R \times C$ in a capacitive circuit. It is best to calculate the time constant at the onset of the problem. Example 17-7 demonstrates this process.

EXAMPLE 17-7

Demonstrate how to calculate the resistor voltage at 2.5 s, the circuit current at 1.25 s, and the capacitor voltage at 3.75 s for the circuit shown in Figure 17-5.

Solution: Use the decay and rise formulas to find the exact voltages and currents.

Resistor voltage after 2.5 s:

$$v_R = 20 \times e^{-2.5\,\text{s}/1\,\text{s}}$$
$$= 20 \times e^{-2.5}$$
$$= 20 \times \text{antiln}(-2.5)$$
$$= 1.6417\ \text{V}$$

[2] [.] [5] [+/−] [÷] [1] [=]
[2nd] [ln] [×] [2] [0] [=]   display   1.6416   00

Circuit current after 1.25 s:

$$i_C = 1 \text{ mA} \times e^{-1.25\text{s}/1\text{s}}$$
$$= 1 \text{ mA} \times e^{-1.25}$$
$$= 0.001 \times \text{antiln}(-1.25)$$
$$= 0.2865 \, \mu\text{A}$$

[1] [.] [2] [5] [+/−] [÷] [1] [=]
[2nd] [ln] [×] [1] [EE] [3] [+/−] [=]   display   2.8650   −04

Capacitive voltage after 3.75 s:

$$v_C = 20 \times (1 - e^{-3.75\text{s}/1\text{s}})$$
$$= 20 \times (1 - e^{-3.75})$$
$$= 20 \times [1 - \text{antiln}(-3.75)]$$
$$= 20 \times 0.9765$$
$$= 19.53 \text{ V}$$

[3] [.] [7] [5] [+/−] [÷] [1] [=] [2nd] [ln] [mem]
[1] [−] [rcl] [=] [×] [2] [0] [=]   display   1.9529   01   ■

A quick comparison will show that the graphic method is surprisingly accurate. It is best to use the graphic approach when it is necessary to do several calculations on the same circuit.

SELF-TEST

Use the mathematic approach to find the missing values. The following sets of data refer to Figure 17-8.

1. $R = 1.2$ kΩ
   $C = 4.7 \, \mu$F
   $v_R = $ _____
   $V = 250$ V
   $t = 15.228$ ms

2. $R = 15$ kΩ
   $C = 330$ pF
   $v_C = $ _____
   $V = 250$ V
   $t = 1.122 \, \mu$s

3. $R = 1.2$ kΩ
   $L = 300 \, \mu$H
   $v_R = $ _____
   $V = 250$ V
   $t = 15.228$ ms

4. $R = 15$ kΩ
   $L = 35$ mH
   $v_L = $ _____
   $V = 250$ V
   $t = 8.122$ s

ANSWERS TO SELF-TEST

1. $v_R = 16.8$ V
2. $v_C = 50.7$ V
3. $v_R = 250$ V
4. $v_L = 0$ V

PULSE AND WAVE SHAPING

Differentiator The circuit shown in Figure 17-10 is called a *differentiator*. Its prime function is to provide an immediate output for any change in input. When it is driven by a triangle wave, the result, as shown in Figure 17-11, will be a square wave. The differentiator produces an output wave that indicates the rate of change in input voltage. A triangle wave has a constant rate of voltage change. In Figure 17-11 the change in voltage is 2 V and the change in time is 0.5 s. So the rate of change (this is listed as dv/dt) is 4 V/s.

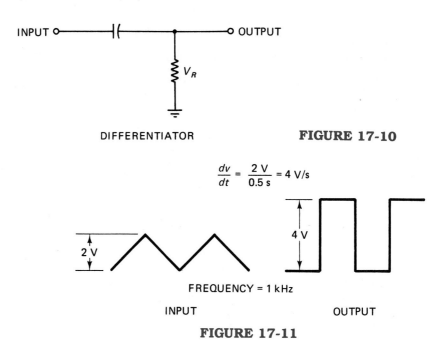

DIFFERENTIATOR   **FIGURE 17-10**

FIGURE 17-11

The time constant plays an important part in the design of a differentiator. The smaller the value of resistor and capacitor, the faster the resulting output will be. For this reason it is necessary to have a short time constant. The frequency of the applied voltage determines whether the time constant is considered long or short. In general, the time constant should be 10 to 20% of the time period of the applied voltage or less. The differentiator is used to trigger a particular response to a change in the input voltage.

Integrator The circuit shown in Figure 17-12 is called an integrator. Its prime function is to provide a slowly changing output for a quick-changing input signal. When it is driven by a square wave, the result is as shown in Figure 17-13, a trian-

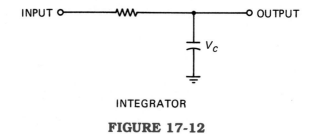

INTEGRATOR

FIGURE 17-12

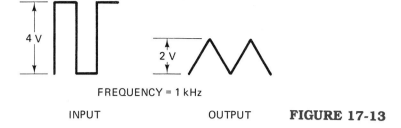

FIGURE 17-13

gle wave. The integrator provides an indication of the total charge (amount of electrons or holes) contained in the capacitor.

The time constant also plays an important role in the design of an integrator. The larger the value of resistor and capacitor, the slower the resulting change in the output will be. Again it is the frequency of the input signal that determines whether the time constant is long or short. In general, the time constant should be 5 to 10 times the time period of the input signal or more.

PRACTICAL APPLICATIONS

RC Time Constants *RC* time constants affect circuit operation primarily when the circuit is first activated. Older electronic devices (TV sets, radios, etc.) took a long time to warm up. This was due to the slow charging of the capacitors in their circuitry. Today, many electric appliances contain digital circuitry. One advantage of digital circuitry is its ability to remember its last setting. *RC* networks are used at times to provide this memory. (They provide power to the circuit when the device is turned off.). *RC* networks can also be used to clear this memory. Example 17.8 demonstrates how *RC* time constants are used to clear this memory.

EXAMPLE 17-8

Find the length of time it will take for the 330-pF capacitor of Figure 17-14 to charge to 3.6 V through the 82-kΩ resistor.

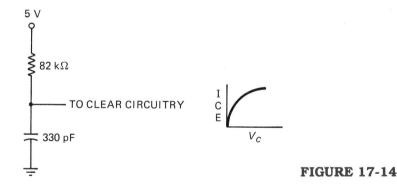

FIGURE 17-14

Solution: First use ICE to determine that there will be a rise in capacitor voltage. Next find the time constant.

$$TC = R \times C$$
$$= 82{,}000 \times 330 \times 10^{-12}$$
$$= 27.06 \ \mu s$$

Then use the formula:

$$v = V \times (1 - e^{-t/\text{TC}})$$

$$3.6 = 5 \times (1 - e^{-t/0.00002706})$$

$$\frac{3.6}{5} = 1 - e^{-t/0.00002706}$$

$$0.72 = 1 - e^{-t/0.00002706}$$

$$0.28 = e^{-t/0.00002706}$$

$$\ln 0.28 = -\frac{t}{0.00002706}$$

$$-1.273 = -\frac{t}{0.00002706}$$

$$t = 34.45 \ \mu s$$

LR Time Constants *LR* networks primarily affect circuit operation when the power is disconnected. The popping sound that sometimes occurs when your stereo is turned off is cause by the induced voltage of the speaker windings. This induced voltage is sometimes called *inductive kickback*. Example 17.9 will demonstrate how to calculate the inductive kickback voltage as a switch is opened.

EXAMPLE 17-9

Given that the switch, S1, in Figure 17-15 has an off resistance of 1 MΩ, calculate the voltage kickback of the inductor and the length of time it will take to deplete.

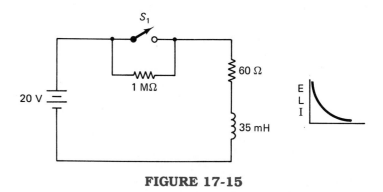

FIGURE 17-15

Solution: Begin by calculating the steady-state current level of the circuit when S1 is closed.

$$I_{\text{steadystate}} = \frac{20}{60}$$

$$= 0.3333 \ A$$

At the instant the switch is opened, the inductor tries to maintain (and does momentarily) this current level. So the inductive voltage kickback is

$$v_{\text{kickback}} = 0.3333 \times 1{,}000{,}000$$

$$= 333.3 \ kV$$

It is important to remember that this voltage is only a momentary value.
Next calculate the time constant for this circuit.

$$TC = \frac{L}{R}$$

$$= \frac{0.035}{1,000,000}$$

$$= 35 \text{ ns}$$

It takes five time constants for the voltage to deplete. So the kickback voltage will be zero after

$$35 \text{ ns} \times 5 = 175 \text{ ns} \qquad \blacksquare$$

SUMMARY

The number e predicts the rate of increase or decrease of a voltage or current transient. The use of natural logarithms makes working with powers of e simple.

ICE and ELI can be used to predict whether a quantity is an increase (rise) or a decrease (decay). The time constant indicates the length of time that the quantity will be transient. After five time constants a circuit reaches its steady-state value. The formulas

$$TC = R \times C$$

and

$$TC = \frac{L}{R}$$

are used to find the time constant, TC.

There are two methods commonly used to predict instantaneous values of voltage and current during the transient period: one is graphical, the other is mathematical. In general, if several instantaneous values are needed, the graphic methods are more practical. When only one or two values are needed, the mathematic approach may be faster.

RC networks are used in two methods of pulse shaping. This is due to their constant time transitions. A differentiator is created by the transient voltage of *R* in the network. An integrator is created by the transient voltage of *C* in the network.

A differentiator transforms a triangle wave into a square wave. It effectively describes the rate of change in voltage and requires a relatively short time constant.

An integrator transforms a square wave into a triangle wave. It requires a long time constant and provides a description of the summing of the charge on the capacitor.

Transients are most important in *RC* networks when the circuit is first activated. Transients do occur when the *RC* network is deactivated, but this is not so crucial. *RL* networks produce transients that affect circuit operation primarily when the circuit is deactivated. The transients produced when *RL* networks are activated do not usually cause a difficulty.

QUESTIONS

17-1. What is the value of *e*?

17-2. Define the term *logarithm*.

17-3. Define the term *arithmetic sequence*. Include an example.

17-4. Define the term *geometric sequence*. Include an example.

17-5. What arithmetic function is accomplished by adding two logarithms?

17-6. What arithmetic function is accomplished by subtracting two logarithms?

17-7. Describe how the logarithm function is used on your calculator.

17-8. What is the inverse operation of the mathematical function logarithm?

17-9. How is your calculator used to provide the antilog function?

17-10. Describe the manner in which logarithms can be used to find powers and roots of numbers.

17-11. What is meant by the term *common logarithm*?

17-12. How are natural logarithms different from common logarithms?

17-13. Define *characteristic* as it pertains to logarithms.

17-14. What is the mantissa of a logarithm?

17-15. Describe how ICE and ELI are used to predict the voltage and current transient in a circuit. Include both circuit activation and deactivation.

17-16. What is meant by the term *time constant*?

17-17. How is the time constant calculated in an *RL* circuit?

17-18. How is the time constant calculated in an *RC* circuit?

17-19. What is meant by steady-state current? By steady-state voltage?

17-20. How does the *time constant* pertain to the use of IC chips?

17-21. How long will a circuit be in a transient state?

17-22. Describe the use of graphs to find voltage and current levels in a circuit during its transient state.

17-23. What is the percentage associated with a voltage rise at the second time constant?

17-24. What is the percentage associated with a current decay at the third time constant?

17-25. Explain how the formula $v = V \times e^{-t/\text{TC}}$ can be used to predict voltage transients. Include whether this is a rise or decay formula.

17-26. When do capacitors have their biggest effect on circuit transients? Explain.

17-27. When do inductors have their biggest effect on circuit transients? Explain.

17-28. What function does a differentiator provide?

17-29. Briefly explain the operation of a differentiator.

17-30. Draw a simple differentiator circuit.

17-31. What function does an integrator provide?

17-32. Briefly explain the operation of an integrator.

17-33. Draw a simple integrator circuit.

17-34. What is meant by the term *pulse shaping*?

17-35. What is meant by *inductive kickback*?

17-36. Briefly describe the manner in which the kickback voltage can be calculated.

PRACTICE PROBLEMS

17-1. Find the common logarithm for each of the following.

log 12 = _____
log 2.7 = _____
log 0.79 = _____
log 250 = _____
log 3500 = _____

17-2. Find the natural logarithm for each of the following.

ln 12 = _____
ln 2.7 = _____
ln 0.79 = _____
ln 250 = _____
ln 3500 = _____

17-3. Find the antilogarithm for each of the following.

antilog 3.4563 = _____
antilog 1.3257 = _____
antilog (−2.4689) = _____
antilog 2.3467 = _____
antilog −0.9765 = _____

17-4. Find the antilogarithm for each of the following.

antiln 3.4563 = _____
antiln 1.3257 = _____
antiln (−2.4689) = _____
antiln 2.3467 = _____
antiln −0.9765 = _____

17-5. Refer to Figure 17-8. Find the missing values.

V = 30 V
v_R = 6 V
R = 2200 Ω
C = _____
t = 3.575 ms

17-6. Refer to Figure 17-8. Find the missing values.

V = 50 V
v_C = 16 V
R = 2200 Ω

$C = 0.022\ \mu\text{F}$
$t = \underline{\qquad}$

17-7. Refer to Figure 17-9. Find the missing values.

$V = 30\ \text{V}$
$v_R = 6\ \text{V}$
$R = 2200\ \Omega$
$L = 1\ \text{mH}$
$t = \underline{\qquad}$

17-8. Refer to Figure 17-9. Find the missing values.

$V = 50\ \text{V}$
$v_R = 16\ \text{V}$
$R = 2200\ \Omega$
$L = \underline{\qquad}$
$t = 51.1\ \text{ns}$

17-9. Refer to Figure 17-8. Find the missing values.

$V = 30\ \text{V}$
$v_R = 6\ \text{V}$
$R = 100\ \text{k}\Omega$
$C = 2200\ \text{pF}$
$t = \underline{\qquad}$

17-10. Refer to Figure 17-8. Find the missing values.

$V = 50\ \text{V}$
$v_C = 16\ \text{V}$
$R = \underline{\qquad}$
$C = 2200\ \text{pF}$
$t = 0.12375\ \text{ms}$

17-11. Refer to Figure 17-9. Find the missing values.

$V = 30\ \text{V}$
$v_R = 6\ \text{V}$
$R = \underline{\qquad}$
$L = 35\ \text{mH}$
$t = 0.875\ \mu\text{s}$

17-12. Refer to Figure 17-9. Find the missing values.

$V = 50\ \text{V}$
$v_L = 16\ \text{V}$
$R = 1200\ \Omega$

$L = 35$ mH
$t =$ _____

17-13. Refer to Figure 17-16. This power-up reset circuit requires 8.4 V to prevent clearing. What-value capacitor is needed to provide a stable operation after 0.3 ms?

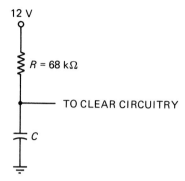

FIGURE 17-16

17-14. Refer to Figure 17-16. Suppose that a 660-pF capacitor were used. How long would it take for the capacitor voltage to reach 8.4 V?

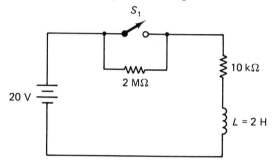

FIGURE 17-17

17-15. Calculate the kickback voltage generated when the switch in Figure 17-17 is opened.

17-16. What would be the *off resistance* of the switch in Figure 17-17 if the kickback voltage were known to be 20 kV?

18

Batteries

Chapter objectives

After reading this chapter and answering the questions, you should be able to:

- Describe the construction and operation of several types of batteries.
- Differentiate between primary and secondary cells.
- List several types of secondary cells, including advantages and disadvantages of each.
- Describe various methods of testing a battery.
- List several types of primary cells, including advantages and disadvantages of each.
- Define capacity and give its most common unit of measure.
- List several considerations in battery charging.
- Given loaded and unloaded terminal voltage and the amount of load current, calculate the internal resistance of a battery.
- Predict the effect that internal battery resistance will have on a circuit.

This chapter is intended to give you an understanding of the nature of voltage sources, particularly batteries. The concept of chemical production of electricity, the concept of internal resistance, and how it affects power delivery, as well as basic ideas concerning battery charging and rating methods are topics explained in this chapter.

BATTERY CONSTRUCTION

All batteries are constructed in the same basic manner (Figure 18-1). They all have two dissimilar metals that are separated by a chemical mixture. One plate is called the *anode*. This is the terminal where the electrons are released. The other is the *cathode*, which is the terminal where the electrons are collected. When you examine the operation of a circuit, it is easy to understand that the electrons (opposite with holes in conventional flow) flow into the negative side of any component and out of its positive side. By definition the side that has the electrons flowing into it is the cathode. For most of the components in an electronic circuit the electrons flow into the negative side; however, the electrons flow into the positive side of the battery and out of its negative side. That is why a diode's (or other circuit component's) negative terminal is called a cathode, whereas a battery's positive terminal is called a cathode. The chemical mixture used to produce the current is called an *electrolyte*.

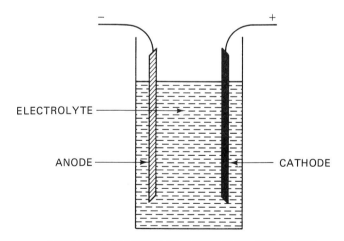

SIMPLIFIED BATTERY CONSTRUCTION. THE ANODE AND CATHODE ARE DISSIMILAR METALS. THE ELECTROLYTE IS A CHEMICAL MIXTURE THAT IS USED TO PRODUCE THE CURRENT

FIGURE 18-1

The electrolyte is the heart of the battery. It can be an acidic or a basic (alkaline) mixture. The particular chemical used determines the effectiveness of the battery (the amount of current and voltage that can be produced). Batteries themselves are not a new development. Records have been found which indicate that batteries were used in ancient times to electroplate jewelry. In recent times it has been found that potatoes, plants, and soil can be used to provide a weak source of electricity, because their internal chemistry is an electrolyte. A mixture of salt water can be used as an electrolyte, as can urine, soda, lemon juice, and aspirin.

BATTERY TYPES

A battery is a device that changes chemical energy into electrical energy. Actually, a battery is divided up into independent sections called *cells*. Each cell produces a voltage. When the cells are connected together, they form a battery. In other words, when we go to the store to buy a flashlight battery (D cell), we are not really buying a battery, we are buying a cell. If the flashlight contains more than one cell, these cells, grouped together, form a battery. When we purchase a battery for an automobile, we are actually purchasing a battery. An auto battery is comprised of a group of cells.

By definition there are two types of cells: primary cells, which are disposable units and cannot be recharged, and secondary cells, units that are not generally disposable and can be recharged. In the following discussion we examine some of the more common materials and battery types. These are profiled in Table 18-1.

TABLE 18-1 Battery Profile

Type	No-Load Voltage	Class
Lead–acid	2.10	Secondary
Nickel–cadmium	1.25	Secondary
Nickel–iron	1.20	Secondary
Silver–zinc	1.50	Secondary
Carbon–zinc	1.50	Primary
Manganese–alkaline	1.50	Primary
Mercury	1.35	Primary
Metal–air	Determined by metal	Primary

Secondary Cells

1. *Lead–acid battery*. (No-load voltage is 2.1 V/cell.) One of the most common secondary cell batteries is the lead–acid battery (Figure 18-2). The chemical reaction that produces the electricity is the result of sulfuric acid and its reaction with lead (negative terminal—anode) and lead peroxide (positive terminal—cathode). The reaction will produce electricity, with by-products of lead sulfate and water. The electrolyte is the main difference between a fully charged and a discharged battery. One way of testing the condition of this type of battery is by checking the ratio of electrolyte (charged) to water (discharged).

   a. *Specific gravity testing*. A tool called a *hydrometer* (Figure 18-3) is used to perform this test. The hydrometer contains a glass bulb which floats when the hydrometer is filled with electrolyte. Acid weighs more than water; as a result, the bulb will float higher in an electrolyte that is mostly acid than it will in an electrolyte that is mostly water. The bulb itself has a marking which indicates the specific gravity (weight) of the electrolyte (Figure 18-4 and 18-5).

FIGURE 18-2 Cut-Away Lead Acid Battery.

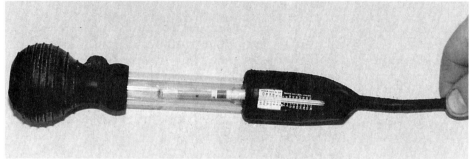

FIGURE 18-3 Hydrometer. Used to measure specific gravity of battery acid.

FIGURE 18-4 Testing a lead-acid battery.

A fully charged battery will have a specific gravity of 1.280 to 1.260 (read "twelve eighty to twelve sixty"). The actual specific gravity varies by location. For instance, the electrolyte used in car batteries in northern climates is a stronger acid (1.380) than that used in southern climates. Testing the battery using a hydrometer requires two basic checks. One is the actual reading. This is sometimes difficult be-

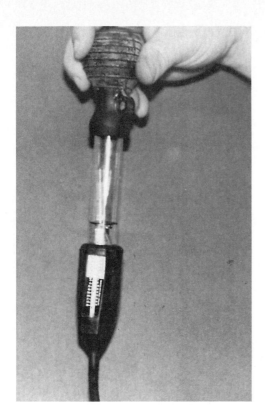 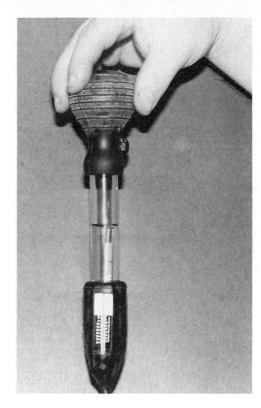

FIGURE 18-5 (A) High (Good) Hydrometer Reading; (B) Low (Bad) Hydrometer Reading.

cause the weight of the liquids depends greatly on their temperature. Good hydrometers have a thermometer and a table. The process requires taking a basic reading of the specific gravity and temperature. The temperature is then found on the table. The table will indicate the correction factor needed (about 0.004 per °F from 70°). Colder liquids are heavier than warmer liquids. If an electrolyte is cooler than 70°, a correction factor will have to be subtracted from the specific gravity reading originally obtained.

The second portion of the testing process requires a comparison between the readings of individual cells. The maximum variance between the highest and the lowest cells is 0.025 to 0.050 (25 to 50 points). It is agreed that a difference between cells exceeding 50 points indicates a faulty battery.

b. *Load testing*. The specific gravity test is probably the most accurate test ever developed for testing a wet cell battery; however, modern battery construction has made this test all but impossible to perform. Most manufacturers state that a load test is more desirable. A load test will check the battery's performance directly as it relates to circuit operation. In short, the test consists of placing a resistance across the terminals of the battery. After a specific length of time, the battery voltage is read. Normally, on a secondary cell, the load is listed as three times the batteries' capacity (ampere-hour rating, which will be discussed later). And the ending voltage is at least 75% of the battery's rating.

EXAMPLE 18-1

Suppose that a car (12 V) battery that has a capacity of 40 ampere-hours (A-h) needs to be tested. Calculate the current load needed for testing and the minimum acceptable voltage level.

Solution: The battery would be tested by placing a resistor across the terminals. The resistor would be sufficiently small to allow 120 A.

$$\text{load current} = 3 \times \text{A-h rating}$$
$$= 3 \times 40$$
$$= 120 \text{ A}$$

of current to flow.

The voltage would be checked after 15 s. At this time the voltage would have to be 9 V.

$$\text{minimum load voltage} = 75\% \times \text{no-load voltage}$$
$$= 0.75 \times 12$$
$$= 9 \text{ V}$$

or higher. ∎

If the battery did not pass this test, it would need to be replaced or recharged. The load test is the testing procedure most often recommended by manufacturers today.

2. *Nickel–cadmium battery.* (No-load voltage is 1.25 V/cell.) The nickel–cadmium (Ni-Cad) battery is used extensively in applications where small size and light weight are essential. It uses potassium hydroxide as an electrolyte, cadmium as an anode, and nickel hydroxide as a cathode.

3. *Nickel–iron battery.* (No-load voltage is 1.2 V/cell.) The nickel–iron or Edison cell is valued for its long life. In fact, some of the batteries built by Edison himself are still in operation. It uses potassium hydroxide as the electrolyte. The electrolyte is laced with lithium hydroxide to provide a higher capacity. The anode is a mixture of ferrous oxide and mercurous oxide, while the cathode is a nickel-plated steel, formed into a pocket, which contains nickel hydroxide and graphite.

4. *Silver–zinc battery.* (No-load voltage is 1.5 V/cell.) The silver–zinc battery is used where weight is a factor. It is extremely light in weight; however, it is very costly and has a short cycle life (limited number of times it can be recharged). The electrolyte is a mixture of potassium hydroxide and zinc hydroxide. The anode is porous zinc with a coating of mercury. The mercury suppresses the corrosion of the anode and thus increases the life expectancy of the battery. The cathode is silver peroxide.

Primary Cells

1. *Carbon–zinc cell.* (No-load voltage is 1.5 V/cell.) The most common type of primary cell is the carbon zinc cell (Figure 18-6). The carbon zinc or Leclanche cell uses ammonium chloride as an electrolyte. The positive terminal (cathode) is a carbon rod that extends through the center of the cell. The zinc canister not only holds the major components of the battery but also forms the negative terminal (anode). The strength and capacity of this type of cell is determined by the purity of the mixture of ammonium chloride used in its construction.

    a. *Local action.* Impurities in the mixture cause small opposing voltages to be produced on the inside of the battery. These opposing voltages are referred to as local action. *Local action* impedes the production of current and voltage. Some manufacturers advertise that their particular batteries outlast those of other manufacturers. The main reason for this claim is a very pure mixture.

    b. *Polarization.* The Leclanche cell wears out when a collection of hydrogen forms around the carbon rod. This grouping of hydrogen molecules is called *polar-*

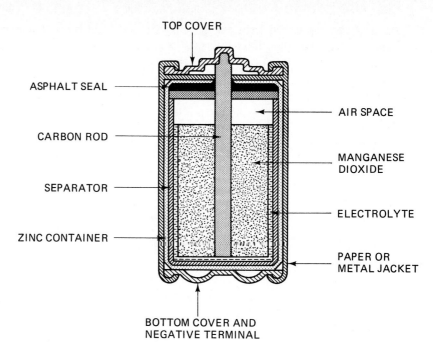

FIGURE 18-6 A Dry Cell consists of a zinc container filled with substances that produce an electric current by reacting chemically with one another. The container itself is the negative terminal of the cell. A carbon rod in the center serves as the positive terminal.

ization. Polarization occurs in a lead–acid battery when the lead plates become coated with sulfate. There is no coating action in a carbon zinc polarization; however, when hydrogen collects around the carbon rod, the amount (surface area) of carbon and ammonium chloride is greatly reduced. If the reacting chemicals cannot contact each other, there cannot be an electricity-producing reaction. Secondary batteries are capable of reversing the polarizing reaction. Primary batteries are not capable of reversing polarization.

2. *Manganese–alkaline cell.* (No-load voltage is 1.5 V/cell.) Commonly called *alkaline batteries,* the manganese–alkaline battery provides a high current capacity for relatively the same size and weight as the carbon zinc cell. It uses potassium hydroxide as an electrolyte, porous zinc as an anode, and manganese dioxide as a cathode.

3. *Mercury cell.* (No-load voltage is 1.35 V/cell.) The mercury cell is valued for its extremely small size and high current capacity. It uses potassium hydroxide as the electrolyte, zinc as the anode, and mercuric oxide with graphite as the cathode.

4. *Metal–air cells.* (No-load voltage is determined by metal.) The metal–air cells are of great importance because they use air as the cathode. In all previous battery designs, the cathode has determined the life expectancy of the battery. The metal–air design uses atmospheric oxygen to eliminate this factor. They are valued because of their unlimited life expectancy. The electrolyte is again potassium hydroxide. The anode can be any of several metals: zinc, magnesium, aluminum, or iron. Presently the aluminum air cell is the most promising.

Modern technology has developed batteries that are relatively small in size but can produce fairly large amounts of current. Also experimentation with other sources (other than chemical reactions) have brought about the use of solar cells. Solar cells convert light into electricity. At present, these cells produce only small amounts of current per square inch. They have their largest application in providing power for hand-held calculators.

A load test is the best manner to test the majority of dry cells. This test is performed in a manner described previously. In all cases remember that the loaded and unloaded voltage of a battery is different. There must be a load attached for an accurate test of a battery's condition.

BATTERY CAPACITY

Ampere-Hour Rating Any given battery can only produce electricity at a certain rate. This rate is referred to as the battery's capacity. A simple explanation of capacity is how many hours a battery can produce a current. For instance, if a battery can produce 2 A of current for 12 hours, it has an ampere-hour capacity of 24 A-h ($2 \times 12 = 24$). A 24 A-h battery can generally be considered to produce 4 A for 6 hours, 3 A for 8 hours, or any other current for a corresponding amount of time, provided that the current multiplied by the length of time equals 24 A-h. This is not always true. A battery can produce a smaller current for a longer time than it can a larger current; however, this is seldom taken into consideration when calculating battery capacity. Methods of finding what capacity battery is needed for a particular job are given in the following examples.

EXAMPLE 18-2

Suppose that an emergency system were designed to operate for 12 hours in time of danger. The system consists of a radio (which requires 3 A), an exhaust fan (which requires 10 A), and a lighting system (which requires 15 A). What is the minimum capacity needed for the battery that powers this system?

Solution: Begin by making a table of the loads and calculating the total required current (Table 18-2). Then find the needed battery capacity by multiplying the total current by the desired length of time of operation.

Battery Capacity = 28 A × 12 hours = 336 A-h ∎

TABLE 18-2 Require Current Table From Example 18-2.

Load	Required Current (A)
Radio	3
Fan	10
Lights	15
Total	28 amps

EXAMPLE 18-3

Find the length of time a 56-A-h battery will provide a 7.25-A current to a circuit.

Solution: Begin by remembering that the ampere-hour rating is found by

$$A\text{-}h = I \times T$$

$$56 \text{ A-h} = 7.25 \, A \times T$$

$$T = \frac{56}{7.25}$$

$$= 7.724 \text{ hours} \qquad \blacksquare$$

Cold Cranking Amps The manufacturers of lead–acid batteries have stopped using the ampere-hour capacity rating. Instead, they use a cold cranking amperes rating. This rating is specially designed for automobiles. The cold cranking amperes rating needed for a particular automobile is equal to the cubic-inch displacement of the auto's engine (cubic-inch displacement = 61 × the liter displacement) In other words, when purchasing a battery for your car, you should be concerned with the CCA rating of the battery. If the car in question has a 350-in.^{3} engine, the minimum requirement for capacity is 350 CCA. The cold cranking amperes rating can be converted to an approximate ampere-hour rating by dividing by 6.

EXAMPLE 18-4

Find the approximate ampere-hour rating of a 350-CCA battery?

Solution: Divide the cold cranking ampere rating by 6 to find the approximate ampere-hour rating.

$$\frac{350}{6} = 58.3 \text{ A-h}$$

Manufacturers will list this as a 59-A-h battery. $\qquad \blacksquare$

Increasing Battery Capacity The capacity of a battery can be raised by placing the batteries together in a parallel fashion (Figure 18-7). In this arrangement the current capacity of each battery is added to that of the other to provide the total current capacity. Understand that when you are connecting the batteries of cars together to use one to jump the other, the connections place the two batteries in parallel. This allows the capacity of one (the charged one) to be combined with that of the other (the discharged one).

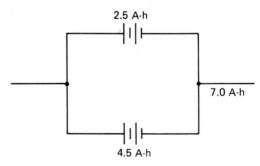

IN PARALLEL CAPACITY IS ADDITIVE FIGURE 18-7

BATTERY CHARGING

When a battery is charged, the battery itself becomes the load. This allows the current to flow through the battery in a direction opposite the norm. That is, a charging battery will have current flowing into its negative terminal and out of its positive terminal. This action is like that of a resistor (or any other component) in a normal circuit. There must be a second voltage source. This is the source that does the actual charging (provides the current) of the battery.

To limit the speed at which the battery is charged, a current-limiting resistance must be used (Figure 18-8). This resistor effectively keeps the battery from overheating. Charging too quickly may cause the battery to explode. How fast a battery can be recharged depends on the type and size of the battery. In general, check the specifications for the battery before charging. Usually, a current of less than 2 A is desired.

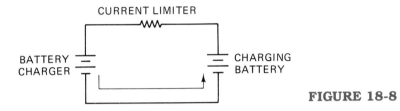

FIGURE 18-8

EXAMPLE 18-5

Suppose that a 100-V source is used to charge a 12-V battery. What resistance should the current limiter have to limit the charge rate to 1.5 A?

Solution: Begin by sketching the charging circuit (Figure 18-9). Use the known values of the charging circuit to find:

Potential difference:
$$V = 100 - 12$$
$$= 88 \text{ V}$$

Current limiter:
$$R = 88 \text{ V}/1.5 \text{ A}$$
$$= 58.666 \text{ }\Omega$$

Resistor power rating:
$$P = I^2 \times R$$
$$= (1.5)^2 \times 58.6$$
$$= 131.85 \text{ W}$$

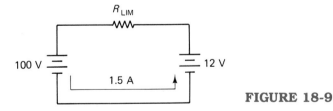

FIGURE 18-9

It is not unusual for the current limiter to need a high power rating. As a rule, this current limiter is constructed as a carbon pile (Figure 18-10). A carbon pile can dissipate hundreds of watts easily. The tension (amount of pressure) exerted on this pile controls the charge rate.

BATTERY CHARGING

FIGURE 18-10 Carbon Pile. Provides adjustable low resistance/high current and high power rating for battery load testing.

INTERNAL RESISTANCE

Several factors combine to limit the rate of current production of a battery. Among these are the purity of the electrolyte, the age of the battery, and the temperature of the battery. Rather than consider all of these factors separately, engineers consider them as a unit. They all have the effect of opposing the current flow in the circuit. So it follows that they can be thought of as a resistance. The internal resistance of a battery affects the way that the circuit operates. In the next few paragraphs we attempt to examine the effect that internal resistance has on the circuit as a whole.

Remember that the no-load voltage and the load voltage of a battery are different values. That was why a battery had to be placed under a load to be tested.

EXAMPLE 18-6

The circuit shown in Figure 18-11 is a typical normally operating starting system for an automobile. As a rule, the terminal voltage of a battery will drop from 13.2 V (no load) to 9.5 V (load) when cranking. It is not unusual for a starting system to require a current of 95 A. Calculate the internal resistance of this battery.

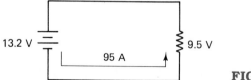

FIGURE 18-11

Solution: Begin by calculating the internal voltage drop. This is the difference between the no-load and the loaded voltage drop.

Internal voltage = 13.2 V − 9.5 V = 3.7 V

Use Ohm's law to find the internal resistance of the battery.

Internal resistance = 3.7 V/95 A = 0.03895 Ω ∎

Don't be confused by these calculations. Think of the circuit as a simple two-resistor series circuit. The voltage across the load (9.5 V) and the voltage across the internal resistance (3.7 V) must total the supply voltage (13.2 V). Once you know the internal voltage drop (3.7 V), you can use this and the current (same throughout a series circuit) to find the internal resistance (0.03895 Ω).

Impedance Matching A microphone, an amplifier, a power supply, and even an TV antenna can be thought of as a voltage source. Every voltage source has an internal resistance. Sometimes we want to achieve a good voltage transfer from one component to another, at other times we want to provide maximum current from one component to another, and at still other times we want to transfer maximum power from one component to another. All of these conditions require a different ratio of internal (source) resistance to load resistance.

When maximum voltage transfer is desirable, the load resistance must be quite large compared to the source resistance (Figure 18-12). When maximum current transfer is wanted, the opposite is true. We must have the source impedance large compared to the load resistance (Figure 18-13). When maximum power transfer is desired, it is necessary for the internal and load resistance to be equal (Figure 18-14).

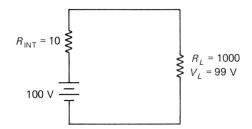

FIGURE 18-12 Maximum voltage transfer occurs when R_L is larger than R_{INT}.

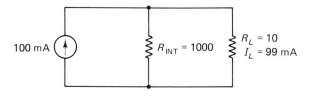

FIGURE 18-13 Maximum current is transferred when load resistance is small.

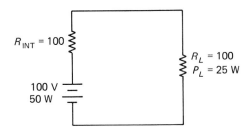

FIGURE 18-14 Maximum power transfer occurs when R_L is equal to R_{INT}.

SUMMARY

There are many, many different types of battery chemical design, but they all have approximately the same construction. They all need to have some material to use as an electrolyte (the electron-carrying medium). Potassium hydroxide is the most commonly used electrolyte. The negative terminal is called the anode because the circuit electrons enter there. The positive terminal is called the cathode because the electrons exit there. The anode and cathode are always two dissimilar metals.

Batteries are made up of groups of cells. Cells are classified by their ability to be recharged. Secondary cells are commonly called storage batteries because they can be recharged. Primary cells are called voltaic batteries and cannot be recharged.

Batteries are rated by their current-producing capacity. This rating is usually an ampere-hour rating, but a cold cranking ampere rating may also be used. Dividing the CCA rating by 6 will provide an approximate ampere-hour rating. The choice of chemicals used in the production of electricity determines the battery's capacity. This choice is offset by the expense, the size, the weight, and the life expectancy desired. Placing cells in parallel will increase the battery's capacity.

The rate at which a battery can be charged is limited by the manufacturer. In general, this rate is low (less than 2 A). A current-limiting resistor is used to control the charging rate.

Battery testing is done in one of two manners. One is by checking the specific gravity of the electrolyte. A hydrometer is used to make this test. A load test is performed when the electrolyte cannot be checked by a hydrometer reading (the electrolyte is a paste or in a sealed container). Here a load is placed on the battery by a resistor (a carbon pile may be used when high power dissipation is necessary) and the voltage reading taken.

The internal resistance of the battery limits its current output. This internal resistance is responsible for the battery's condition (charged or discharged), as well as current, voltage, and power transfer. In general, for good current transfer the internal resistance should be high compared to the load resistance. For good voltage transfer the internal resistance should be low compared to the load resistance. For good power transfer the internal resistance should be equal to the load resistance.

QUESTIONS

18-1. Briefly describe the construction of a battery.
18-2. Define the term *anode* as it pertains to batteries.
18-3. Define the term *cathode* as it pertains to batteries.
18-4. What is an electrolyte? Give an example.
18-5. What determines the amount of current and voltage that a battery can produce?
18-6. Define the term *battery*.
18-7. What is the difference between a voltage cell and a battery?
18-8. Explain what is meant by the term *secondary cell*?
18-9. Explain what is meant by the term *primary cell*?
18-10. List several types of secondary cells. Include their voltage rating and the material used as the anode, cathode, and electrolyte for each.
18-11. What is meant by *no-load voltage*?
18-12. Why are the no-load and load voltage different? Be specific.

18-13. What is a hydrometer? How is it used? Describe its construction and operation.

18-14. Explain what is meant by *specific gravity*. What material is used as a basis for specific gravity?

18-15. How does temperature affect specific gravity?

18-16. What is the normal specific gravity reading for a lead–acid battery?

18-17. What is the maximum allowable difference in specific gravity reading between cells of a lead–acid battery?

18-18. Describe how a load test is performed on a battery.

18-19. List several types of primary cells. Include the voltage rating and the material used as the anode, cathode, and electrolyte for each.

18-20. What is meant by the term *local action?*

18-21. How does local action affect battery performance?

18-22. What is meant by *polarization?*

18-23. Describe the manner in which a Leclanche cell polarizes.

18-24. Describe the manner in which a lead–acid battery polarizes.

18-25. Name a few applications for each of the battery types you have listed.

18-26. What characteristics would you want in a battery?

18-27. If you were given a choice of batteries, which would you choose? Why?

18-28. What is a solar cell? What is it used for?

18-29. What is considered the best way to test a battery's condition? Why?

18-30. Define *battery capacity*.

18-31. If you had a choice between a 36-A-h battery and a 52-A-h battery, which would you choose? Why?

18-32. Under the same loading conditions, which would provide current for the longest period of time?

18-33. What is a cold cranking ampere?

18-34. How can cold cranking amperes be converted to ampere-hours?

18-35. Describe the manner in which you would increase battery capacity.

18-36. What happens when batteries are connected in series?

18-37. What happens when batteries are connected in parallel?

18-38. List some of the precautions that need to be taken when charging a battery.

18-39. What do you think would happen if you tried to charge a primary cell? Why?

18-40. What do you think would happen if you tried to charge a secondary cell? Why?

18-41. What is meant by the term *trickle charge?* (Some research may be necessary.)

18-42. Why is a current limiter needed when charging a battery?

18-43. What is a carbon pile?

18-44. What is meant by *internal resistance* of a battery? How does this resistance affect battery performance?

18-45. Which is higher, no-load or load voltage? Why?

18-46. Which would be better, a battery with an internal resistance of 0.33 Ω or one with an internal resistance of 1.5 Ω? Why?

18-47. A perfect voltage source would have an infinite resistance. True or false?

18-48. What is meant by *impedance matching?*

QUESTIONS

18-49. If you wanted to transfer voltage, would you want high or low internal resistance? To transfer current?

18-50. When would you want the internal resistance and the load resistance to be equal? What advantage would this provide?

PRACTICE PROBLEMS

18-1. 350 CCA is equivalent to _____ A-h.

18-2. 550 CCA is equivalent to _____ A-h.

18-3. 42 A-h is equivalent to _____ CCA.

18-4. 36 A-h is equivalent to _____ CCA.

18-5. How much current draw should be placed on a 59-A-h battery?

18-6. How much current draw should be placed on a 36-A-h battery?

18-7. How much current draw should be placed on a 12-V 350-CCA battery?

18-8. How much current draw should be placed on a 12-V 550-CCA battery?

18-9. How much current draw should be placed on a 1.5-V 0.33-A-h battery?

18-10. How much current draw should be placed on a 1.25-V 1.1-A-h battery?

18-11. How much current draw should be placed on a 1.25-V 0.45-A-h battery?

18-12. How much current draw should be placed on a 4.8-V 0.25-A-h battery?

18-13. Calculate the minimum load voltage for the batteries in Problems 18-7 to 18-12.

18-14. Find the resistance value needed for each load test in Problems 18-7 to 18-12.

18-15. Calculate the power dissipation for each resistor used in Problems 18-8 to 18-12.

18-16. The no-load voltage of a particular lead–acid battery is found to be 13.6 V. Its cold cranking ampere rating is 440 CCA. What is its approximate ampere-hour rating? How much current draw is required to perform a load test? What is the acceptable minimum load voltage? What is the value of the resistor needed to perform the load test? What is the resistor's power rating?

18-17. How long can a 3.6-V 280 mA-h battery deliver 18 mA of current? 1 A of current? (Give answers in minutes.)

18-18. Refer to Table 18-3. Design an emergency system that will provide 3.75 V and 0.8 A for a minimum of 30 minutes. Include original expense and operating expense. (*Note:* There are at least three different design possibilities.)

TABLE 18-3

Battery	Volts	Ampere-Hours	Price
1	1.25	1.10	2/$6.29
2	1.25	0.45	2/$4.39
3	3.75	0.50	$9.95

18-19. Calculate the capacity battery needed to provide 0.65 A of current to a system for 6 hours.

18-20. Given that an emergency system consisting of three lights (each requiring 0.87 A), a radio (requiring 1.2 A), and a fan (requiring 2.5 A) must operate for 340 minutes. Find the ampere-hour capacity needed for the battery.

18-21. The no-load voltage for a 3-V battery is 3.1 V. This voltage drops to 2.9 V when a 250-mA load is attached. Find the internal resistance of the battery.

18-22. The no-load voltage for a 12-V battery is 13.1 V. This voltage drops to 9.2 V when a 250-A load is attached. Find the internal resistance of the battery.

18-23. The internal resistance of a 12-V battery is 0.022 Ω. Its no-load voltage is 13.2 V. What current load will pull its voltage down to 10 V?

18-24. The internal resistance of a 1.5-V battery is 0.35 Ω. Its no-load voltage is 1.7 V. What current load will pull its voltage down to 1.0 V?

18-25. The no-load voltage of a battery is 6.9 V. Its internal resistance is 0.24 Ω. What will be its terminal voltage when a 3-A load is attached? What is the resistance of the load?

18-26. The no-load voltage of a battery is 6.9 V. Its internal resistance is 0.42 Ω. What will be its terminal voltage when a 3-A load is attached? What is the resistance of the load?

18-27. The no-load voltage of a battery is 1.3 V. Its internal resistance is 0.34 Ω. What will be its terminal voltage when a 30-mA load is attached? What is the resistance of the load?

18-28. Find the needed current-limiting resistance when a 12-V battery is used to charge a 3-V battery. The current is to be limited to 100 mA.

18-29. A 100-Ω current-limiting resistance is used when a 12-V battery is used to charge a 3-V battery. To what is the current limited?

18-30. Which would provide better current transfer to a 150-Ω load, a current source that has an internal resistance of 1500, 150, or 15 Ω? Why? Refer to Figure 18-15.

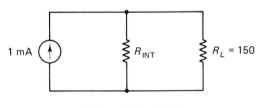

FIGURE 18-15

18-31. Which would provide better voltage transfer to a 150-Ω load, a voltage source that has an internal resistance of 1500, 150, or 15 Ω? Why? Refer to Figure 18-16.

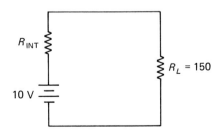

FIGURE 18-16

18-32. Which would provide better power transfer to a 150-Ω load, a power source that has an internal resistance of 1500, 150, or 15 Ω? Why? Refer to Figure 18-17.

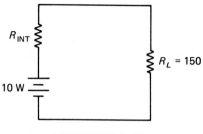

FIGURE 18-17

RCL Circuit Analysis

19

Chapter objectives

After reading this chapter and answering the questions and problems, you should be able to:

- State the interrelation between right triangles and rectangular and polar notation.
- State the effect of capacitive and inductive reactance when they occur in the same circuit.
- Given the source voltage, the inductive reactance, the capacitive reactance, and the resistance of either a series or a parallel circuit, find the total impedance, phase angle, and current and voltage for each component.
- Define *power factor* and state how it relates to power consumption and dissipation.
- Define *apparent power*, *real power*, and *reactive power*. Give units for each.
- Use rectangular and polar notation to describe total impedance on an impedance block.
- Use series–parallel combination rules to find the total impedance of a network of impedance blocks.
- Define *resonance* and state where and why it occurs. Give its effect on series and parallel circuits.

Most circuits, even semiconductor (transistorized and IC) circuits, can be reduced down to an equivalent circuit that contains only resistors, capacitors, inductors, and voltage or current sources. At high frequencies (microwave and satellite communications), even circuits that contain only resistors must be considered as if they were an *RCL* combination type of circuit. In this chapter we describe some of the methods used to analyze such circuits.

CAPACITIVE REACTANCE VERSUS INDUCTIVE REACTANCE

Compared to resistive voltages, the capacitive voltage is lagging behind by 90°. Inductive voltage, on the other hand, leads the resistive voltage by 90°. This means that the inductive voltage in a circuit leads the capacitive voltage by 180°. Understand that a 180° phase inversion has a canceling effect on these two voltages. Examine the waves shown in Figure 19-1.

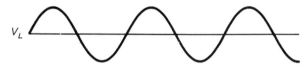

WAVES ARE 180° OUT OF PHASE

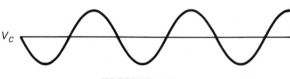

FIGURE 19-1

One way to analyze a circuit that contains both inductors and capacitors is just to subtract the capacitive reactance value and the inductive reactance value. It is important to remember the individual rules that apply to the circuit type being analyzed. That is, a series circuit has two appropriate triangles, a voltage triangle and an impedance triangle. With a series circuit it is appropriate to begin to analyze the circuit by subtracting the inductive and capacitive voltages (use the triangle that is correct for the largest of these two values) or by subtracting the inductive reactance and the capacitive reactance (again use the largest for the triangle).

A parallel circuit can be analyzed in a similar manner; however, a current triangle is the most appropriate triangle for this analysis. Here the inductive and capacitive currents are subtracted. The triangle direction is determined by the larger of these two quantities. Examples of the analysis of these circuit types follow.

EXAMPLE 19-1

Suppose that a resistance of 1500 Ω is connected in series with an inductive reactance of 2500 Ω and a capacitive reactance of 3750 Ω. The circuit is powered by a 250-mV signal (Figure 19-2). Find the total impedance, the current, and the voltages across each component.

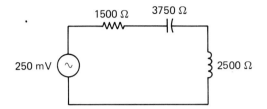

FIGURE 19-2

Solution: Begin by finding the net reactance. This is done by subtracting the inductive and capacitive reactance (Figure 19-3)

$$X_C - X_L = 3750 - 2500$$
$$= 1250 \; \Omega$$

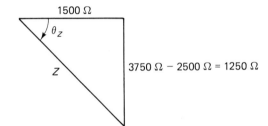

FIGURE 19-3

Then use the Pythagorean theorem to find the impedance of the circuit.

$$Z = \sqrt{1500^2 + 1250^2}$$
$$= 1953 \; \Omega$$

Use trigonometry to find the phase angle. Since there is more capacitive reactance than inductive reactance, the phase angle is lagging and negative.

$$\theta = \tan^{-1} \frac{1250}{1500}$$
$$= -39.8°$$

Ohm's law can be used to find the current and remaining voltages.

$$I = \frac{V}{Z}$$
$$= \frac{0.250}{1953}$$
$$= 0.1280 \; \text{mA}$$
$$V_R = I \times R$$
$$= 0.1280 \; \text{mA} \times 1500$$
$$= 192.1 \; \text{mV}$$

$$V_C = I \times X_C$$
$$= 0.1280 \text{ mA} \times 3750$$
$$= 480.1 \text{ mV}$$
$$V_L = I \times X_L$$
$$= 0.1280 \text{ mA} \times 2500$$
$$= 320.1 \text{ mV}$$

See Figure 19-4.

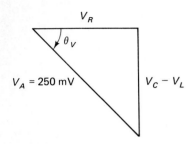

FIGURE 19-4

Note: The net reactive voltage is equal to the difference between the capacitive and inductive voltage. ∎

EXAMPLE 19-2

Suppose that a parallel circuit consists of a 200-Ω resistor, an inductive reactance of 40 Ω, and a 80-Ω capacitive reactance. The circuit is powered by a 10-V source (Figure 19-5). Find the individual current through the components, the total current, and the total impedance of the circuit.

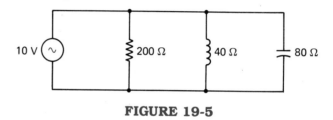

FIGURE 19-5

Solution: Begin by finding the individual currents through each component.

$$I_R = \frac{V}{R}$$
$$= \frac{10}{200}$$
$$= 50 \text{ mA}$$
$$I_L = \frac{V}{X_L}$$
$$= \frac{10}{40}$$
$$= 250 \text{ mA}$$
$$I_C = \frac{V}{X_L}$$
$$= \frac{10}{80}$$
$$= 125 \text{ mA}$$

Next find the net current through the reactive components. This is done by subtracting the capacitive and inductive currents (Figure 19-6).

$$I_L - I_C = 250 \text{ mA} - 125 \text{ mA}$$
$$= 125 \text{ mA}$$

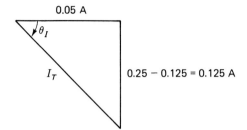

FIGURE 19-6

Use the Pythagorean theorem to find the total current,

$$I_T = \sqrt{50^2 + 125^2}$$
$$= 134.6 \text{ mA}$$

and trignometry to find the phase angle,

$$\theta = \tan^{-1} \frac{0.125}{0.05}$$
$$= -68.2°$$

Use Ohm's law to find the impedance of the circuit.

$$Z = \frac{V}{I}$$
$$= \frac{10}{134.6 \text{ mA}}$$
$$= 74.28 \text{ } \Omega \qquad \blacksquare$$

SELF-TEST

1. Given that a 400-Ω resistance, a 900-Ω inductive reactance, and a 200-Ω capacitive reactance are connected in series to a 20-V ac source, find the current through the circuit, the voltage across each component, and the phase angle of the circuit.
2. A series circuit containing a 2.5-kΩ inductive reactance, a 3.9-kΩ capacitive reactance, and a 720-Ω resistance are powered by a 15-V ac source. Calculate the circuit current, the voltage across each component, and the phase angle of the circuit.
3. Given that a 400-Ω resistance, a 900-Ω inductive reactance, a a 200-Ω capacitive reactance are connected in parallel to a 20-V ac source, find the current through each component, the total current supplied by the voltage source, and the phase angle of the circuit.
4. A parallel circuit containing a 2.5-kΩ inductive reactance, a 3.9-kΩ capacitive reactance, and a 720-Ω resistance are powered by a 15-V ac source. Calculate the current through each component, the total current supplied by the voltage source, and the phase angle of the circuit.

ANSWERS TO SELF-TEST

1. $I = 24.8$ mA, $V_R = 9.923$ V, $V_L = 22.33$ V, $V_C = 4.961$ V
2. $I = 9.528$ mA, $V_R = 6.86$ V, $V_L = 23.82$ V, $V_C = 37.16$ V
3. $I_R = 50$ mA, $I_L = 22.22$ mA, $I_C = 100$ mA, $I_{net} = 77.78$ mA, $I_T = 92.46$ mA, $\theta = 57.3°$
4. $I_R = 20.83$ mA, $I_L = 6$ mA, $I_C = 3.846$ mA, $I_{net} = 2.154$ mA, $I_T = 20.94$ mA, $\theta = -5.9°$

POWER FACTOR

In earlier chapters the power dissipated by the circuit (usable power—power to do work) was defined as I^2R. That definition does not change with *RCL* circuitry. However, the power delivered to an *RCL* circuit is not always equal to the usable power of the circuit. Note that it is always necessary to work in rms quantities. The process for calculating usable power and delivered power is outlined in the following example.

EXAMPLE 19-3

Examine the series circuit shown in Figure 19-7, which contains a 2500-Ω resistor and a 1500-Ω capacitive reactance. The circuit is powered by a 20-V rms source.

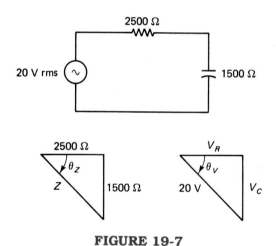

FIGURE 19-7

Solution: Begin by calculating the impedance and phase angle for the circuit.

$$Z = \sqrt{2500^2 + 1500^2}$$
$$= 2915 \, \Omega$$
$$\theta = \tan^{-1} \frac{1500}{2500}$$
$$= -31°$$

Next, find the voltage across the resistor and the capacitor.

$$V_R = 20 \times \cos(-31°)$$
$$= 17.15 \text{ V rms}$$
$$V_C = 20 \times \sin(-31°)$$
$$= 10.29 \text{ V rms}$$

Next, find the current flow of the circuit.

$$I = \frac{20}{2915}$$
$$= 6.86 \text{ mA} \qquad \blacksquare$$

Apparent Power Referring to Example 19.3, at first glance you would think that the circuit power would be $I \times V_A$. It is not.

$$6.86 \text{ mA} \times 20 \text{ V rms} = 137.2 \text{ mVA}$$

This value is called the *apparent power* of the circuit. Notice that the unit is the volt-ampere, not the milliwatt. In other words, if the current is multiplied by the total voltage of the circuit, the result is the apparent power of the circuit, which is measured in volt-amperes (VA). Apparent power can be thought of as the amount of power that is actually being provided by the generator. In terms of efficiency, this is the amount we are charged for (what we must pay for).

$$P_{\text{apparent}} = V_A \times I$$
$$= 20 \times 6.86 \text{ mA}$$
$$= 137.2 \text{ mVA}$$

Real Power Referring to Example 19.3, there are a number of ways in which the real power can be found. Multiplying the resistance voltage by the current is one way.

$$P_{\text{real}} = I \times V_R$$
$$= 6.86 \text{ mA} \times 17.15$$
$$= 117.6 \text{ mW}$$

Real power is measured in watts. Another way to calculate real power is by using the I^2R formula.

$$P_{\text{real}} = I^2R$$
$$= (6.86 \text{ mA})^2 \times 2500 \text{ }\Omega$$
$$= 117.6 \text{ mW}$$

Real power can also be found by multiplying the apparent power by the cosine of the phase angle.

POWER FACTOR

$$P_{real} = P_{apparent} \times \cos\theta$$
$$= 137.2 \times \cos(-31°)$$
$$= 117.6 \text{ mW}$$

The cosine of the phase angle is called the *power factor,* because it can be used in the manner described above to provide the real power. The real power can be though of as the amount of power actually used by the circuit. That is, the real power is the amount of power that is used to do work (run motors, light lights, etc.) in the circuit. Also, the power factor gives a statement of the circuit efficiency. Our circuit is 85.75% efficient because the cos of $-31°$ is 0.8575.

Reactive Power Referring to Example 19-3, the reactive power can be thought of as the amount of power that is wasted in the circuit's operation. The reactive power can be calculated in the same basic manner as the real power. Also, the reactive power is measured in volt-amperes reactive (VAR). For the circuit described previously, reactive power could be found by multiplying the reactive voltage by the circuit current. (*Note:* The process is the same whether the reactive component is a capacitor or an inductor.)

$$P_{reactive} = I \times V_C$$
$$= 6.86 \text{ mA} \times 10.29$$
$$= 70.58 \text{ mVAR}$$

Reactive power can be calculated by multiplying the square of the current times the capacitive reactance.

$$P_{reactive} = I^2 \times X_C$$
$$= (6.86 \text{ mA})^2 \times 1500$$
$$= 70.58 \text{ mVAR}$$

Or like real power, the phase angle can be used to find the reactive power.

$$P_{reactive} = P_{apparent} \times \sin\theta$$
$$= 137.2 \text{ mVA} \times \sin(-31°)$$
$$= 70.58 \text{ mVAR}$$

You will notice that the sign of the phase angle is not used when calculating the power levels. In an ac circuit the sign is an indication of the direction of the current with respect to the reference. In other words, negative power does not indicate a loss and positive power does not indicate a power gain. This is a common misconception and should be avoided. The presence of a capacitor or an inductor in any ac circuit will draw extra power from the source.

APPLIED VOLTAGE AS A REFERENCE

From a calculation standpoint it is easier to use current as a reference on a series circuit and voltage as a reference on a parallel circuit. However, when actually testing or working on a circuit, it is best to use the applied voltage as a reference. It is easiest for the technician to connect the applied voltage to the external trigger of an oscilloscope and read the resulting phase angle. When used to trigger the oscilloscope, the applied voltage will cause the circuit phase to appear to shift. All of the phase relations are shifted an amount equal to what has until now been referred to as the phase angle. This is explained best by example.

EXAMPLE 19-4

Refer to Figure 19-8. Find the total impedance, the current, and voltages across each component. Also examine the phase of each of these as referenced from the applied voltage.

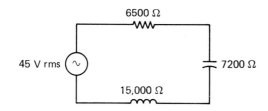

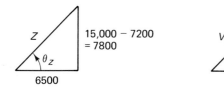

FIGURE 19-8

Solution: Begin the analysis in the normal manner by finding the net reactance.

$$X_L - X_C = 15{,}000 - 7200$$
$$= 7800 \ \Omega$$

Next, find the circuit impedance and phase.

$$Z = \sqrt{6500^2 + 7800^2}$$
$$= 10.15 \ k\Omega$$

$$\theta = \tan^{-1} \frac{7800}{6500}$$
$$= 50.19°$$

(*Note:* This is the phase provided that the current is used as a reference.) Then find the circuit current.

$$I = \frac{V_A}{Z}$$
$$= \frac{45}{10.15 \ k\Omega}$$
$$= 4.432 \ mA$$

APPLIED VOLTAGE AS A REFERENCE

Use Ohm's law to find the voltage across each component.

$$V_R = V_A \times \cos \theta$$
$$= 45 \times \cos 50.19°$$
$$= 28.81 \text{ V}$$
$$V_C = I \times X_C$$
$$= 4.432 \text{ mA} \times 7200$$
$$= 31.91 \text{ V}$$
$$V_L = I \times X_L$$
$$= 4.432 \text{ mA} \times 15,000$$
$$= 66.48 \text{ V}$$

∎

Examine Figures 19-9 and 19-10. It is necessary that you understand that these two phasor diagrams are equivalent. Figure 19-9 uses current as the reference for each of the resulting phasors. Figure 19-10 uses the applied voltage as the reference.

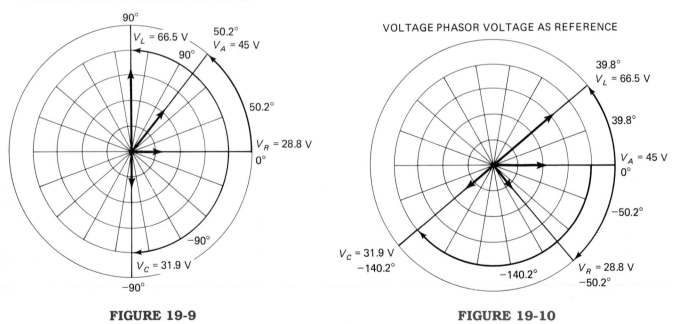

FIGURE 19-9 **FIGURE 19-10**

The onset of using applied voltage as a reference can cause considerable confusion. It is necessary that you come to understand this procedure. It is the standard for working with *RCL* circuits. Polar notation and rectangular notation are two methods of making this transition easier.

RECTANGULAR NOTATION

At first glance rectangular notation seems complicated. After you understand this method of writing impedances and how it relates to the triangle, rectangular notation is not quite as formidable. Rectangular notation uses a j in front of any reactance. The j is called a j-factor and it represents an imaginary number, the square root of -1. Don't let this bother you. We simply use the j as a method of keeping the resistance value and the reactance values separate. An impedance written in rectangular notation would look like this:

$$1500 + j2500$$

A positive j indicates a positive 90°, while a negative j refers to a negative 90°. The impedance, $1500 + j2500$, describes a circuit containing 1500 Ω of resistance and 2500 Ω of inductive reactance. If we relate this notation to the triangle, the real term (the number without the j, in this case the 1500) represents the horizontal side of the triangle. The imaginary component (the j-factor, the 2500 here) corresponds to the vertical axis. The sign ($+$ or $-$) in the center indicates whether the triangle points upward (positive) or downward (negative). Circuits containing several resistors, inductors, and capacitors can be analyzed quickly using rectangular notation.

EXAMPLE 19-5

Use rectangular notation to find the impedance of the circuit shown in Figure 19-11.

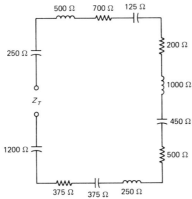

FIGURE 19-11

Solution: Using methods described in earlier chapters would have us combine the resistances, then the inductive reactances, and then the capacitive reactances. The capacitive and inductive reactances would be subtracted to find the net reactance. Then the triangle would be used to solve for total impedance. Rectangular notation accomplishes the same objectives: however, it allows us to use algebra to solve the circuit faster. Begin by examining the circuit (go in one direction from the positive terminal of the source to the negative source). Place a $(+j)$ in front of all inductive reactances and a $(-j)$ in front of all capacitive reactances. Resistances have no j operators. Then combine like terms.

$$Z_{rect} = -j250 + j500 + 700 - j125 + 200 + j1000 - j450 + 500 + j250 - j375 + 375 - j1200$$

$$= 1775 + j1750 - j2400$$

$$= 1775 - j650 \qquad \blacksquare$$

The circuit in Example 19.5 was analyzed by moving around the circuit in a clockwise direction (the direction is a matter of choice) and writing down the particular value of the component. Resistance values were not given a prefix. Inductive reactances were given a prefix of $+j$. Capacitive reactance were given a prefix of $-j$. In this manner, different reactances types can be distinguished from each other and from resistances.

POLAR NOTATION

Like rectangular notation, polar notation seems complicated at first. Again learning to relate this notation to the triangle will make it meaningful and less complicated. Polar notation looks like:

$$1450 \, \underline{/-68.23°} \, \Omega$$

This would represent 1450 Ω of impedance with a phase angle of $-68.23°$. The first quantity in polar notation is related to the hypotenuse of the triangle associated with the circuit. The second quantity is preceded by an angle symbol and it corresponds to the phase of the circuit.

Performing circuit calculations using rectangular and polar notation boil down to nothing other than converting from polar to rectangular notation or from rectangular to polar notation. A large number of modern calculators have this function as a built-in command. In this chapter we discuss when and where each of these notations is most appropriate; however, each calculator manufacturer uses a different method for activating these functions, so the operator's manual should be consulted for further information. It is suggested that you learn how to perform these functions yourself before you learn shortcuts on your calculator. The following examples demonstrate these conversions.

EXAMPLE 19-6

Suppose that the rectangular impedance for a circuit is $1200 + j750$. Find the equivalent polar impedance.

Solution: Understand that $1200 + j750$ represent the adjacent and opposite sides, respectively, of a triangle that points upward (the j-factor is positive; Figure 19-12). The polar representation of this impedance utilizes the hypotenuse and the angle. Use the Pythagorean theorem to find the hypotenuse and the phase angle.

$$Z = \sqrt{1200^2 + 750^2}$$
$$= 1415 \, \Omega$$

$$\theta = \tan^{-1} \frac{750}{1200}$$
$$= 32°$$

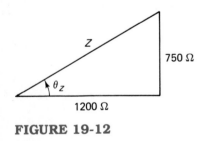

FIGURE 19-12

The rectangular impedance is

$$Z_{\text{rect}} = 1200 + j750 \text{ }\Omega$$

The polar impedance is

$$Z_{\text{polar}} = 1415 \text{ }\underline{/32°} \text{ }\Omega$$ ∎

It is necessary also to change from the polar form of notation to the rectangular form of notation. Here, again, learning to relate this process to the triangle will simplify this transaction. The following example will demonstrate conversion from polar to rectangular notation.

EXAMPLE 19-7

Suppose that the polar impedance of a circuit is 2575 $\underline{/-54°}$ Ω. Find the rectangular impedance.

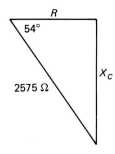

FIGURE 19-13

Solution: Understand that 2575 $\underline{/-54°}$ Ω represents the hypotenuse and the angle, respectively. The rectangular representation of this impedance will consist of the adjacent and opposite sides of a downward-pointing triangle (the angle is negative).

Begin by sketching the triangle as shown in Figure 19-13. The find the adjacent (resistance) and opposite (reactance) sides.

$R = 2575 \times \cos(-54°)$

$ = 1514 \text{ }\Omega$

$X_C = 2575 \times \sin(-54°)$

$ = -2083 \text{ }\Omega$ (the $-$ means that this is a capacitive reactance)

The polar impedance is

$$Z_{\text{polar}} = 2575 \text{ }\underline{/-54°} \text{ }\Omega$$

The rectangular impedance is

$$Z_{\text{rect}} = 1514 - j2083 \text{ }\Omega$$ ∎

IMPEDANCE BLOCKS

Large and complicated circuits can be broken down into smaller sections. These sections are called *impedance blocks*. Any combination of resistors, inductors, and capacitors can be viewed as a block of impedance. It is important to remember that impedances must be in rectangular form in order to perform addition and subtraction. Multiplication and division require that the impedances be in polar form. The following examples demonstrate this process.

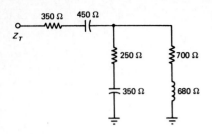

FIGURE 19-14

EXAMPLE 19-8

Consider the circuit shown in Figure 19-14. Analyze the circuit and break it into convenient impedance blocks. Write the impedance of each block.

Solution: This circuit can be viewed as three impedance blocks. One block would be the 250 Ω resistor and the 350 Ω capacitive reactance. Another would be the 700 Ω resistor and the 680 Ω inductive reactance. The third impedance block would be the 350 Ω resistor and the 450 Ω capacitive reactance. These impedances would be written as

$$Z_1 = 250 - j350 \text{ } \Omega = 430.1 \text{ /} {-54.5°} \text{ } \Omega$$
$$Z_2 = 700 + j680 \text{ } \Omega = 975.9 \text{ /} {44.2°} \text{ } \Omega$$
$$Z_3 = 350 - j450 \text{ } \Omega = 570.1 \text{ /} {-52.1°} \text{ } \Omega$$

The circuit of Figure 19-14 can be further analyzed by examining the positioning of these impedance blocks as shown in Figure 19-15. Also, these impedance blocks can be treated as we would treat individual resistors to find the total impedance of the circuit of Figure 19-14. The total impedance of the circuit can be found by first finding the parallel impedance of Z_1 and Z_2. This impedance ($Z_1 \parallel Z_2$) can be considered in series with Z_3. Several ideas must be remembered. One of these is that the impedances must be in rectangular form for addition and subtraction. Also the impedances must be in polar form for multiplication and division. Typically, the Z_{block} values can be treated as you would treat a resistance in a dc circuit analysis. All of the resistance formulas can be used to calculate these impedances. This process is illustrated in Example 19.9.

EXAMPLE 19-9

Find the total impedance of the impedance blocks of Figure 19-15.

Solution: Begin by finding the parallel impedance of Z_1 and Z_2. Remember that impedance should be in polar form to multiply and in rectangular form to add.

$$Z_1 \parallel Z_2 = \frac{Z_1 \times Z_2}{Z_1 + Z_2}$$

$$Z_1 \times Z_2 = (430.1 \text{ /} {-54.5°}) \times (975.9 \text{ /} {44.2°})$$
$$= 430.1 \times 975.9 \text{ /} {-54.5° + 44.2°}$$
$$= 419{,}800 \text{ /} {-10.3} \text{ } \Omega^2$$

Notice that the hypotenuses were multiplied and the angles were added.

$$Z_1 + Z_2 = (250 - j350) + (700 + j680)$$
$$= 950 + j330$$
$$= 1006 \text{ /} {19.2°} \text{ } \Omega$$

$$Z_1 \parallel Z_2 = \frac{419{,}800 \text{ /} {-10.3°}}{1006 \text{ /} {19.2°}}$$
$$= 417.4 \text{ /} {-29.5°} \text{ } \Omega$$

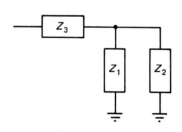

FIGURE 19-15

Notice that the hypotenuses were divided and the angles were subtracted.
This is the impedance of the two parallel blocks. This impedance must be placed in series with Z_3. Impedance values are additive in series. Complex numbers must be in rectangular form to add, so the next step is to convert the 417.4 $\underline{/-29.45°}$ Ω into its rectangular counterpart.

$$Z_1 \| Z_2 = 417.4 \underline{/-29.5°}$$
$$= 363.5 - j205.2 \text{ Ω}$$

Then

$$Z_T = (Z_1 \| Z_2) + Z_3$$
$$= (363.5 - j205.2) + (350 - j450)$$
$$= 713.5 - j655.2$$
$$= 968.7 \underline{/-42.6°} \text{ Ω} \qquad \blacksquare$$

Another application of this procedure follows.

EXAMPLE 19-10

Suppose that we needed to find the voltage across the impedance block made up of the 650-Ω resistor and the 430-Ω inductance of Figure 19-16.

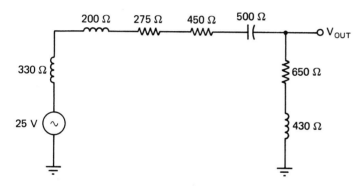

FIGURE 19-16

Solution: Begin analysis by dividing the circuit into two impedance blocks. One block should consist of the 650-Ω resistor and the 430-Ω inductor ($Z_1 = 650 + j430$). The other should consist of all of the other impedances in the circuit ($Z_2 = -j500 + 450 + 275 + j200 + j330$; see Figure 19-17).

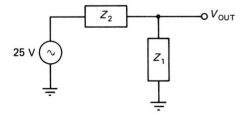

FIGURE 19-17

The standard voltage divider formula can then be used to find the voltage $[V = Z_1/(Z_1 + Z_2) \times V_T]$. The arithmetic is outlined in the following section.

$$Z_1 = 650 + j430$$
$$= 779.4 \underline{/33.5°} \text{ Ω}$$

IMPEDANCE BLOCKS

$$Z_2 = -j500 + 450 + 275 + j200 + j330$$
$$= 725 + j30$$
$$Z_1 + Z_2 = (650 + j430) + (725 + j30)$$
$$= 1375 + j460$$
$$= 1450 \underline{/18.5°} \, \Omega$$
$$V = \frac{Z_1}{Z_1 + Z_2} \times V_T$$
$$= \frac{779.4 \underline{/33.5°}}{1450 \underline{/18.5°}} \times 25$$
$$= (0.5375 \underline{/15°}) \times 25$$
$$= 13.44 \underline{/15°} \, V$$

The voltage across the impedance block is 13.44 V. This voltage leads the applied voltage by 15°. ∎

The impedance of an impedance block can be substituted into any resistance formula and not affect the validity of the equation. This is true for transistor circuits, operational amplifier circuits, Millman's theorem, delta and wye conversions, and so on.

RESONANCE

A special case of *RCL* circuitry occurs any time that X_C and X_L are equal to each other. The equal and opposite effects of these two quantities cause the circuit current to reduce (parallel) or the circuit (V_C and V_L) voltage to peak (series). When resonance occurs in a series circuit containing both an inductor and a capacitor, the two components provide the effect of magnifying the circuit voltage. This mode of operation is called *series resonance* (see Figure 19-18) and is used in communication circuitry (radios and television) for tuning purposes (selecting the proper station).

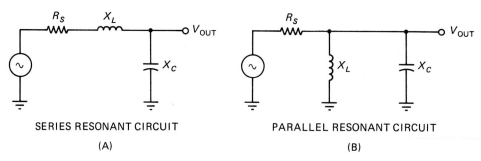

SERIES RESONANT CIRCUIT
(A)

PARALLEL RESONANT CIRCUIT
(B)

FIGURE 19-18

Parallel resonance magnifies the impedance of the circuit. When parallel resonance is reached, the impedance of both parallel branches (the capacitor and the inductor) is at an equal but high value (see Figure 19-18), which reduces the total current of the circuit. This mode of resonance is also used for tuning and selecting purposes in communication circuits. The formula for calculating the resonant frequency is

$$f_{res} = \frac{1}{2 \times \pi \times \sqrt{L \times C}}$$

EXAMPLE 19-11

Suppose that a 25-pF capacitor and a 50-μH inductor were connected together in the same circuit. At what frequency would the circuit resonate? Also, what would be the reactance values at this frequency?

Solution: Begin by using the resonant frequency formula to find f_{res}.

$$f_{res} = \frac{1}{2 \times \pi \times \sqrt{L \times C}}$$

$$= \frac{1}{2 \times \pi \times \sqrt{25 \text{ pF} \times 50 \text{ } \mu\text{H}}}$$

$$= \frac{1}{2 \times \pi \times \sqrt{1.25 \times 10^{-15}}}$$

$$= \frac{1}{2 \times \pi \times 3.536 \times 10^{-8}}$$

$$= 4.502 \text{ MHz}$$

Then use the reasonance frequency to find the capacitive and inductive reactance.

$$X_C = \frac{1}{2 \times \pi \times f \times C}$$

$$= \frac{1}{2 \times \pi \times 4.502 \text{ MHz} \times 25 \text{ pF}}$$

$$= 1414 \text{ }\Omega$$

$$X_L = 2 \times \pi \times f \times L$$

$$= 2 \times \pi \times 4.502 \text{ MHz} \times 50 \text{ }\mu\text{H}$$

$$= 1414 \text{ }\Omega \qquad \blacksquare$$

Notice that the reactance values are equal. This is a phenomenon that is always associated with resonance ($X_C = X_L$). Also, the resonant frequency is the same for a series or a parallel combination of these two components. More details regarding resonance will be discussed in Chapter 21.

SUMMARY

Capacitive reactance and inductive reactance effectively cancel each other. It is the net reactance (the difference between these two) that is used as the reactive component in circuit analysis. The j-operator is positive when working with inductive reactance values and negative when working with capacitive reactance values. In rectangular form the j-operators will cancel capacitive and inductive quantities.

The cosine of the phase angle is referred to as the power factor. The power factor is an indication of the usable amount of power dissipated by the circuit and is measured in watts. The reactive power is measured in volt-amperes-reactance (VAR) and represents the unusable power dissipated by the circuit. The volt-ampere is the unit of delivered power to the circuit and is calculated by $V_A \times I_A$. All power calculations should be done using rms values.

When using the oscilloscope, the applied voltage is used as a reference. This is done by connecting the trigger input to the applied voltage. The applied voltage when used as a reference has a phase angle of zero. All subsequent calculations have both a magnitude and a phase angle and are done in polar form.

Resonance is a special *RCL* circuit combination. It occurs when X_C and X_L are equal. Series resonance will cause the circuit impedance to minimize and results in a high current and voltage drop across the reactive components. In short, the input voltage is magnified at series resonance. Parallel resonance, on the other hand, results in a high impedance and reduces the total current flow of the circuit.

QUESTIONS

19-1. In your own words, define *rectangular notation*.

19-2. Relate rectangular notation to a right triangle.

19-3. In your own words, define *polar notation*.

19-4. Relate polar notation to a right triangle.

19-5. What is the phase difference between capacitive and inductive reactance? How do they affect each other?

19-6. If you have 80 Ω of inductive reactance and 70 Ω of capacitive reactance, what is the value of the equivalent reactance? Is it inductive or capacitive? Why?

19-7. What is the power factor? How does it affect the circuit?

19-8. Use your own words to define *apparent power*, *real power*, and *reactive power*. Include symbols and units. Which is most important? Why?

19-9. Mathematically, what needs to be done to use applied voltage as a reference?

19-10. Technically, what needs to be done to use applied voltage as a reference?

19-11. Why do we use the *j*-operator on reactances and not on resistances?

19-12. When are $+j$ and $-j$ operators used? Why?

19-13. In the polar form, what does a positive angle indicate? A negative angle?

19-14. In your own words, describe the manner used to convert from the polar form of a number to the rectangular form.

19-15. In your own words, describe the manner used to convert from the rectangular form of a number to the polar form.

19-16. What is an impedance block?

19-17. How can complex arithmetic be used to analyze an impedance block?

19-18. Compare $V = R/R_T \times V_T$ with $V = Z/Z_T \times V_T$. Are they equivalent? Why?

19-19. What is resonance? When does resonance occur? What is meant by *resonant frequency*?

19-20. What does parallel resonance magnify? Series resonance?

PRACTICE PROBLEMS

19-1. A series circuit is powered by a 15-V ac supply. The circuit contains a 2.7-kΩ resistor and a 2-kΩ inductive reactance. Find the apparent, real, and reactant powers.

19-2. A circuit consists of a resistance of 200 Ω, a capacitive reactance of 300 Ω and an inductive reactance of 400 Ω that are connected in parallel. The circuit is powered by a 200-mV ac source. Find the apparent and real power level for this circuit.

Problems 19-3 and 19-4 refer to the circuit shown in Figure 19-19.

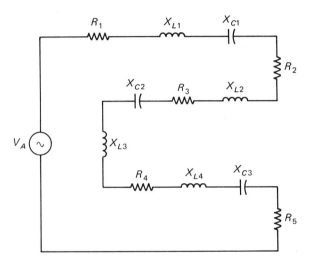

FIGURE 19-19

19-3. Given: $R_1 = 470\ \Omega$ Find: $Z_T =$ _____
$R_2 = 680\ \Omega$ $V_{R4} =$ _____
$R_3 = 270\ \Omega$ $V_{XL3} =$ _____
$R_4 = 820\ \Omega$ $V_{XC2} =$ _____
$R_5 = 220\ \Omega$ $I =$ _____
$X_{L1} = 700\ \Omega$
$X_{L2} = 340\ \Omega$
$X_{L3} = 1200\ \Omega$
$X_{L4} = 290\ \Omega$
$X_{C1} = 500\ \Omega$
$X_{C2} = 120\ \Omega$
$X_{C3} = 490\ \Omega$
$V_A = 25$ V

19-4. Given: $R_1 = 560\ \Omega$ Find: $Z_T =$ _____
$R_2 = 330\ \Omega$ $V_{R2} =$ _____
$R_3 = 820\ \Omega$ $V_{XL4} =$ _____
$R_4 = 120\ \Omega$ $V_{XC1} =$ _____
$R_5 = 200\ \Omega$ $I =$ _____
$X_{L1} = 800\ \Omega$
$X_{L2} = 540\ \Omega$
$X_{L3} = 1500\ \Omega$

$X_{L4} = 320\ \Omega$
$X_{C1} = 500\ \Omega$
$X_{C2} = 720\ \Omega$
$X_{C3} = 490\ \Omega$
$V_A = 15\ V$

19-5. Given that a 600-Ω resistance, a 800-Ω inductive reactance, and a 200-Ω capacitive reactance are connected in series to a 150-V ac source, find the current through the circuit, the voltage across each component, and the phase angle of the circuit.

19-6. A series circuit containing a 3.3-kΩ inductive reactance, a 2.5-kΩ capacitive reactance, and a 680-Ω resistance are powered by a 20-V ac source. Calculate the circuit current, the voltage across each component, and the phase angle of the circuit.

19-7. Given that a 2700-Ω resistance, a 1200-Ω inductive reactance, and a 2000-Ω capacitive reactance are connected in parallel to a 20-V ac source, find the current through each component, the total current supplied by the voltage source, and the phase angle of the circuit.

19-8. A parallel circuit containing a 2.5-kΩ inductive reactance, a 3.9-kΩ capacitive reactance, and a 2700-Ω resistance are powered by a 15-V ac source. Calculate the current through each component, the total current supplied by the voltage source, and the phase angle of the circuit.

19-9. Refer to Figure 19-20. Use the times plus parallel resistance formula to find the resistance of $Z_1 = R \parallel X_L$. Then repeat the process to find $Z_T = Z_1 \parallel X_C$. Given that $R = 2.2$ kΩ, $X_L = 3.3$ kΩ, and $X_C = 1.2$ kΩ.

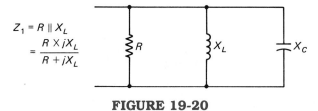

FIGURE 19-20

Problems 19-10 and 19-11 refer to Figure 19-21.

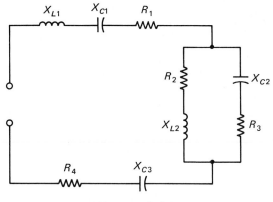

FIGURE 19-21

19-10. Given that $R_1 = 330\ \Omega$, $R_2 = 270\ \Omega$, $R_3 = 680\ \Omega$, $R_4 = 820\ \Omega$, $X_{L1} = 650\ \Omega$, $X_{L2} = 1250\ \Omega$, $X_{C1} = 400\ \Omega$, $X_{C2} = 530\ \Omega$, and $X_{C3} = 290\ \Omega$, use impedance blocks to find Z_T.

19-11. Given that $R_1 = 560\ \Omega$, $R_2 = 330\ \Omega$, $R_3 = 1.5\ k\Omega$, $R_4 = 470\ \Omega$, $X_{L1} = 1650\ \Omega$, $X_{L2} = 1750\ \Omega$, $X_{C1} = 1400\ \Omega$, $X_{C2} = 830\ \Omega$, and $X_{C3} = 590\ \Omega$, use impedance blocks to find Z_T.

19-12. Refer to Figure 19-22. Find f_{res}.

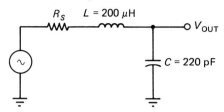

FIGURE 19-22

19-13. Refer to Figure 19-23. Find f_{res}.

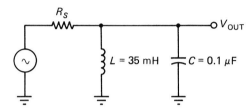

FIGURE 19-23

19-14. Refer to Figure 19-24. Find V_{out}.

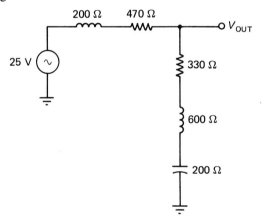

FIGURE 19-24

20
Filters

Chapter objectives

After reading this chapter and answering the questions and problems, you should be able to:

- Given any two of input, output, or decibel level, find the missing quantity.
- Define *decade* and *octave* in terms of frequency.
- List and describe the five basic types of filter response.
- Predict which type of filter response a circuit will exhibit by examining the placement of inductors and capacitors.
- List and describe the four filter families grouped by response rolloff.
- Describe the construction of simple, L-network, T-network and Π-network filters. Give advantages and disadvantages of each.
- Describe how ac and dc voltages can be separated using transformers and coupling and bypass capacitors.
- Given cutoff frequency, source resistance, and load resistance. Find the proper value capacitor for coupling and bypass circuits.
- Describe the concept of constant k and m-derived filter construction.

As the frequency of a signal changes, the voltage and power output of that circuit will also change. This is true for all circuits. Some circuits are specifically designed to exhibit this property. They are called *filters*. Filters are used to separate dc and ac components of a signal. They are also used to allow only certain frequencies to be transmitted. (Some cable TV systems use filters to control which pay stations are received). There are two major categories of filters, active and passive. Active filters require the use of an amplifier and although they are beyond the scope of this book, the concepts discussed in this chapter do apply. Passive filters are combinations of resistors, capacitors, and inductors. They will be discussed in depth. Topics in this chapter are the different types of filter responses, the major filter families, and methods used to separate dc and ac signals. It is important to understand that the terms and principles of active and passive filter are identical.

DECIBELS, POWER, AND VOLTAGE

Decibels are the unit used to measure power, voltage, and current changes. The half-power point (the frequency that cuts power from maximum to one-half maximum) and the 70.7% voltage point (the frequency that reduces the voltage level from maximum to 70.7% of maximum) are of particular importance to filter design. This frequency point is referred to as the 3-dB (decibel) down point. That is, the half-power point and the 70.7% voltage point both equate to 3 dB.

Decibels provide a quick method for measuring losses (power, voltage, and current) because the decibel level is the same for all parameters (power, voltage, and current). The formulas for calculating the decibel levels are

$$dB_{level} = 20 \times \log \frac{V_{out}}{V_{in}} \quad \text{for voltage}$$

$$dB_{level} = 20 \times \log \frac{I_{out}}{I_{in}} \quad \text{for current}$$

$$dB_{level} = 10 \times \log \frac{P_{out}}{P_{in}} \quad \text{for power}$$

For our purposes the voltage formula will be most applicable.

Any time the output signal is greater than the input signal, we say that the signal has gain and the dB_{level} will be positive. When there is a loss in voltage or power from the input to the output, we say that the signal is attenuated and the dB_{level} will be negative. Passive filters will generally have an attenuated output.

EXAMPLE 20-1

Given that the input voltage of a particular filter is 100 mV at a frequency of 200 kHz and the output voltage is 60 mV, calculate the decibel attenuation of this circuit.

Solution: Use the decibel formula to find the decibel level.

$$dB_{level} = 20 \times \log \frac{0.06}{0.1}$$

$$= 20 \times \log 0.6$$

$$= 20 \times -0.2218$$

$$= -4.4369 \text{ dB}$$

Note. The negative value indicates that there has been a loss of voltage through the circuit.

[6] [0] [EE] [3] [+/−] [÷]

[1] [0] [0] [EE] [3] [+/−] [=]

[log] [×] [2] [0] [=]   display   −4.4368   ∎

SELF-TEST

1. V_{in} = 200 mV   attenuation or gain?
   V_{out} = 150 mV
   dB_{level} = _____

2. P_{in} = 400 mW   attenuation or gain?
   P_{out} = 150 mW
   dB_{level} = _____

3. I_{in} = 50 μA   attenuation or gain?
   I_{out} = 10 μA
   dB_{level} = _____

4. P_{in} = 0.045 W   attenuation or gain?
   P_{out} = 0.1 W
   dB_{level} = _____

5. P_{in} = 400 mW   attenuation or gain?
   P_{out} = 650 mW
   dB_{level} = _____

6. I_{in} = 50 mA   attenuation or gain?
   I_{out} = 35 mA
   dB_{level} = _____

7. V_{in} = 0.2 V   attenuation or gain?
   V_{out} = 1 V
   dB_{level} = _____

8. V_{in} = 200 mV   attenuation or gain?
   V_{out} = 450 mV
   dB_{level} = _____

9. V_{in} = 10 V   attenuation or gain?
   V_{out} = 7 V
   dB_{level} = _____

10. V_{in} = 2.5 V attenuation or gain?
    V_{out} = 1.25 V
    dB_{level} = _____

ANSWERS TO SELF-TEST

1. Attenuation $dB_{level} = -2.499$ dB
2. Attenuation $dB_{level} = -4.26$ dB
3. Attenuation $dB_{level} = -13.98$ dB
4. Gain $dB_{level} = 3.468$ dB
5. Gain $dB_{level} = 2.109$ dB
6. Attenuation $dB_{level} = -3.098$ dB
7. Gain $dB_{level} = 13.98$ dB
8. Gain $dB_{level} = 7.044$ dB
9. Attenuation $dB_{level} = -3.098$ dB
10. Attenuation $dB_{level} = -6.021$ dB

When the dB output level is plotted on semilog graph paper (see Figure 20-1) as a function of frequency, a Bode (pronounced Bo-dé) plot is generated. The Bode plot is used to demonstrate the filter's (or other circuit's) frequency response. The frequency is plotted on the horizontal axis and the voltage level (in decibels) is plotted on the vertical axis. Notice that the lines indicating the frequency are not linear; instead, they are exponential. This allows for a shortened graph.

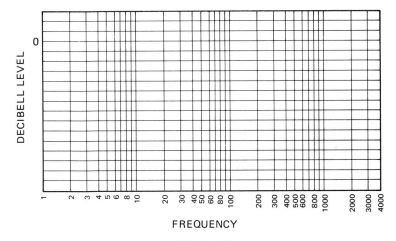

FIGURE 20-1

When frequency response is discussed, rather than state frequency changes in units, it is stated in terms of octaves or decades. An *octave* is any doubling (or halving) of the reference frequency. A *decade* is an increase (or decrease) of the reference frequency by a factor of 10. The following examples will demonstrate octaves and decades.

EXAMPLE 20-2

Suppose that the reference frequency is 2.2 kHz. Find the frequencies that are one octave higher and one octave lower than this reference frequency.

Solution: The term *octave* means that the reference frequency is doubled. So one octave higher than 2.2 kHz is

$$2200 \times 2 = 4400 \text{ Hz}$$

and one octave lower is

$$\frac{2200}{2} = 1100 \text{ Hz}$$

EXAMPLE 20-3

Suppose that the reference frequency is 5 kHz. Find the frequency that is three octaves higher than this reference frequency.

Solution: Three octaves higher would mean that the reference frequency is doubled three times.

$$5000 \times 2 = 10,000 \text{ H} \quad \text{one octave}$$
$$10,000 \times 2 = 20,000 \text{ Hz} \quad \text{two octaves}$$
$$20,000 \times 2 = 40,000 \text{ Hz} \quad \text{three octaves}$$

EXAMPLE 20-4

Given that the reference frequency is 20 kHz, find the frequencies that correspond with the first three decades above and below this reference frequency.

Solution: The term *decade* corresponds to a change by a factor of 10.

Above:

$$20 \text{ kHz} \times 10 = 200 \text{ kHz} \quad \text{one decade}$$
$$200 \text{ kHz} \times 10 = 2 \text{ MHz} \quad \text{two decades}$$
$$2 \text{ MHz} \times 10 = 20 \text{ MHz} \quad \text{three decades}$$

Below:

$$\frac{20 \text{ kHz}}{10} = 2 \text{ kHz} \quad \text{one decade}$$
$$\frac{2 \text{ kHz}}{10} = 200 \text{ Hz} \quad \text{two decades}$$
$$\frac{200 \text{ Hz}}{10} = 20 \text{ Hz} \quad \text{three decades}$$

Normal frequency response for the rolloff (the rate at which attenuation occurs) of a single component filter is 20 dB/decade (read Decibels per decade) or 6 dB/octave. Frequency response which pertains to Bode plots and semi-log graphs are usually rated in decades. Frequency changes which pertain to music or music reproduction are usually rated in octaves. You may encounter both.

FILTER TYPES

As stated earlier the first distinction made between filters is whether they are active and passive. After that distinction is made, filters are classified by the type of frequency response they demonstrate. There are five basic types of responses. They are high pass, low pass, band pass, band stop, and all pass. Each will be discussed in detail, but first it is important that you learn some principles behind their construction and response identification.

One skill that you must develop is circuit identification. It is important for you to be able to determine the circuit function by examining the position of the components. The following figures and paragraphs will help you to do this.

Capacitors pass high frequencies. When you encounter a capacitor in series as in Figure 20-2, that circuit will pass an ac signal or a high frequency signal. When a capacitor is placed in parallel as in Figure 20-3, that circuit will shunt a high frequency signal to ground preventing its transfer.

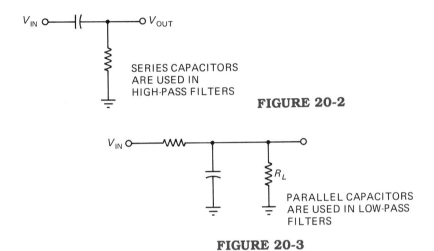

FIGURE 20-2

FIGURE 20-3

Inductors pass low frequencies. A series inductor, like the one in Figure 20-4, will pass a DC voltage or a low frequency signal. When an inductor is placed in parallel as the one in Figure 20-5, that circuit will shunt a low frequency signal to ground. This prevents its transfer.

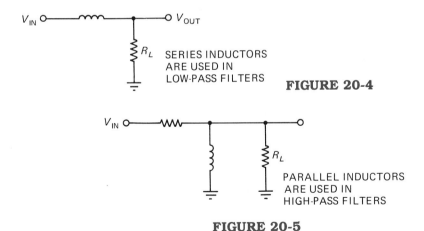

FIGURE 20-4

FIGURE 20-5

A tank circuit (a capacitor and inductor in parallel) in series (see Figure 20-6) with a circuit will allow all but a band of frequencies to pass. Thus passing some frequencies and rejecting others. A tank circuit in parallel with a circuit will allow only a band of frequencies to pass.

FILTER TYPES

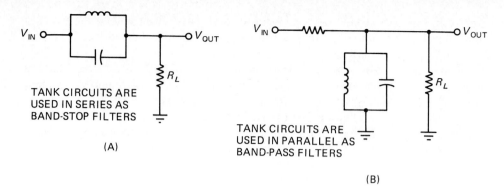

FIGURE 20-6

High-Pass Response A high-pass filter will allow only frequencies higher than a specified frequency to pass. This specified frequency is called the *cutoff frequency*. The cutoff frequency is the point where the dB level is 3 dB down from the maximum level. A 45° phase shift between the input and the output voltage results in the 3-dB level associated with the cutoff frequency. Some high-pass filters and their general response curve are shown in Figure 20-7. Notice that a capacitor in series corresponds to a high-pass filter. A parallel inductor also equates to this type of response.

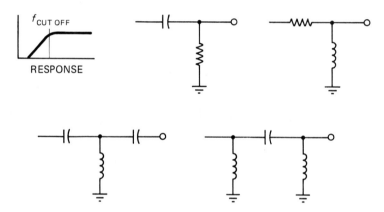

TYPICAL HIGH-PASS FILTERS

FIGURE 20-7

Stereos have a tone control, which provides control over the bass level of the music. This tone control is a high-pass filter. When you make an adjustment using this control, you are changing the cutoff frequency. Lowering this cutoff frequency allows a lower-frequency tone to pass. Raising this cutoff frequency reduces the amount of bass in the music.

Low-Pass Response A low-pass filter will allow only frequencies lower than a specified frequency to pass. Again this specified frequency is referred to as the cutoff frequency and is 3 dB down from the maximum level and shifted 45° from input to output. Several low-pass filters and their general response curve are shown in Figure 20-8. Notice that a parallel capacitor will give a low-frequency response, as can a series inductor.

Automobile radios sometimes pick up spark plug noise from the engine. A device called a *noise suppressor* can be used to reduce or eliminate this noise. This

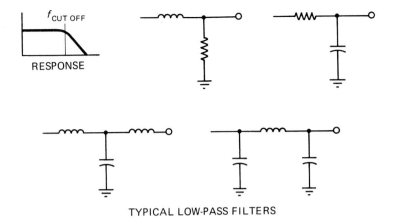

TYPICAL LOW-PASS FILTERS

FIGURE 20-8

noise suppressor is a low-pass filter that prevents the engine noise (high frequency) from entering the circuit through the auto's battery and power system.

Band-Pass Response When a high-pass and a low-pass filter are used in tandem (cascaded), the result is a band-pass filter. A band-pass filter will pass any frequency from the lower cutoff frequency to the upper cutoff frequency. The difference between the upper and lower cutoff frequencies is called the bandwidth. The cutoff frequencies are again at the 3 dB down level and correspond to a 45° phase shift. Several band-pass filters are shown in Figure 20-9.

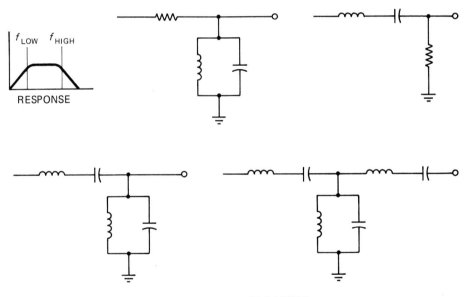

TYPICAL BAND-PASS FILTERS

FIGURE 20-9

Your AM/FM radio has two separate band-pass filter networks incorporated in its circuitry. One filter network allows frequencies between 535 and 1605 kHz to pass and rejects all other frequencies. This is the AM broadcast band. The FM band filter network allows frequencies from 88 to 108 MHz to pass. Remember that your AM/FM radio has a switch that allows you to use the lower band-pass filter (AM band) or the higher band-pass filter (FM band).

FILTER TYPES

Band-Stop Response When a high-pass and a low-pass filter are connected in parallel, the result is a band-stop filter. A band-stop filter will pass all but a specified group of frequencies. The band-stop filter exhibits a response like the one shown in Figure 20-10. A 45° phase shift and 3-dB down point are also associated with band-stop filters.

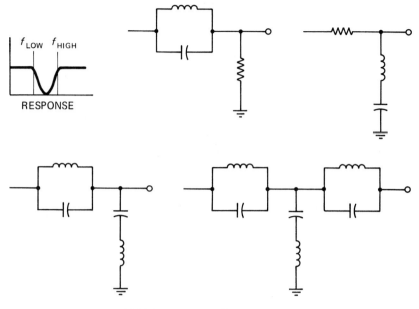

TYPICAL BAND-STOP FILTERS

FIGURE 20-10

Some cable TV systems insert band-stop filters that prevent the reception of one or more stations. These are stations to which the customer does not subscribe.

All-Pass Response As the name implies, an all-pass filter will pass all frequencies with little or no attenuation; however, this type of filter will cause a phase shift between the input and output voltages.

FILTER FAMILIES

Modern filters are classified by the steepness of the rolloff and the flatness of the pass band. It is important that you learn the names of these filter families and that you can describe each.

The modern filter families are grouped together by the mathematic formulas used to describe their response curves. They are Bessel, Butterworth, Chebyshev (also Tchebycheff), and Cauer. Each of these names refers to the mathematician who developed the mathematical function that describes the response curve. Understand that these formulas were, in most cases, developed long before electronic filters were thought of. Although the mathematics used to describe these functions is beyond the scope of this book, a brief description of the characteristics and applications of each of these filter types follows.

Bessel Bessel function filters are known for their shallow rolloff (less than 20 dB/decade; Figure 20-11). They also exhibit a linear phase shift (one that varies

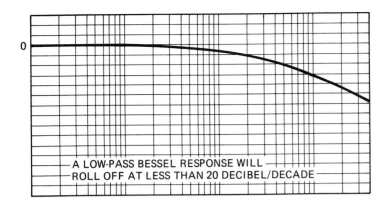

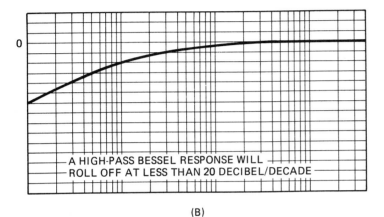

(B)

FIGURE 20-11

directly and proportionally to changes in frequency), which provides a transient response with minimal distortion. They are used in some frequency to voltage meters to provide a voltage that is proportional to frequency.

Butterworth Butterworth filters are probably the most common of the modern filter types. They have a rolloff of 20 dB/decade and a very flat pass-band response (Figure 20-12). When a steeper rolloff is desired, several Butterworth filter sections

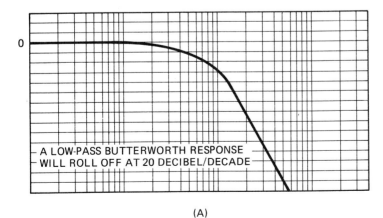

(A)

FIGURE 20-12

FILTER FAMILIES

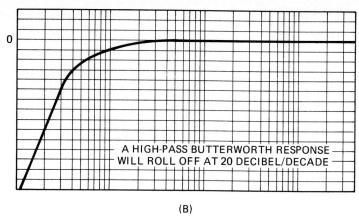

(B)

FIGURE 20-12 (continued)

(each section is called a *pole* or an *order*) are cascaded. This provides an additional 20 dB/decade for each additional section. For example, a two-pole Butterworth filter would provide a (2 × 20 dB =) 40-dB/decade rolloff.

Chebyshev It is possible to construct single-section filters that provide a rolloff greater than 20 dB/decade. Doing so causes at least some distortion in the pass band. The Chebyshev filter is such a filter type. It is characterized by a rippled pass band. Typically, a Chebyshev response curve will start the pass band with a gain rather than an attenuation (Figure 20-13). This gain is the result of an inductor ringing ac-

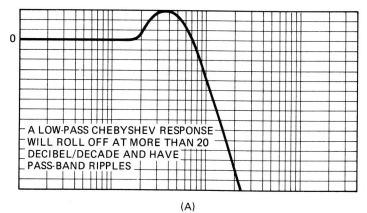

(A)

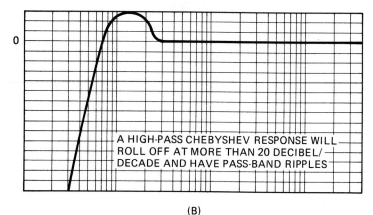

(B)

FIGURE 20-13

tion provided by its construction. The Chebyshev is used when a steep rolloff is desired and pass-band distortion is not a factor.

Cauer The filter that provides the steepest rolloff for a given order is the Cauer (also called elipptic) filter. Although the rolloff is extremely steep, it is filled with ripples, as are both the pass and stop bands (Figure 20-14). It is used when the bandwidth is already limited at about the cutoff frequency and frequencies beyond this point are not expected. Telephone circuitry is the area where the Cauer filter finds its most extensive use.

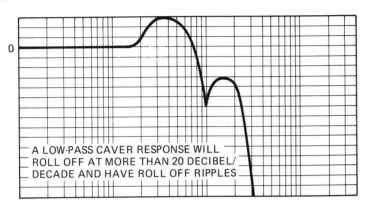

(A)

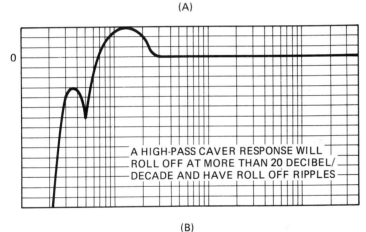

(B)

FIGURE 20-14

Classic Classic filters are grouped by the methods used to calculate their components. There are two classic filter designs: the constant k and the m derived. A brief description of each of these filter designs follows.

The concept of a constant k filter specifies that the input impedance and output impedance should remain equal for all frequencies (Figure 20-15). Understand that even though a constant k filter can provide any of the modern filter responses (Bessel, Butterworth, Chebyshev, and Cauer), they typically provide a Butterworth or Chebyshev response. The idea behind their design is for the product of the filter series impedance (Z_1) and the filter parallel impedance (Z_2) to be constant for all frequencies, thus keeping the input and output impedance equal. The reflective coefficient, k, indicates that the impedance is (or would be) matched for any number of cascaded filter sections.

The m-derived filter contains an extra section (or element) that provides an infinite attenuation at a given frequency, $f_{infinite}$. Again this type of filter may provide

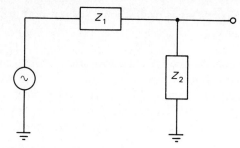

THE CONSTANT k DESIGN REQUIRES THAT $Z_1 \times Z_2$ BE A CONSTANT VALUE FOR ALL INPUT FREQUENCIES

FIGURE 20-15

any of the modern filter responses; however, typically it will provide a Chebyshev or Cauer response. The design fraction, *m*, is based on a ratio between the cutoff frequency and the frequency at which infinite attenuation occurs. This decimal can be any value between 0 and 1, but is usually 0.6 to 0.8.

FILTER CONSTRUCTION

Simple Any single reactive component or tank circuit can provide a 20-dB/decade filter response. Capacitors are used most often because of their light weight, small size, and relative low cost. When designing this type of filter, the reactive component is set to equal the load at the cutoff frequency. This process is demonstrated in the following example.

EXAMPLE 20-5

Calculate the value of the capacitor needed to provide a high-frequency response having a cutoff frequency of 10 kHz when connected to a 2-kΩ load resistance.

Solution: Begin by examining the high-pass filters shown in Figure 20-7. Notice that the one containing a single capacitor, like Figure 20-16, must be used.

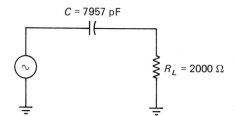

FIGURE 20-16

Next set $X_C = 2000\ \Omega$ and calculate C for a frequency of 10 kHz.

$$X_C = \frac{1}{2 \times \pi \times C \times f}$$

$$2000 = \frac{1}{2 \times \pi \times C \times 10{,}000}$$

$$C = \frac{1}{2 \times \pi \times 2000 \times 10{,}000}$$

$$= 7958\ \text{pF}$$

The filter design procedure (Example 20.5) is similar for inductor and tank circuits. Again the reactance is set at a value equal to the load resistance. Filters of this type have simplicity as their main advantage. The disadvantage of using this type of filter is that the frequency must vary a full decade before the output signal is low enough to disregard. When the reactive component is placed in parallel (Figure 20-17), this circuit also has the disadvantage of causing generator loading.

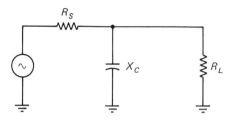

THE CAPACITOR IS SHORTED TO HIGH
FREQUENCIES AND MAY DAMAGE
THE GENERATOR

FIGURE 20-17

L Network Dual-element filters use a combination of an inductor and a capacitor (see Figure 20-18). This design is called an *L network* or sometimes an inverted-*L* arrangement. It has the advantage of producing a 40-dB/decade rolloff and prevents the generator loading effect that is associated with the single-element parallel filter design. The disadvantage of using the L network as a filter is that it results in a ripple in the pass band, usually at the cutoff frequency. This ripple may provide as much as a 3-dB gain at cutoff.

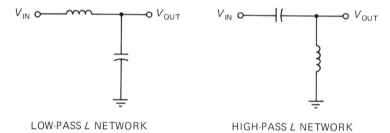

LOW-PASS *L* NETWORK   HIGH-PASS *L* NETWORK

FIGURE 20-18

Refer to the high-pass filter in Figure 20-18. When the frequency is very low, the capacitor has a very high impedance and the inductor has a very low impedance. The inductor then shorts the load resistance and prevents the signal from reaching it. The capacitor, on the other hand, provides a high impedance for the generator, thus preventing it from becoming loaded down.

When the frequency is very high, the capacitor has a very low impedance and the inductor has a very high impedance. The capacitor then shorts, allowing the full signal from the generator to reach the output of the circuit. The high impedance of the inductor assures that the parallel impedance is no lower than that of the output impedance.

T and Π Networks Triple element filters are sometimes used. The T (also wye) and Π (also delta) networks are examples of the triple-element arrangement. These two circuit types were developed in an effort to match input and output impedances. Understand that for each capacitor there is a high frequency which causes that component to appear to be short (low frequency for an inductor). In the L network the output is shorted for low frequencies on a high-pass filter. This shorting may result in damage to the output device. The Π and T networks are methods of reducing this and similar problems.

The high-pass Π and T filter networks are shown in Figure 20-19. The generator source resistance has the least effect on the Π network output. That is, it maintains a higher output than the T network. The T network, on the other hand, has a steeper rolloff and a flatter pass-band response. When maximum output is needed for a wide range of input impedances, an Π network works well. When a steep rolloff and flat response are desired, a T network is the better of the two.

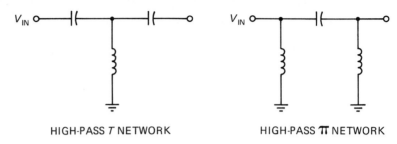

FIGURE 20-19

The low-pass Π and T filter networks are shown in Figure 20-20. Here the T is least dependent on generator source resistance and the Π network shows the steeper rolloff and flatter pass band. When maximum output is desired on a low-pass filter, a T network will provide the best response for a wide range of source impedances. A Π network, on the other hand, will provide a sharper rolloff and flatter pass band.

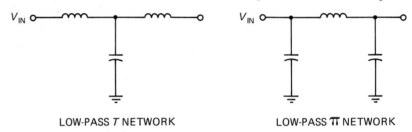

FIGURE 20-20

FLUCTUATING DC VOLTAGES

In most electronic circuits ac and dc voltage signals are combined. This combination is called by any of several names: fluctuating dc, pulsating dc, and sometimes ac riding on dc. Any of these is correct; however, *fluctuating dc* is most often used. One important aspect of filtering is the ability to separate the ac and dc signals.

AC and DC Voltage Separation Transformers offer the most complete separation of ac and dc. Remember that only changes in current (results in a pulsating magnetic field) are passed from the primary to the secondary windings of a transformer. Refer to Figure 20-21. The dc current in the primary winding provides a steady magnetic field, which cannot create any secondary voltage. The ac portion of the primary signal provides a changing magnetic field, which does induce a secondary voltage. The result is a pulsating dc signal (ac and dc combined) on the primary, which induces an ac output on the secondary.

A phase shift (between primary and secondary) may be associated with this voltage transfer. Schematic symbols for transformers will sometimes have dots by

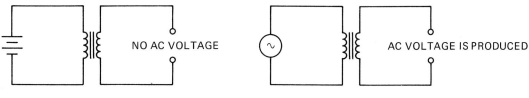

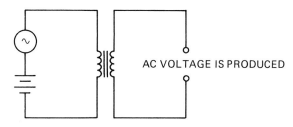

FIGURE 20-21

the primary and secondary windings (see Figure 20-22). When the dots are at the same side (top or bottom), the primary and secondary voltage are in phase. Dots at opposite sides (one at the top and the other at the bottom) indicate that the voltages are 180° out of phase.

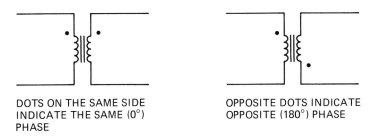

FIGURE 20-22

EXAMPLE 20-6

Given that the primary of a 1 : 2 transformer is connected to a dc source that fluctuates between 9.5 and 10.5 V, find the secondary voltage.

Solution: Begin by finding the dc or average voltage.

$$V_{dc} = \frac{9.5 + 10.5}{2}$$
$$= 10 \text{ V}$$

Next find the ac voltage. The amount of change in dc voltage.

$$V_{ac} = 10.5 - 9.5$$
$$= 1 \text{ V peak to peak}$$

The primary ac voltage is 1 V. The transformer has a 1 : 2 ratio, so the output is twice the input, 2 V peak to peak. ∎

FLUCTUATING DC VOLTAGE

The disadvantage of using transformers in this manner is their expense and size. Today, transformers (as filters) find their most extensive use in radio receivers. In earlier times transformers were used in many filter applications, especially when quality was a factor. Miniaturization has reduced their use as filters.

Coupling Capacitors Another method of separating ac and dc voltage is using a coupling capacitor (see Figure 20-23). The coupling capacitor transfers the majority of the ac signal from one point to another. Its disadvantage is that it does not allow the entire ac signal to be transferred. Size and inexpense are the advantages of using a coupling capacitor.

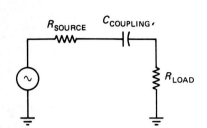

COUPLING CAPACITORS ARE USED TO TRANSFER (LINK) AN AC SIGNAL FROM ONE POINT TO ANOTHER

FIGURE 20-23

Refer to Figure 20-23. Seventy percent of the source voltage is transferred to the load, when the reactance is set to equal the resistance at the desired frequency. When the reactance is set to equal one-tenth of the resistance (FCC standard), the ac transfer is greater than 90% of the source signal above a specified frequency. The coupling capacitor does provide complete dc filtering; however, a portion of the ac signal is lost. A simple thing to remember is that unless a particular frequency is to be considered, a larger capacitor value will provide better coupling. The following examples will demonstrate this process.

EXAMPLE 20-7

Refer to Figure 20-24. Calculate the minimum-value capacitor needed to provide 90% transfer of the ac signal for frequencies higher that 1.5 kHz.

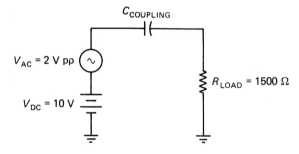

FIGURE 20-24

Solution: Begin by Thévenizing the circuit around the capacitor to find the circuit resistance. In this case the circuit resistance is 1500 Ω. Next, set X_C equal to one-tenth of this value.

$$X_C = \frac{1500}{10}$$

$$= 150 \ \Omega$$

Then calculate C for 1.5 kHz.

$$C = \frac{1}{2 \times \pi \times 1500 \times 150}$$

$$= 0.7074 \ \mu F$$

Any value of capacitor greater than 0.7074 μF will provide adequate coupling at this frequency. ∎

EXAMPLE 20-8

Refer to Figure 20-25. Use the FCC standard to find the value of capacitor that will provide good ac coupling at 300 Hz.

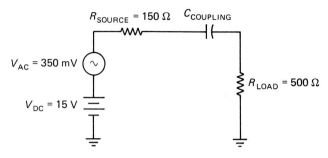

FIGURE 20-25

Solution: Begin by Thévenizing the circuit around the capacitor. Here the circuit resistance is the sum of the source and the load resistance, 650 Ω. Then set X_C equal to one-tenth of the circuit resistance:

$$X_C = \frac{650}{10}$$
$$= 65 \ \Omega$$

Next calculate the capacitor needed at 300 Hz.

$$C = \frac{1}{2 \times \pi \times 300 \times 65}$$
$$= 8.162 \ \mu F$$

■

Bypass Capacitors Sometimes it is necessary to remove the ac portion of a fluctuating dc voltage without disturbing the dc component of that voltage. This is done with a bypass capacitor (see Figure 20-26). Like the coupling capacitor, the bypass capacitor does not remove all of the ac signal. This is the disadvantage of its use. It is small, lightweight, and inexpensive, so it is used very often. The FCC again requires that the reactance be one-tenth of the resistance. An example of how to determine the value of the capacitor needed follows.

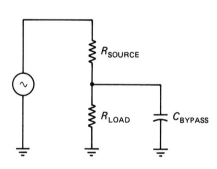

FIGURE 20-26

EXAMPLE 20-9

Refer to Figure 20-27. Use the FCC standard to find the value needed to shunt the majority of the ac signal to ground at frequencies above 500 Hz.

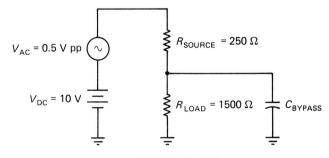

FIGURE 20-27

FLUCTUATING DC VOLTAGE

Solution: Begin by finding the Thévenin resistance of the circuit, looking in from the capacitor. In this case the circuit resistance is 250 Ω in parallel with 1500 Ω.

$$250 \parallel 1500 = 214.3 \ \Omega$$

Set X_C equal to one-tenth of the circuit resistance.

$$X_C = 21.43 \ \Omega$$

Then calculate C at 500 Hz.

$$C = \frac{1}{2 \times \pi \times 500 \times 21.43}$$
$$= 14.85 \ \mu F$$
■

SUMMARY

Filters are used to separate ac and dc signals or to allow only certain frequency signals to pass. In the latter case some frequencies are accepted (passed); others are rejected (shunted to ground). The amount or percentage of signal that is transmitted is measured in decibels. A positive decibel level indicates a voltage (or power) gain. A negative decibel level indicates a voltage (or power) loss. A voltage or power loss is referred to as an attenuation.

The formulas for calculating the decibel levels are

$$dB_{level} = 20 \times \log \frac{V_{out}}{V_{in}} \quad \text{for voltage}$$

$$dB_{level} = 20 \times \log \frac{I_{out}}{I_{in}} \quad \text{for current}$$

$$dB_{level} = 10 \times \log \frac{P_{out}}{P_{in}} \quad \text{for power}$$

Bode plots are a graph of frequency versus decibel level. They are made on semilog graph paper in an effort to produce a shorter graph. Frequency changes are measured in decades (multiples of 10) or octaves (multiples of 2). Most work pertaining to Bode plots uses frequencies measured in decades. Work pertaining to music or music reproduction uses frequencies measured in octaves.

There are five types of filters. They are rated by the group of frequencies which they allow to pass. The cutoff frequency (3-dB down point) is the specific frequency at which acceptance or rejection occurs. A low-pass filter will accept all frequencies lower than cutoff. A high-pass filter will accept all frequencies higher than cutoff. A band-pass filter has two cutoff frequencies, a lower and an upper. The band-pass filter will pass all frequencies higher than the lower cutoff but lower than the upper cutoff. A band-stop filter has two cutoff frequencies and rejects all frequencies higher than the lower cutoff but lower than the upper cutoff.

Circuit identification is made by looking for series and parallel components. A capacitor in series is associated with a high-pass filter. An inductor in series is associated with a low-pass filter. Capacitors in parallel relate to low-pass filters. Inductors in parallel relate to high pass filters. Band-pass and band-stop filters contain tank circuits. Parallel tanks are associated with band-pass filters, while series tanks are associated with band-stop filters.

Filters are subdivided into four groups, depending on the mathematical function that describes their response. It is important to know the names of the functions and to be able to describe the appearance of their response curves.

The Bessel filter has a smooth pass band and rolloff. Its rolloff is classified as shallow (less than 20 dB/decade).

The Butterworth also has a smooth pass band and rolloff. It is characterised by a 20-dB/decade rolloff.

The Chebyshev filter has a rippled pass band and a rolloff greater than 20 dB/decade.

The Cauer filter has the steepest rolloff, but has a rippled pass band and rolloff. At times the stop band is also rippled.

The order or number of sections in any of these filters will increase the rate of rolloff. A two-section Butterworth, for example, would rolloff at 40 dB/decade.

There are two classic design methods. The constant k filter design maintains equal input and output impedances for all frequencies. It also provides impedance matching for any number of cascaded filter section.

The m-derived filter uses an extra reactive element which provides an infinite attenuation at a given frequency. The design fraction, m, is based on a ratio between the cutoff frequency and the frequency that infinite attenuation occurs.

Filters can be configured in single-, dual-, and triple-element arrangements. Each has it own advantages. Single elements are simple and use equal reactance and resistance as design criteria, but they can cause generator loading.

Dual-element L networks reduce the possibility of generator loading but increase the probability that the pass band will be distorted. Triple-element Π and T networks also reduce generator loading, but may result in pass band and rolloff distortion.

With high-pass filters a Π configuration provides maximum output for a wide range of input impedances. A T network provides a steeper rolloff and flatter response. The reverse is true of triple-element low-pass filters.

One major aspect of filtering is the separating of the ac and dc portions of a fluctuating dc voltage. Transformers provide the most complete separation; however, their size and cost prevent extended use.

Coupling capacitors are the most used method of transferring an ac signal from one point to another. They have the advantage of simplicity and inexpense. Typically, reactance is set to equal one-tenth of the resistance, at this value 90% or more of the signal is transferred.

Bypass capacitors remove the ac portion of the signal, leaving only the dc level. The one-tenth rule is used as a design criterion.

QUESTIONS

20-1. The text says that the decibel level is the same for all parameters. What does that mean to you?

20-2. Use your own words to define *gain*.

20-3. What is the opposite of gain?

20-4. By examining the decibel level, how can you tell if there is an attenuation or gain.

20-5. By examining the input and output level (voltage, current, or power), how can you tell if there is an attenuation or gain?

20-6. Use your own words to describe a Bode plot.

20-7. The text says that semilog graph paper is used on a Bode plot to shorten the graph. What does that mean to you?

20-8. Define the term *octave*. Give an example and an application.

20-9. Define the term *decade*. Give an example and an application.

20-10. Why are 20 dB/decade and 6 dB/octave so important?

20-11. List each of the filter types. Include methods for circuit identification.

20-12. List as many applications for each filter type as you can.

20-13. List each of the filter families and give the characteristics that describe their response.

20-14. What are the advantages of using a single-element filter?

20-15. What are the disadvantages of using a single-element filter?

20-16. When designing a filter for a particular cutoff frequency you are to set X_C equal to the circuit resistance. When designing a coupling capacitor circuit you are to set X_C to one-tenth of the circuit resistance. Can you explain?

20-17. What is the advantage of using an L network compared to a simple filter?

20-18. Which construction types are most likely to produce a Chebyshev response?

20-19. Which construction types are most likely to produce a Cauer response?

20-20. Discuss Π and T network filters. Include the advantages and disadvantages of each.

20-21. When will a T-network filter produce a steeper rolloff than that of a Π filter.

20-22. When are transformers used to separate ac and dc signals?

20-23. Briefly describe the theory behind using a transformer as a coupling device.

20-24. List the advantages of using a capacitor as a coupling device.

20-25. What does FCC stand for?

20-26. Describe the function of a bypass capacitor.

20-27. Describe the function of a coupling capacitor.

20-28. Refer to Figure 20-28. Briefly state how you would find the Thévenin resistance as seen from the capacitor.

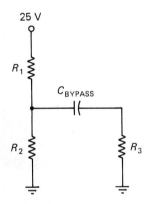

FIGURE 20-28

PRACTICE PROBLEMS

20-1. V_{in} = _____ attenuation or gain?
V_{out} = 150 mV
dB_{level} = 10.2 dB

20-2. P_{in} = 400 mW attenuation or gain?
P_{out} = _____
dB_{level} = −20.3 dB

20-3. I_{in} = _____ attenuation or gain?
I_{out} = 10 μA
dB_{level} = −5.6 dB

20-4. P_{in} = _____ attenuation or gain?
P_{out} = 0.1 W
dB_{level} = 12 dB

20-5. I_{in} = 50 mA attenuation or gain?
I_{out} = _____
dB_{level} = 3 dB

20-6. V_{in} = 0.2 V attenuation or gain?
V_{out} = _____
dB_{level} = 6.2 dB

20-7. V_{in} = 200 mV attenuation or gain?
V_{out} = 450 mV
dB_{level} = _____

20-8. V_{in} = _____ attenuation or gain?
V_{out} = 7 V
dB_{level} = −30 dB

20-9. V_{in} = 2.5 V attenuation or gain?
V_{out} = _____
dB_{level} = −40 dB

20-10. Refer to Table 20-1. Use the data from the table to create a Bode plot of the given frequencies. Describe its response.

TABLE 20-1

Frequency	V_{in}	V_{out}
10,000	180	177.84
20,000	180	176.59
30,000	180	175.34
40,000	180	174.10
50,000	180	172.88
60,000	180	171.66
70,000	180	170.45
80,000	180	169.25
90,000	180	168.07
100,000	180	166.89
200,000	180	155.73
300,000	180	145.56
400,000	180	136.46
500,000	180	128.18
600,000	180	120.83
700,000	180	114.22
800,000	180	108.78
900,000	180	102.78
1,000,000	180	97.72
2,000,000	180	50.82
3,000,000	180	19.12
4,000,000	180	8.17
5,000,000	180	4.15
6,000,000	180	2.39
7,000,000	180	1.49
8,000,000	180	1.00
9,000,000	180	0.70
10,000,000	180	0.51
20,000,000	180	0.06
30,000,000	180	0.02
40,000,000	180	0.01
50,000,000	180	0.00

20-11. Refer to Table 20-2. Use the data from the table to create a Bode plot of the given frequencies. Describe its response.

20-12. Given that you are to couple a source (impedance = 100 Ω) to a load of 1000 Ω for any frequency higher than 3000 Hz, find the value of the coupling capacitor needed.

PRACTICE PROBLEMS

TABLE 20-2

Frequency	V_{in}	V_{out}
10,000	180	0.10
20,000	180	0.63
30,000	180	1.67
40,000	180	3.16
50,000	180	5.05
60,000	180	7.29
70,000	180	9.84
80,000	180	12.63
90,000	180	15.62
100,000	180	18.75
200,000	180	49.45
300,000	180	71.31
400,000	180	86.30
500,000	180	97.28
600,000	180	105.76
700,000	180	112.57
800,000	180	118.17
900,000	180	122.88
1,000,000	180	126.90
2,000,000	180	148.55
3,000,000	180	157.48
4,000,000	180	162.37
5,000,000	180	165.45
6,000,000	180	167.58
7,000,000	180	169.13
8,000,000	180	170.31
9,000,000	180	171.25
10,000,000	180	172.00
20,000,000	180	175.48
30,000,000	180	176.67
40,000,000	180	177.27
50,000,000	180	177.64

20-13. Refer to Figure 20-28. Given that $R_1 = 3$ kΩ, $R_2 = 5$ kΩ, and $R_3 = 200$ Ω, calculate the Thévenin circuit resistance and the value of capacitor needed to couple this circuit for frequencies above 25 kHz.

20-14. Show how to couple a fluctuating dc voltage (5.9 to 8.2 V) to a load. Use a transformer (turns ratio = 3 : 1) and provide a one-third attenuation of the input signal. Include the value of the output voltage.

20-15. Design a single-element high-pass filter that will pass frequencies above 20 kHz. Given that the load resistance is 600 Ω, use a capacitor as the reactive element.

20-16. Design a single-element high-pass filter that will pass frequencies above 20 kHz. Given that the load resistance is 75 Ω, use an inductor as the reactive element.

Resonance

21

Chapter objectives

After reading this chapter and answering the questions and problems, you should be able to:

- List the three series resonant characteristics.
- Describe the flywheel effect of a tank circuit.
- List the three parallel resonant characteristics.
- Describe the effect that the tank coil quality has on v_{out}, z_{tank}, and f_{res}.
- Calculate bandwidth when given inductance and capacitance, resonant frequency and quality, or upper and lower cutoff frequencies.
- Given a series resonant circuit and values for L, r_{coil}, C, and v_{gen}, find f_{res}, I, Q, v_{out}, f_{high}, f_{low}, and BW.
- Given a parallel resonant circuit and values for L, r_{coil}, C, and v_{gen}, find f_{res}, Q, Z_{tank}, I_T, BW, f_{high}, and f_{low}.
- Calculate the effective quality of a damped resonant circuit.
- Calculate the value of damped resistance needed to produce a specific bandwidth.
- Discuss several possible causes of problems in resonant circuits.

A special case of *RCL* circuitry occurs any time that X_C and X_L are equal to each other. The equal and opposite effects of these two quantities cause the circuit current to reduce (parallel) or the circuit (V_C and V_L) voltage to peak (series). Resonance (both series and parallel) is used in communication circuitry (radios and television) for tuning purposes (selecting the proper station). The major topics of this chapter are series and parallel resonant circuit concepts, the quality factor and how it relates to bandwidth and resonance, the bandwidth and resonant frequency, and resonant dampening resistance. A section is also devoted to circuit analysis, tank circuit design, and troubleshooting.

SERIES RESONANT CONCEPTS

When an inductor and capacitor are placed in series, there is a particular frequency (f_{res}) where X_C and X_L are equal. At this point, capacitive and inductive reactance effectively cancel each other (equal and opposite phase angles; see Table 21-1). This places circuit impedance at a minimum and circuit current at a maximum. Also, the voltage drop across both the inductor and capacitor are at a maximum.

Again, there are three circuit conditions that occur when we have a circuit at series resonance.

1. $X_C = X_L$, so there is no phase shift and impedance is at a minimum.
2. Minimum impedance results in maximum current flow.
3. Maximum current flow results in maximum voltage drop across the reactive components.

THE TANK CIRCUIT

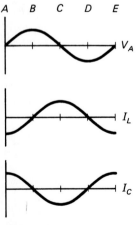

FIGURE 21-1

Any time that an inductor and capacitor are placed in parallel that circuit is referred to as a *tank circuit*. At resonance the tank circuit provides a maximum impedance. Also, the mainline (or generator) current is at a minimum; however, the current through the inductor and capacitor are at high levels. This is due primarily to the 180° phase difference between the inductor and the capacitor currents (see Figure 21-1). In essence, most of the current in the tank revolves from the inductor to the capacitor. This is referred to as the *flywheel effect*.

The flywheel effect begins as the tank voltage builds (goes positive). This results in a current flow through the inductor. See point *A* in Figure 21-1. Remember that the inductor resists any change in rate of current flow through it. As the tank voltage builds to a maximum (reaches the positive peak) and begins to drop, the inductor starts to discharge (in opposition to the change in current) and in doing so forces current into the capacitor. Refer to Figure 21-1, points *B* and *C*. Most of the inductive discharge current flows into the capacitor, causing the tank voltage to drop

TABLE 21-1 Series Resonant Values

X_L (Ω)	X_C (Ω)	f (Hz)	I (mA)	V_A (V)	Z (Ω)	V_L (V)	V_C (V)	R_{coil} (Ω)	Phase (deg)
9,000	11,111	180,000	4.731	10	2113	42	52	100	−88
9,049	11,049	181,000	4.994	10	2002	45	55	100	−88
9,100	10,989	182,000	5.286	10	1831	48	58	100	−87
9,149	10,928	183,000	5.612	10	1781	51	61	100	−87
9,199	10,869	184,000	5.378	10	1672	55	64	100	−87
9,249	10,810	185,000	6.393	10	1564	59	69	100	−87
9,299	10,752	186,000	6.867	10	1456	63	73	100	−87
9,349	10,695	187,000	7.413	10	1348	69	79	100	−86
9,399	10,638	188,000	8.049	10	1242	75	85	100	−86
9,449	10,582	189,000	8.799	10	1136	83	93	100	−85
9,499	10,526	190,000	9.697	10	1031	92	102	100	−85
9,549	10,471	191,000	10.791	10	926	103	113	100	−84
9,599	10,416	192,000	12.154	10	822	116	126	100	−84
9,649	10,362	193,000	13.895	10	719	134	143	100	−83
9,699	10,309	194,000	16.196	10	617	157	166	100	−81
9,749	10,256	195,000	19.372	10	516	188	198	100	−79
9,799	10,204	196,000	24.022	10	416	235	245	100	−77
9,849	10,152	197,000	31.407	10	318	309	318	100	−72
9,899	10,101	198,000	44.541	10	224	440	449	100	−64
9,949	10,050	199,000	70.621	10	141	702	709	100	−46
9,999	10,000	200,000	99.999	10	100	999	999	100	−1
10,049	9,950	201,000	70.798	10	141	711	704	100	44
10,099	9,900	202,000	44.899	10	222	453	444	100	63
10,149	9,852	203,000	31.834	10	314	323	313	100	71
10,199	9,803	204,000	24.479	10	408	249	239	100	75
10,249	9,756	205,000	19.844	10	503	203	193	100	78
10,299	9,708	206,000	16.676	10	599	171	161	100	80
10,349	9,661	207,000	14.38	10	635	148	138	100	81
10,399	9,615	208,000	12.642	10	790	131	121	100	82
10,450	9,563	209,000	11.283	10	886	117	107	100	83
10,499	9,523	210,000	10.19	10	981	107	97	100	84
10,550	9,478	211,000	9.293	10	1075	98	88	100	84
10,599	9,433	212,000	8.544	10	1170	90	80	100	85
10,650	9,389	213,000	7.909	10	1264	84	74	100	85
10,699	9,345	214,000	7.364	10	1357	78	68	100	85
10,750	9,302	215,000	6.891	10	1451	74	64	100	86
10,799	9,259	216,000	6.476	10	1543	69	59	100	86
10,850	9,216	217,000	6.11	10	1636	66	56	100	86
10,899	9,174	218,000	5.785	10	1728	63	53	100	86
10,950	9,132	219,000	5.493	10	1820	60	50	100	86

to zero. See point C. At this point, the inductor can no longer maintain the current flow into the capacitor and, as a result, the capacitor begins to discharge, allowing current to flow back into the inductor (point D). The process then repeats. The overall effect is that most of the current oscillates between the inductor and the capacitor while mainline current is very low.

PARALLEL RESONANT CONCEPTS

The tank circuit is the basis for a parallel resonant circuit. Parallel resonance provides a maximum impedance, a maximum voltage, and minimum current for the circuit. At resonance X_C and X_L are equal. When they are in parallel, both provide an impedance for the generator. At the frequencies below resonance, the inductor dominates the circuit and effectively shorts out the capacitor. At frequencies above the resonant frequency, the capacitor dominates the circuit and effectively shorts out the inductor. But at resonance both elements provide an opposition to current flow (see Table 21-2).

TABLE 21-2 Parallel Resonant Values

X_L (Ω)	X_C (Ω)	f (Hz)	I_T (mA)	V_A (V)	Z (Ω)	I_L (mA)	I_C (mA)	R_{coil} (Ω)	Phase (deg)
9,000	11,111	180,000	0.21	10	47,318	1.11	0.9	100	86
9,049	11,049	181,000	0.2	10	49,947	1.1	0.9	100	86
9,100	10,989	182,000	0.18	10	52,866	1.09	0.91	100	86
9,149	10,928	183,000	0.17	10	56,127	1.09	0.91	100	86
9,199	10,869	184,000	0.16	10	59,792	1.08	0.92	100	85
9,249	10,810	185,000	0.15	10	63,941	1.08	0.92	100	85
9,299	10,752	186,000	0.14	10	68,679	1.07	0.93	100	85
9,349	10,695	187,000	0.13	10	74,138	1.06	0.93	100	85
9,399	10,638	188,000	0.12	10	80,498	1.06	0.94	100	84
9,449	10,582	189,000	0.11	10	88,000	1.05	0.94	100	84
9,499	10,526	190,000	0.1	10	96,982	1.05	0.95	100	83
9,549	10,471	191,000	0.09	10	107,925	1.04	0.95	100	83
9,599	10,416	192,000	0.08	10	121,547	1.04	0.95	100	82
9,649	10,362	193,000	0.07	10	138,958	1.03	0.96	100	81
9,699	10,309	194,000	0.06	10	161,970	1.03	0.97	100	80
9,749	10,256	195,000	0.05	10	193,737	1.02	0.97	100	78
9,799	10,204	196,000	0.04	10	240,240	1.02	0.97	100	75
9,849	10,152	197,000	0.03	10	314,090	1.01	0.98	100	71
9,899	10,101	198,000	0.02	10	445,435	1.01	0.99	100	62
9,949	10,050	199,000	0.01	10	706,254	1	0.99	100	44
9,999	10,000	200,000	0	10	1,000,043	0.99	1	100	−1
10,049	9,950	201,000	0.01	10	708,021	0.99	1	100	−46
10,099	9,900	202,000	0.02	10	449,012	0.99	1.01	100	−64
10,149	9,852	203,000	0.03	10	318,359	0.98	1.01	100	−73
10,199	9,803	204,000	0.04	10	244,805	0.98	1.01	100	−77
10,249	9,756	205,000	0.05	10	198,451	0.97	1.02	100	−80
10,299	9,708	206,000	0.05	10	166,769	0.97	1.03	100	−81
10,349	9,661	207,000	0.06	10	143,810	0.96	1.03	100	−83
10,399	9,615	208,000	0.07	10	126,434	0.96	1.03	100	−84
10,450	9,569	209,000	0.08	10	112,836	0.95	1.04	100	−85
10,499	9,523	210,000	0.09	10	101,910	0.95	1.04	100	−85
10,550	9,478	211,000	0.1	10	92,942	0.94	1.05	100	−86
10,599	9,433	212,000	0.11	10	85,450	0.94	1.06	100	−86
10,650	9,389	213,000	0.12	10	79,099	0.93	1.06	100	−87
10,699	9,345	214,000	0.13	10	73,646	0.93	1.06	100	−87
10,750	9,302	215,000	0.14	10	68,915	0.93	1.07	100	−87
10,799	9,259	216,000	0.15	10	64,770	0.92	1.07	100	−87
10,850	9,216	217,000	0.16	10	61,109	0.92	1.08	100	−88
10,899	9,174	218,000	0.17	10	57,853	0.91	1.08	100	−88
10,950	9,132	219,000	0.18	10	54,937	0.91	1.09	100	−88

Again there are three circuit conditions that occur when we have a circuit at parallel resonance. They are:

1. At frequencies below resonance X_L is low and shorts X_C. At frequencies above resonance X_C is low and shorts X_L. At resonance both provide equal impedance, resulting in maximum impedance.
2. Maximum impedance results in minimum current flow (through the circuit).
3. The flywheel effect of the tank provides maximum voltage across the reactive components.

THE QUALITY FACTOR

The ratio between inductive reactance and ac resistance is the quality of the inductor. The quality of the inductor plays a large role in analyzing resonant circuits. The formula for calculating the quality of the inductor is

$$Q = \frac{X_L}{r_{coil}}$$

Typically, the quality of an inductor is between 10 and 250. The higher the number, the better the inductor. Qualities over 250 are possible but are considered very high. Qualities less than 10 are very low.

It is important to understand that the ac resistance (r_{coil} in the quality formula) is not the same for all frequencies. It is determined by a combination of the dc resistance, the skin effect of the inductor at the frequency, and the amount of eddy currents produced in the inductor core at the frequency. At low frequencies (below 10 kHz), the ac the dc resistance of the coil are approximately equal, but at higher frequencies (over 100 kHz) the ac resistance and dc resistance of the inductor may differ greatly. So when manufacturers rate the Q of their inductor, they will also give a specific frequency which corresponds to that Q.

The Q of the coil is used to predict the circuit conditions that will occur at resonance. Series resonance results in an increase in voltage. The output voltage (V_C or V_L) of the resonant circuit is Q times the input voltage.

$$v_{out} = Q \times v_{in}$$

Parallel resonance results in an increase in circuit impedance. The circuit impedance is Q times the inductive (or capacitive) reactance.

$$Z_{tank} = Q \times X_L$$

The quality of the inductor also determines the bandwidth of the frequency response. The resonant frequency of the circuit is Q times the bandwidth.

$$f_{res} = Q \times BW$$

or

$$BW = \frac{f_{res}}{Q}$$

The bandwidth, remember, is the amount of frequencies between the lower cutoff frequency (f_{low}) and the upper cutoff frequency (f_{high}). This will be explained in more detail when we discuss circuit analysis. But these are the formulas (relationships) which need to be memorized (and understood).

RESONANT FREQUENCY AND BANDWIDTH

As stated earlier, the quality of the coil determines the bandwidth of the resonant circuit. The response curve of a resonant circuit is similar to that of a band-pass filter. Examine the response curve shown in Figure 21-2. Notice the lower cutoff

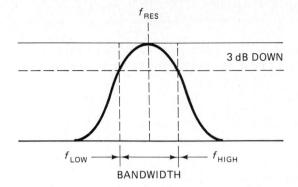

FIGURE 21-2

frequency, f_{low}. The circuit begins resonance at this point. Notice, also, the upper cutoff frequency, f_{high}. The circuit ends resonance at this point. The bandwidth is made up of the frequencies between f_{low} and f_{high}. At the center of the bandwidth is f_{res}, the resonant frequency. The resonant frequency is determined by the product of the inductance and capacitance. The formula follows:

$$f_{res} = \frac{1}{2 \times \pi \times \sqrt{L \times C}}$$

The distance between f_{low} and f_{res} (amount of frequencies) is approximately equal to the distance between f_{res} and f_{high}. There are several formulas that can be used to find the bandwidth of the circuit. Some follow:

$$BW = \frac{f_{res}}{Q}$$

$$BW = f_{high} - f_{low}$$

$$BW = 2 \times (f_{res} - f_{low})$$

$$BW = 2 \times (f_{high} - f_{res})$$

The width of the frequency band is determined almost entirely by the quality of the coil. So the first of these formulas is the most used; however, the word *width* in bandwidth should make the others self-evident. The last two formulas are usually rewritten to provide formulas for f_{low} and f_{high}.

$$f_{low} = f_{res} - \frac{BW}{2}$$

$$f_{high} = f_{res} + \frac{BW}{2}$$

CIRCUIT ANALYSIS

Circuit analysis uses Ohm's law and ac series and parallel circuit concepts. In general, a slow, step-by-step analysis of any resonant circuit is no more complicated than that of other ac circuits. The following examples will demonstrate the analysis of both series and parallel resonance.

Series Resonant Analysis

EXAMPLE 21-1

Refer to Figure 21-3. Find the resonant frequency, X_L and X_C, the circuit current, the quality factor, V_L and V_C, the bandwidth, and f_{low} and f_{high}. Also sketch a rough Bode plot of the resonant response for this circuit.

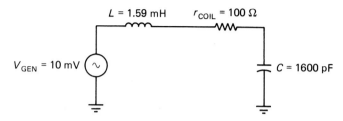

FIGURE 21-3

Solution: Begin by finding the resonant frequency.

$$f_{res} = \frac{1}{2 \times \pi \times \sqrt{L \times C}}$$

$$= \frac{1}{2 \times \pi \times \sqrt{1.59 \times 10^{-3} \times 1600 \times 10^{-12}}}$$

$$= 99.78 \text{ kHz}$$

Use this frequency to find X_C and X_L. Remember that they are equal at resonance, so only one calculation is needed.

$$X_L = 2 \times \pi \times f \times L$$

$$= 2 \times \pi \times 99{,}780 \times 1.59 \times 10^{-3}$$

$$= 996.9 \text{ } \Omega$$

Next find the circuit current. Notice that the circuit impedance is

$$Z = r_{coil} + jX_L - jX_C$$

Since X_C and X_L are equal, they cancel and

$$Z = r_{coil}$$

Ohm's law then gives

$$I = \frac{v_{gen}}{Z}$$

$$= \frac{0.01}{100}$$

$$= 0.0001 \text{ A}$$

Then use Ohm's law to find the reactive voltages.

$$V_C = I \times X_C$$

$$= 0.0001 \times 996.9$$

$$= 99.69 \text{ mV}$$

CIRCUIT ANALYSIS

V_L approximately equals V_C. In circuits, the output voltage is usually taken from across the capacitor.

Next find the quality of the coil.

$$Q = \frac{X_L}{r_{coil}}$$

$$= \frac{996.9}{100}$$

$$= 9.969$$

Notice that if you multiply v_{gen} (v_{in}) by Q you get V_C (v_{out}).

$$v_{out} = Q \times v_{in}$$

$$= 9.969 \times 0.01$$

$$= 99.69 \text{ mV}$$

This is a quick way to find the output voltage.

The quality factor can now be used to find the bandwidth of the circuit.

$$BW = \frac{f_{res}}{Q}$$

$$= \frac{99,780}{9.969}$$

$$= 10.01 \text{ kHz}$$

Next find the lower cutoff frequency.

$$f_{low} = f_{res} - \frac{BW}{2}$$

$$= 99.78 \text{ kHz} - \frac{10.01 \text{ kHz}}{2}$$

$$= 94.78 \text{ kHz}$$

Then the upper cutoff frequency.

$$f_{high} = f_{res} + \frac{BW}{2}$$

$$= 99.78 \text{ kHz} + \frac{10.01 \text{ kHz}}{2}$$

$$= 104.8 \text{ Hz}$$

See Figure 21-4. ∎

The source impedance affects the operation of the circuit. In essence, the source impedance lowers the quality of the circuit, widens the bandwidth, and reduces the output voltage. Calculations are very similar to Example 21.1, with the exception of finding the quality of the circuit. Note that we refer here to the quality of the circuit (also called effective quality), while before we referred to the quality of the coil. They are not the same. The following example will demonstrate the differences.

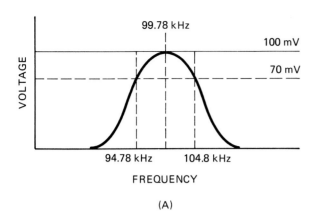
(A)

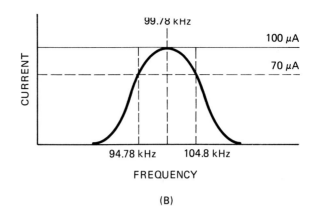
(B)

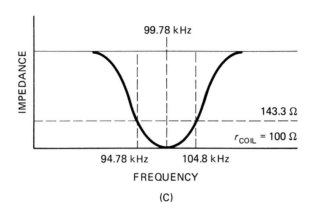
(C)

FIGURE 21-4

EXAMPLE 21-2

Refer to Figure 21-5. Find the resonant frequency, X_L and X_C, the circuit current, the quality factor, V_L and V_C, the bandwidth, and f_low and f_high. Also sketch a rough Bode plot of the resonant response for this circuit.

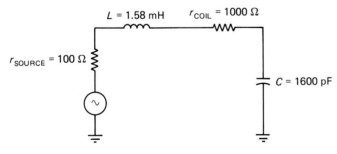

FIGURE 21-5

Solution: Begin by finding the resonant frequency. This is the same as Example 21.1.

$$f_\text{res} = \frac{1}{2 \times \pi \times \sqrt{L \times C}}$$

CIRCUIT ANALYSIS

$$= \frac{1}{2 \times \pi \times \sqrt{1.59 \times 10^{-3} \times 1600 \times 10^{-12}}}$$

$$= 99.78 \text{ kHz}$$

Use this frequency to find X_C and X_L. Remember that they are equal at resonance, so only one calculation is needed (no change in procedure).

$$X_L = 2 \times \pi \times f \times L$$

$$= 2 \times \pi \times 99.78 \text{ kHz} \times 1.59 \text{ mH}$$

$$= 996.9 \, \Omega$$

Next find the circuit current. Here we begin to see a difference in procedure. Notice that the circuit impedance is

$$Z = r_{\text{source}} + r_{\text{coil}} + jX_L - jX_C$$

Since X_C and X_L are equal, they cancel and

$$Z = r_{\text{source}} + r_{\text{coil}}$$

$$= 100 + 100$$

$$= 200 \, \Omega$$

Ohm's law then gives

$$I = \frac{v_{\text{gen}}}{Z}$$

$$= \frac{0.01}{200}$$

$$= 0.05 \text{ mA}$$

Then use Ohm's law to find the reactive voltages.

$$V_C = I \times X_C$$

$$= 0.00005 \times 996.9$$

$$= 49.84 \text{ mV}$$

V_L approximately equals V_C.

Next find the quality of the circuit. Understand that it is the series resistance in the circuit which determines the quality of the circuit.

$$Q_{\text{effective}} = \frac{X_L}{(r_{\text{source}} + r_{\text{coil}})}$$

$$= \frac{996.9}{200}$$

$$= 4.984$$

Notice that if you multiply v_{gen} (v_{in}) by Q you get V_C (v_{out}).

$$v_{\text{out}} = Q_{\text{effective}} \times v_{\text{in}}$$

$$= 4.984 \times 0.01$$

$$= 49.84 \text{ mV}$$

This is a quick way to find the output voltage.

The quality factor can now be used to find the bandwidth of the circuit.

$$BW = \frac{f_{res}}{Q_{effective}}$$

$$= \frac{99.78 \text{ kHz}}{4.984}$$

$$= 20.02 \text{ kHz}$$

Next find the lower cutoff frequency,

$$f_{low} = f_{res} - \frac{BW}{2}$$

$$= 99.78 \text{ kHz} - 20.02 \text{ kHz}/2$$

$$= 89.77 \text{ kHz}$$

then the upper cutoff frequency,

$$f_{high} = f_{res} + \frac{BW}{2}$$

$$= 99.78 \text{ kHz} + \frac{20.02 \text{ kHz}}{2}$$

$$= 109.8 \text{ kHz}$$

See Figure 21-6. ■

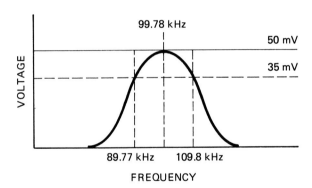

(A)

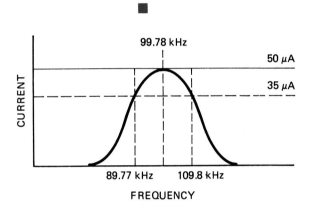

(B)

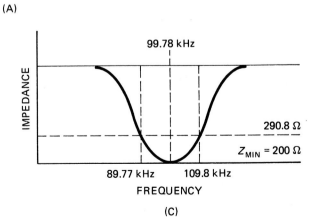

(C)

FIGURE 21-6

Parallel Resonant Analysis

EXAMPLE 21-3

Refer to Figure 21-7. Find the resonant frequency, X_L and X_C, the circuit impedance at resonance, the quality factor, I_L and I_C, the bandwidth, and f_{low} and f_{high}. Also sketch a rough Bode plot of the resonant response for this circuit.

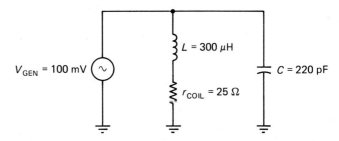

FIGURE 21-7

Solution: Begin by calculating the resonant frequency.

$$f_{res} = \frac{1}{2 \times \pi \times \sqrt{L \times C}}$$

$$= \frac{1}{2 \times \pi \times \sqrt{0.3 \text{ mH} \times 220 \text{ pF}}}$$

$$= 619.5 \text{ kHz}$$

Use this frequency to find X_L and X_C.

$$X_L = 2 \times \pi \times f \times L$$

$$= 2 \times \pi \times 619.5 \text{ kHz} \times 0.3 \text{ mH}$$

$$= 1168 \text{ }\Omega$$

Remember that X_L and X_C are equal at resonance. Use this value to find the quality of the coil.

$$Q = \frac{X_L}{r_{coil}}$$

$$= \frac{1168}{25}$$

$$= 46.71$$

Next find the resonant impedance. Understand that the coil resistance is in series with the inductive reactance while the combination is in parallel with the capacitive reactance.

$$Z = (r_{coil} + jX_L) \parallel -jX_C$$

but

$$r_{coil} + jX_L = 25 + j1168$$

$$= 1168 \underline{/+88.77°} \text{ }\Omega$$

so

$$Z = \frac{1168\ /+88.77° \times 1168\ /-90°}{25 + j1168 - j1168}$$

$$= \frac{1{,}364{,}000\ /-1.23°}{25}$$

$$= 54.56\ \text{k}\Omega\ /-1.23°$$

Notice that this impedance is, for practical purposes, equivalent to $Q \times X_L$.

$$Q \times X_L = 46.71 \times 1168$$

$$= 54.55\ \text{k}\Omega$$

Typically, calculations are done by making this approximation. Understand that the higher the quality factor, the closer the approximation is to exact. Also, at resonance the circuit is primarily resistive and as such the phase angle is (or is close to) zero. The remainder of the calculations will be done using 54.55 kΩ (the approximate value) as the circuit impedance.

Next use Ohm's law to find the current through the inductor and the capacitor. Remember that X_L and X_C are equal and r_{coil} is negligible.

$$I_C\ \&\ I_L = \frac{v_{\text{gen}}}{X_L}$$

$$= \frac{0.1}{1168}$$

$$= 85.63\ \mu\text{A}$$

The total current from the generator is much less than this.

$$I_T = \frac{v_{\text{gen}}}{Z}$$

$$= \frac{0.1}{54{,}550}$$

$$= 1.833\ \mu\text{A}$$

Notice that I_L and I_C are equal to $Q \times I_T$.

$$Q \times I_T = 46.71 \times 1.833\ \mu\text{A}$$

$$= 85.63\ \mu\text{A}$$

This relationship can also be used to speed up calculations.

Next use the quality factor to find the bandwidth and cutoff frequencies.

$$\text{BW} = \frac{f_{\text{res}}}{Q}$$

$$= \frac{619.5\ \text{kHz}}{46.71}$$

$$= 13.26\ \text{kHz}$$

CIRCUIT ANALYSIS

$$f_{\text{low}} = f_{\text{res}} - \frac{\text{BW}}{2}$$

$$= 619.5 \text{ kHz} - \frac{13.26 \text{ kHz}}{2}$$

$$= 612.9 \text{ kHz}$$

$$f_{\text{high}} = f_{\text{res}} + \frac{\text{BW}}{2}$$

$$= 619.5 \text{ kHz} + \frac{13.26 \text{ kHz}}{2}$$

$$= 626.1 \text{ kHz}$$

Next sketch the frequency response for this circuit (Figure 21-8). ∎

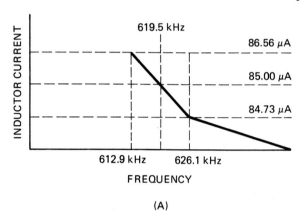
(A)

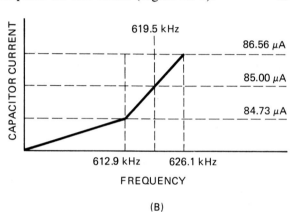
(B)

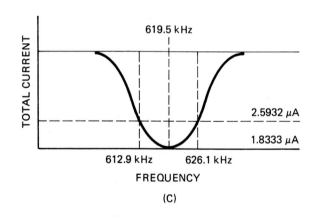
(C)

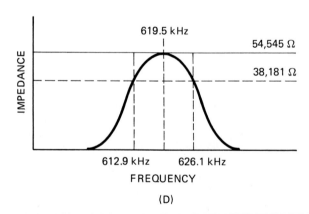
(D)

FIGURE 21-8

Source impedance also has an effect on parallel resonant circuits. It is important to understand that in a resonant circuit the tank is the only source of quality. That is, the circuit quality can be no greater than that of the coil. With that in mind, we can think of the coil as the Q generator and the impedance of the tank as the source impedance of the Q generator.

A Q generator is a matter of concept. Most texts will not describe anything similar; however, treating a parallel resonant circuit in this manner will help your understanding of the theory behind several formulas used in connection with resonance. As in a series resonant circuit, source impedance will lower the circuit quality, widen the bandwidth, and reduce the voltage output of the circuit.

Consider the circuit shown in Figure 21-9. Now think of it from the tank's point of view (see Figure 21-10) and in terms of quality, not voltage. The tank produces the quality. That quality is the quality of the coil.

$$Q_{coil} = \frac{X_L}{r_{coil}}$$

The tank has an impedance. It is that impedance which helps to determine the effective quality of the circuit.

$$Z_{tank} = Q_{coil} \times X_L$$

The voltage divider formula (now a Q divider) can be used to find the effective quality of the circuit. Understand that the tank is distributing the quality across the source.

$$Q_{effective} = Q_{coil} \times \frac{r_{source}}{Z_{tank} + r_{source}}$$

The effective circuit quality is used to make calculations pertaining to bandwidth, as shown in the previous examples.

It is important to understand that the output voltage is found in a very similar manner. Try carefully not to confuse the two. An example of these calculations follow.

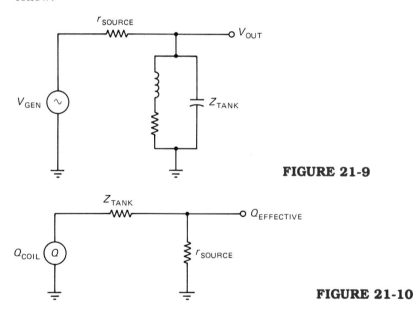

FIGURE 21-9

FIGURE 21-10

EXAMPLE 21-4

Given that a 1-mH choke and a 220-pF capacitor are used to form a tank circuit that is powered by a 200-mV source having a source resistance of 100 kΩ. Find the resonant frequency, bandwidth with cutoff frequencies, the effective quality of the circuit, and the output voltage at parallel resonance. The quality of the coil is 75.

Solution: Use the capacitance and inductance to calculate the resonant frequency.

$$f_{res} = \frac{1}{2 \times \pi \times \sqrt{1 \text{ mH} \times 220 \text{ pF}}}$$

$$= 339.3 \text{ kHz}$$

Use this frequency to find the reactance of each component. Remember that $X_L = X_C$ at resonance.

$$X_L = 2 \times \pi \times 339.3 \text{ kHz} \times 1 \text{ mH}$$

$$= 2132 \text{ }\Omega$$

Next find the impedance of the tank circuit.

$$Z_{tank} = 75 \times 2132$$

$$= 159.9 \text{ k}\Omega$$

Remember that at resonance the circuit is resistive (no, or little, phase shift).

Sketch the equivalent circuit (voltage as a reference; Figure 21-11). Use the voltage divider formula to find the output voltage at resonance. Note that the generator provides the voltage and the tank serves as the load.

$$v_{out} = \frac{Z_{tank}}{Z_{tank} + r_{source}} \times v_{gen}$$

$$= \frac{159.9 \text{ k}\Omega}{159.9 \text{ k}\Omega + 100 \text{ k}\Omega} \times 0.2$$

$$= 123.0 \text{ mV}$$

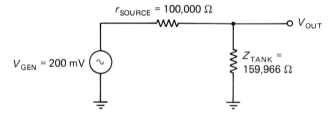

FIGURE 21-11

Next sketch the equivalent circuit (tank quality as a reference; Figure 21-12). Use the voltage divider formula (now a Q divider) to find the effective quality of the circuit. Note that the tank provides the quality and the generator (source resistance) serves as the load for this quality.

$$Q_{effective} = \frac{r_{source}}{Z_{tank} + r_{source}} \times Q_{coil}$$

$$= \frac{100 \text{ k}\Omega}{159.9 \text{ k}\Omega + 100 \text{ k}\Omega} \times 75$$

$$= 28.86$$

Use the effective quality of the circuit to find the bandwidth and cutoff frequencies.

$$BW = \frac{339.3 \text{ kHz}}{28.86}$$

$$= 11.76 \text{ kHz}$$

$$f_{\text{low}} = 339.3 \text{ kHz} - \frac{11.76 \text{ kHz}}{2}$$

$$= 333.4 \text{ kHz}$$

$$f_{\text{high}} = 339.3 \text{ kHz} + \frac{11.76 \text{ kHz}}{2}$$

$$= 345.2 \text{ kHz} \qquad ∎$$

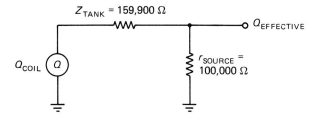

FIGURE 21-12

When source impedance is a considered factor in parallel resonance, it is important to understand when to use r_{source} and Z_{tank} in the calculations. Simply stated: When doing voltage calculations, use Z_{tank} as the load; when doing quality calculations use r_{source} as the load.

DAMPING RESISTANCE

Sometimes it is necessary to widen (or control) the bandwidth of the circuit. When this is necessary, a parallel damping resistance is used (see Figure 21-13). This damping resistor will decrease the voltage output, widen the bandwidth, and lower the quality of the circuit. Understand that the only way to decrease the bandwidth is to improve the quality of the coil or use multiple tank sections (usually done with amplified sections).

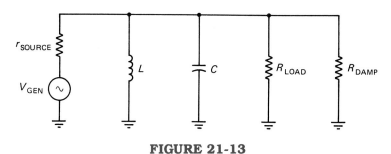

FIGURE 21-13

The resistance of the source is a major contributor to damping (see Example 21-4). The circuit load (the device that is being driven) is another source of damping. An additional parallel resistor may also be used in some designs to provide a specific bandwidth. Each of these has a different effect on the manner in which the circuit is analyzed. A list of steps in this analysis follows.

1. The circuit is Thévenized (from the generator) to find the circuit resistance, R_{THvolts}.
2. R_{THvolts} is used in the voltage divider formula to find the output voltage.
3. Next the circuit is again Thévenized (from the tank) to find the circuit resistance, $R_{\text{THeffective}}$.
4. $R_{\text{THeffective}}$ is used in the voltage divider formula (now a Q divider) to find the effective quality of the circuit.

Example 21-5 demonstrates this process.

EXAMPLE 21-5

Refer to Figure 21-13. Given that $v_{\text{gen}} = 300$ mV, $r_{\text{source}} = 150$ kΩ, $R_{\text{damp}} = 50$ kΩ, $R_{\text{load}} = 200$ kΩ, $L = 3.5$ mH, $Q_{\text{coil}} = 50$, and $C = 35$ pF, find f_{res}, f_{low}, f_{high}, BW, $Q_{\text{effective}}$, and v_{out} before and after R_{damp} has been placed in the circuit.

Solution: Start by finding the resonant frequency. This is the same both with and without R_{damp} in the circuit.

$$f_{\text{res}} = \frac{1}{2 \times \pi \times \sqrt{3.5 \text{ mH} \times 35 \text{ pF}}}$$
$$= 454.7 \text{ kHz}$$

Use this frequency to find X_C and X_L. They also will not change when R_{damp} is added to the circuit.

$$X_L = 2 \times \pi \times 454.7 \text{ kHz} \times 3.5 \text{ mH}$$
$$= 10 \text{ k}\Omega$$

Next use X_L and Q_{coil} to find Z_{tank}. Z_{tank} will not be effected by the addition of R_{damp}.

$$Z_{\text{tank}} = 50 \times 10000$$
$$= 500.0 \text{ k}\Omega$$

Without R_{damp}: Make a quick sketch of the voltage circuit (Figure 21-14). Find R_{THvolts}. Examining the circuit will show that Z_{tank} and R_{load} are in parallel.

$$R_{\text{THvolts}} = Z_{\text{tank}} \| R_{\text{load}}$$
$$= 500.0 \text{ k}\Omega \| 200.0 \text{ k}\Omega$$
$$= 142.9 \text{ k}\Omega$$

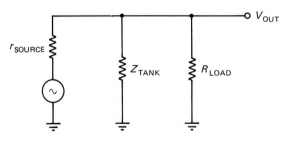

$R_{\text{TH VOLT}} = Z_{\text{TANK}} \| R_{\text{LOAD}}$

FIGURE 21-14

Use this resistance to find v_{out}.

$$v_{out} = \frac{R_{THvolts}}{R_{THvolts} + r_{source}} \times v_{gen}$$

$$= \frac{142.9 \text{ k}\Omega}{142.9 \text{ k}\Omega + 150 \text{ k}\Omega} \times 0.3$$

$$= 146.3 \text{ mV}$$

Make a quick sketch of the quality circuit (Figure 21-15). Find $R_{THeffective}$. Examining the circuit will show that R_{Load} and r_{source} are in parallel.

$$R_{THeffective} = R_{load} \parallel r_{source}$$

$$= 200 \text{ k}\Omega \parallel 150 \text{ k}\Omega$$

$$= 85.71 \text{ k}\Omega$$

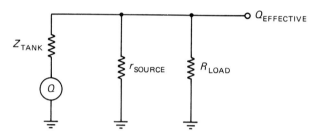

FIGURE 21-15

Use this to find $Q_{effective}$.

$$Q_{effective} = \frac{R_{THeffective}}{R_{THeffective} + Z_{tank}} \times Q_{coil}$$

$$= \frac{85.71 \text{ k}\Omega}{85.71 \text{ k}\Omega + 500 \text{ k}\Omega} \times 50$$

$$= 7.317$$

Next find BW, f_{low}, and f_{high}.

$$\text{BW} = \frac{454.7 \text{ kHz}}{7.317}$$

$$= 62.15 \text{ kHz}$$

$$f_{low} = 454.7 \text{ kHz} - \frac{62.15 \text{ kHz}}{2}$$

$$= 423.7 \text{ kHz}$$

$$f_{high} = 454.7 \text{ kHz} + \frac{62.15 \text{ kHz}}{2}$$

$$= 485.8 \text{ kHz}$$

With R_{Damp}: Make a quick sketch of the voltage circuit (Figure 21-16). Find $R_{THvolts}$. Examining the circuit will show that Z_{tank}, R_{damp}, and R_{load} are in parallel.

$$R_{\text{THvolts}} = Z_{\text{tank}} \| R_{\text{damp}} \| R_{\text{load}}$$
$$= 500{,}000 \| 50{,}000 \| 200{,}000$$
$$= 37.04 \text{ k}\Omega$$

Use this resistance to find v_{out}.

$$v_{\text{out}} = \frac{R_{\text{THvolts}}}{(R_{\text{THvolts}} + r_{\text{source}})} \times v_{\text{gen}}$$
$$= \frac{37.04 \text{ k}\Omega}{37.04 \text{ k}\Omega + 150 \text{ k}\Omega} \times 0.3$$
$$= 59.41 \text{ mV}$$

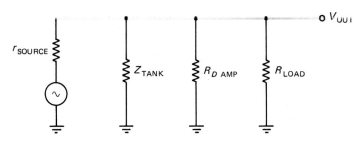

FIGURE 21-16

Make a quick sketch of the quality circuit (Figure 21-17). Find $R_{\text{THeffective}}$. Examining the circuit will show that R_{load}, R_{damp}, and r_{source} are in parallel.

$$R_{\text{THeffective}} = R_{\text{load}} \| R_{\text{damp}} \| r_{\text{source}}$$
$$= 200{,}000 \| 50{,}000 \| 150{,}000$$
$$= 31.58 \text{ k}\Omega$$

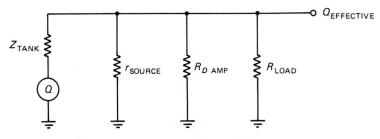

FIGURE 21-17

Use this to find $Q_{\text{effective}}$.

$$Q_{\text{effective}} = \frac{R_{\text{THeffective}}}{R_{\text{THeffective}} + Z_{\text{tank}}} \times Q_{\text{coil}}$$
$$= \frac{31.58 \text{ k}\Omega}{31.58 \text{ k}\Omega + 500 \text{ k}\Omega} \times 50$$
$$= 2.970$$

Next find BW, f_{low}, and f_{high}.

$$BW = \frac{454.7 \text{ kHz}}{2.970}$$

$$= 153.1 \text{ kHz}$$

$$f_{low} = 454.7 \text{ kHz} - \frac{153.1 \text{ kHz}}{2}$$

$$= 378.2 \text{ kHz}$$

$$f_{high} = 454.7 \text{ kHz} + \frac{153.1 \text{ kHz}}{2}$$

$$= 531.3 \text{ kHz} \qquad \blacksquare$$

Notice that the addition of a damping resistor to the circuit in Example 21.5 resulted in a lower effective circuit quality, a wider bandwidth, and lower output voltage. Again there are three sources of circuit damping. They are the internal resistance of the generator, the impedance of the load, and the intentional addition of a parallel damping resistor. It is important to note that when performing tests on these types of circuits, your voltmeter may also be a source of circuit damping. For this reason you should always use a high-impedance instrument for measuring these circuit parameters.

In some texts, you will find $R_{THeffective}$ listed as R_p or $R_{parallel}$. It is important for you to realize that they are the same. Texts usually include the formula

$$Q_{effective} = \frac{X_L}{r_{coil} + X_L^2/R_P}$$

for calculating the effective quality of the circuit. Example 21.6 demonstrates how this formula would be used on the circuit described in Example 21.5. The use of this formula can greatly speed up the calculations.

EXAMPLE 21-6

Refer to Example 21-5. Use the formula

$$Q_{effective} = \frac{X_L}{r_{coil} + X_L^2/R_P}$$

to find the effective quality of the circuit of Figure 21-12, with and without R_{damp} in the circuit.

Solution: Remember that $R_P = R_{THeffective}$. So $R_P = 85.71$ kΩ without R_{damp} and $R_P = 31.58$ kΩ with R_{damp}. Next find the value of r_{coil}.

$$r_{coil} = \frac{X_L}{Q_{coil}}$$

$$= \frac{10,000}{50}$$

$$= 200 \text{ } \Omega$$

Now insert these values into the formula.
Without R_{damp}:

$$Q_{\text{effective}} = \frac{X_L}{r_{\text{coil}} + X_L^2/R_P}$$

$$= \frac{10{,}000}{200 + 10{,}000^2/85.71 \text{ k}\Omega}$$

$$= 7.317$$

With R_{damp}:

$$Q_{\text{effective}} = \frac{X_L}{r_{\text{coil}} + X_L^2/R_P}$$

$$= \frac{10{,}000}{200 + 10{,}000^2/31.58 \text{ k}\Omega}$$

$$= 2.970 \qquad \blacksquare$$

Understand that formulas are wonderful when the same types of calculations must be done over and over again. When you are first learning how to do something, as stated before, they can lead to your destruction. It is better to gain an understanding of what is being accomplished than to memorize formulas blindly. For that reason it is suggested that you master the method outlined in Example 21-5 rather than memorize the formula method outlined in Example 21-6.

TANK DESIGN

The first consideration in tank design is choosing the proper value of L and C to provide the desired resonant frequency. Understand that there is a wider choice of values for capacitors than inductors. So, typically, the tank is designed around the available inductor. The formula

$$C = \frac{1}{4 \times \pi^2 \times f_{\text{res}}^2 \times L}$$

is used to calculate the desired value of capacitance. When the tank is to be designed around a particular capacitor, the formula

$$L = \frac{1}{4 \times \pi^2 \times f_{\text{res}}^2 \times C}$$

is used. In any case, a variable inductor and/or capacitor may be necessary in order to provide the exact frequency desired. When multiple frequencies are desired, a tapped inductor and variable capacitor are needed (Figure 21-18).

EXAMPLE 21-7

Calculate the value of capacitor needed to provide a resonant frequency of 1120 kHz when a 300-μH choke and said capacitor are used as a tank circuit.

Solution: Insert values into the formula and solve.

$$C = \frac{1}{4 \times \pi^2 \times f_{\text{res}}^2 \times L}$$

$$= \frac{1}{4 \times \pi^2 \times 1{,}120{,}000^2 \times 0.0003}$$

$$= 67.31 \text{ pF}$$

The closest standard value would be 68 pF. Use this value and 300 μH to check for the desired resonant frequency.

$$f_{\text{res}} = \frac{1}{2 \times \pi \times \sqrt{300 \ \mu\text{H} \times 68 \text{ pF}}}$$

$$= 1114 \text{ kHz}$$

Finally, you would check to see if 1120 kHz is a frequency between f_{low} and f_{high}. ∎

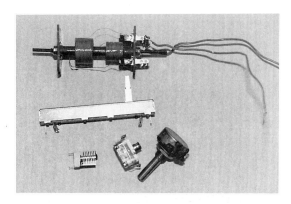

FIGURE 21-18 Adjustable RCL units—Tapped Inductor—top, Potentiometers—center and lower right, Variable Air Capacitors—bottom left and center.

The desired bandwidth is another important consideration. Understand that the source impedance and the load impedance play an important role in determining the quality of the circuit, which in turn determines the bandwidth of the circuit. Also, the narrowness of the bandwidth is determined by the quality of the coil itself. That is, the bandwidth can be no less than

$$\frac{f_{\text{res}}}{Q_{\text{coil}}}$$

so choose a coil that will provide a sufficiently narrow bandwidth. A bandwidth less than one-half the desired (end result) is typically used. Then the design is calculated in a manner that is almost the reverse of the analysis method. Several examples follow. These should clarify any questions you may have.

EXAMPLE 21-8

Design a tank circuit that will have a bandwidth of 25 kHz and operate at a resonant frequency of 1120 kHz. The source impedance is 350 kΩ and the load impedance is 300 kΩ. Use the 68-pF capacitor and 300-μH coil from Example 21-7 and consider the resonant frequency produced to be 1120 kHz (not 1114 kHz, as calculated).

Solution: Begin by finding the needed coil quality. Here begin with a value that will produce approximately one-half the 25-kHz bandwidth desired: say, 10 kHz.

$$\text{BW} = \frac{f_{\text{res}}}{Q_{\text{coil}}}$$

$$10{,}000 = \frac{1{,}120{,}000}{Q_{\text{coil}}}$$

$$Q_{\text{coil}} = \frac{1{,}120{,}000}{10{,}000}$$

$$= 112$$

When choosing a coil for this circuit, choose one that has Q equal to or greater than 112. For our purposes, let's use a coil that will produce a Q of 120.

Now calculate the circuit quality needed to produce the 25-kHz bandwidth desired.

$$\text{BW} = \frac{f_{\text{res}}}{Q_{\text{effective}}}$$

$$25{,}000 = \frac{1{,}120{,}000}{Q_{\text{effective}}}$$

$$Q_{\text{effective}} = \frac{1{,}120{,}000}{25{,}000}$$

$$= 44.8$$

Next make calculations for X_L, X_C, and Z_{tank}.

$$X_L \ \& \ X_C = 2 \times \pi \times 1{,}120{,}000 \times 0.0003$$

$$= 2111 \ \Omega$$

$$Z_{\text{tank}} = 120 \times 2111$$

$$= 253.3 \ \text{k}\Omega$$

Also calculate the parallel impedance of the source and load impedance, $R_{\text{THeffective}}$.

$$R_{\text{THeffective}} = 350{,}000 \ \| \ 300{,}000$$

$$= 161.5 \ \text{k}\Omega$$

Use this value in a voltage divider (now a Q divider) to find the effective quality generated by damping caused by the source and load.

$$Q_{\text{effective1}} = \frac{R_{\text{THeffective1}}}{R_{\text{THeffective1}} + Z_{\text{tank}}} \times Q_{\text{coil}}$$

$$= \frac{161.5 \ \text{k}\Omega}{161.5 \ \text{k}\Omega + 253.3 \ \text{k}\Omega} \times 120$$

$$= 46.72$$

Now compare the desired circuit quality (44.8) to the actual circuit quality (46.72). If the actual value is less than the desired value, you must choose a coil with a higher quality rating. If the actual value is greater than the desired value, a parallel damping resistance may be added to the circuit. At times the actual value may be acceptable. For the purposes of this example, let's add a damping resistor.

Add a damping resistor by first calculating the necessary parallel impedance, $R_{\text{THeffective2}}$. Again use a Q divider and $Q_{\text{effective}} = 44.8$, the desired circuit quality.

$$Q_{\text{effective}} = \frac{R_{\text{THeffective2}}}{R_{\text{THeffective2}} + Z_{\text{tank}}} \times Q_{\text{coil}}$$

$$44.8 = \frac{R_{\text{THeffective2}}}{R_{\text{THeffective2}} + 253.3 \text{ k}\Omega} \times 120$$

$$\frac{44.8}{120} = \frac{R_{\text{THeffective2}}}{R_{\text{THeffective2}} + 253.3 \text{ k}\Omega}$$

$$0.3733 = \frac{R_{\text{THeffective2}}}{R_{\text{THeffective2}} + 253.3 \text{ k}\Omega}$$

$$0.3733 \times (R_{\text{THeffective2}} + 253.3 \text{ k}\Omega) = R_{\text{THeffective2}}$$

$$0.3733 \times R_{\text{THeffective2}} + 94.58 \text{ k}\Omega = R_{\text{THeffective2}}$$

$$0.6267 \times R_{\text{THeffective2}} = 94.58 \text{ k}\Omega$$

$$R_{\text{THeffective2}} = \frac{94.58 \text{ k}\Omega}{0.6267}$$

$$= 150.9 \text{ k}\Omega$$

Understand that $R_{\text{THeffective2}}$ is the parallel impedance of r_{source}, R_{load}, and R_{damp}. Now calculate R_{damp}.

$$R_{\text{THeffective2}} = r_{\text{source}} \parallel R_{\text{load}} \parallel R_{\text{damp}}$$

$$150.9 \text{ k}\Omega = 350{,}000 \parallel 300{,}000 \parallel R_{\text{damp}}$$

$$R_{\text{damp}} = 2.291 \text{ M}\Omega \qquad\blacksquare$$

TROUBLESHOOTING

Troubleshooting resonant and tank circuits is more of an art than a science. Nothing short of experience can make you good at it. Let's begin by talking about standard test procedures. The tank (or resonant) circuit consists of an inductor (coil) and a capacitor. You can make a fair judgment of the condition of the coil by taking a resistance measurement on its windings. Remember that this resistance will probably be a low value (values of 2 or 3 Ω are not unusual), so readings must be taken with a very sensitive ohmmeter. The capacitor can be tested using an ohmmeter also, but since the capacitor is a normal open, the ohmmeter must have a high range (usually 1 MΩ) to make this test accurately. Remember that the ohmmeter should show a low reading (lower one-third of the scale) at first and gradually increase to a high reading (near infinity).

It would be nice if the inductor and capacitor were the only components that could be at fault. That is not the case. There are many, many variables that could result in a resonant circuit not functioning properly. So many, in fact, that it is impossible to discuss all of them here. One of the more common problems is the length of the leads used in building the circuit. Resonant circuits require that all leads be kept to a minimum length (typically $\frac{3}{8}$ in.). If the circuit is being breadboarded (built on a proto-board), all of the jumper wires must also be kept at a minimum length. Resonant circuits do emit electromagnetic waves (radio waves). Any extra-long wire (lead) can pick up these radio waves (like an antenna), which may result in interference with the resonant circuit.

Dc bias voltage is another important factor in resonant circuitry. The dc voltage level should be at (or as close to) zero as possible. The presence of any dc voltage in the resonant circuit has a tendency to charge the capacitor and thus reduce its role in creating the flywheel effect. When working on a tank circuit, it is important to ensure that the proper static (dc) voltage level is present (check the schematic). When using an RF generator or function generator, make sure that there is no dc offset voltage present.

Dc voltage is dependent, in part, on circuit ground. A faulty common or ground circuit in many buildings can create additional problems. Any time you are working with several devices (dc power supply, RF generator, function generator, etc.), try to keep them all plugged into the same power supply. Plugging them in outlets only feet apart can create difficulties. Always work in areas that have a good ground circuitry.

Any device that emits electromagnetic waves can cause interference. Such devices as high-speed motors, television sets, computers, and fluorescent lights may create an interference that can prevent your resonant circuit from operating correctly. Try to avoid working with resonant circuits in areas where these are operating.

SUMMARY

Series resonance produces maximum circuit current, by allowing the equal capacitive and inductive reactances to cancel. Maximum current results in a maximum output voltage (voltage across an reactive component).

Parallel resonance produces maximum circuit impedance and minimum circuit current. The current through the reactive members is quite high, due to the flywheel effect.

The quality of the coil,

$$Q = \frac{X_L}{r_{\text{coil}}}$$

can be used to predict the bandwidth, the tank impedance, and the output voltage.

It is important to understand that series (source) resistance can and will reduce the quality of the circuit (effective quality). The exact effect can be found by Thévenizing the circuit around the tank and then using the voltage divider (now a Q-divider) proportionality to predict the effective quality of the circuit.

A parallel damping resistance can increase the bandwidth (by reducing the quality factor) and is sometimes used for this purpose. The damping resistance and source resistance will, however, reduce the output voltage.

Resonant circuits are subject to electromagnetic interference. One precaution against this is maintaining a short lead length; another is to make sure that adequate grounding is maintained.

QUESTIONS

21-1. What circuit conditions are associated with series resonance?

21-2. In your own words, describe what is meant by the *flywheel effect*.

21-3. What circuit conditions are associated with parallel resonance?

21-4. In your own words, describe what is meant by *coil quality*.

21-5. What would be a good value for coil quality?

21-6. What circuit conditions can the quality of the coil predict?

21-7. Describe the response curve of a resonant circuit.

21-8. How do series resonant response curves differ from parallel resonant response curves?

21-9. There are three frequencies (f_{res}, f_{low}, and f_{high}) associated with resonance. Explain the differences and relationships between them.

21-10. Briefly describe how a series resonant circuit is analyzed.

21-11. Briefly state the effect of source impedance on a series resonant circuit.

21-12. In your own words, state the difference between the quality of the circuit and the quality of the coil.

21-13. In your own words, state what effect source impedance has on parallel resonant circuits.

21-14. What is a Q generator?

21-15. From where in the circuit does Q come?

21-16. What phase shift is associated with a resistive circuit? A capacitive circuit? An inductive circuit?

21-17. Briefly describe how source impedance affects the way you would analyze a parallel resonant circuit.

21-18. Which components in a parallel resonant circuit can cause circuit damping?

21-19. In your own words, describe how to determine the Thévenin resistance controlling output voltage.

21-20. In your own words, describe how to determine the Thévenin resistance controlling circuit quality.

21-21. Which values will not be affected by the addition of a damping resistor?

21-22. What resonant circuit conditions are affected by parallel damping?

21-23. A circuit is constructed using a coil ($Q = 100$). What can be done to the circuit to make the effective quality of the circuit greater than 100?

21-24. Refer to Figures 21-13 and 21-14. In your own words, describe the difference between these sketches. Remember that they are sketches of the same circuit.

21-25. What is another name for $R_{THeffective}$?

21-26. The text states that it is better to gain an understanding of what is being accomplished than to memorize formulas. Do you agree? Explain.

21-27. Get a catalog from a local electronics supplier. List the available values of capacitors and inductors. Which list is longer?

21-28. Get a catalog from a local electronics supplier. What are the values of coil quality listed?

21-29. Sketch a circuit having one inductor, a multiposition switch, and several capacitors. The circuit is to be of a tank design and must resonate at several frequencies.

21-30. In your own words, describe the difference between tank circuit analysis and tank circuit design.

21-31. What is the standard method of testing a tank circuit? Describe.

21-32. What things (other than defective components) can cause resonant circuit malfunction?

PRACTICE PROBLEMS

21-1. Given that a series resonant circuit is powered by a 200-microvolt source and has the following component values: $r_{coil} = 100\ \Omega$, $L = 35$ mH, and $C = 22$ pF, find the resonant frequency, the upper and lower cutoff frequency, the quality of the coil, the bandwidth of the circuit, and the output voltage of the circuit.

21-2. A series resonant circuit resonates at 1120 kHz and has a bandwidth of 25 kHz. The circuit contains a 250-μH choke. Find the quality of the inductor, the ac resistance of the coil, and the value of the capacitor needed to produce this resonant frequency. Also find the upper and lower cutoff frequency.

21-3. An experimenter has found the inductive reactance of a series resonant circuit to be 25 kΩ. He has measured the upper and lower cutoff frequencies to be 325 kHz and 275 kHz, respectively. What are the bandwidth and resonant frequency of this circuit? What are the quality and ac resistance of the circuit?

21-4. A series resonant circuit produces a maximum of 3.75 V across the capacitor when powered by a 15-mV source operating at a frequency of 500 kHz. The circuit contains a 500-μH choke. Find the bandwidth of this circuit. Also find the quality and ac resistance of the coil.

21-5. Given that a series resonant circuit is powered by a voltage source having an internal resistance of 600 Ω. The inductive quality of the coil is 250 and the ac resistance of the coil at resonance is 150 Ω. Find the effective quality of the circuit.

21-6. A 100-mV source has an impedance of 300 Ω and is used to power a series resonant circuit. The circuit contains a coil whose ac resistance is 200 Ω and has a quality of 75. Find the output voltage and effective quality of this circuit.

21-7. A tank circuit consists of a 30-mH choke having an ac resistance of 250 Ω and a 22,000-pF capacitor. Find the resonant frequency, the bandwidth, the upper and lower cutoff frequenies, the impedance of the tank at resonance, and the quality of the coil.

21-8. Given that a tank circuit consists of a 300-μH choke and a 2.2-μF capacitor. At what frequency will the generator current be at a minimum?

21-9. Given that $r_{coil} = 200\ \Omega$, $L = 3.3$ mH, $C = 0.22$ pF, and $v_{gen} = 250$ mV. Find f_{res}, BW, f_{low}, f_{high}, I_T, and Z_{tank}.

21-10. Given that $r_{source} = 250$ kΩ, $r_{coil} = 290$ kΩ, $L = 4.7$ mH, $c = 1.8$ pF and $v_{gen} = 500$ V, find f_{res}, BW, Z_{tank}, $Q_{effective}$, and I_T.

21-11. Given that $r_{source} = 600$ kΩ, $r_{coil} = 250$ kΩ, $R_{load} = 100$ kΩ, $L = 6$ mH, $C = 2.2$ pF, and $v_{gen} = 350$ mV, find $Q_{effective}$, Z_{tank}, BW, v_{out}, f_{low}, and f_{high}.

21-12. Given that $r_{source} = 300$ kΩ, $r_{coil} = 200$ kΩ, $R_{load} = 500$ kΩ, $R_{damp} = 2$ MΩ, $L = 3.9$ mH, $C = 3.9$ pF, and $v_{gen} = 400$ mV, find $Q_{effective}$, Z_{tank}, BW, v_{out}, f_{res}.

21-13. Use a 35-mH choke having a quality rating of 50 to design a tank circuit that will resonate at 500 kHz and has a bandwidth of 15 kHz.

21-14. Use a 0.035-mH choke having a quality rating of 50 to design a tank circuit that will resonate at 500 kHz and has a bandwidth of 15 kHz when powered by a 200-mV generator having an internal resistance of 20 kΩ.

21-15. Use a 0.035-mH choke having a quality rating of 50 to design a tank circuit that will resonate at 500 kHz and has a bandwidth of 15 kHz when powered by a 200-mV generator having an internal resistance of 20 kΩ and driving a 100-kΩ load.

Calculators

For electronics work, there are two types of calculators that are adequate, both are referred to as slide-rule calculators. One uses *algebraic notation,* which means that you enter numbers in the same manner as you would write it down in algebra. The other is called a *RPN* (*reverse Polish notation*) calculator. A comparison of the advantages and disadvantages of each follows. In this book we use algebraic notation throughout. It is suggested that you use a cheap ($20) algebraic calculator that has engineering notation. Experience has shown that learning electronics and how to operate a fancy calculator at the same time is extremely difficult.

TYPES OF CALCULATORS

Algebraic Since the algebraic calculator receives data in the same way as you would normally write it down, it is the easiest for students to learn how to use. Algebraic calculators have two distinct advantages: they are easy to use, and changing from one algebraic calculator to another or even to a computer is not a difficult transition.

Reverse Polish Notation RPN calculators receive data in a manner similar to the way you would work the problem out on paper. They can easily be identified since they do not have parentheses or an equal sign. In complicated calculations the RPN calculator takes fewer strokes to generate the same answer. The main disadvantage of RPN calculators is the degree of difficulty experienced in switching from an RPN to an algebraic calculator. This is a disadvantage since the algebraic type is the most common calculator.

CALCULATOR FUNCTIONS

Standard Arithmetic Functions

1. *Add* [+]: performs addition operations.
2. *Subtract* [−]: performs subtraction operations.
3. *Multiply* [×]: performs multiplication operations.
4. *Divide* [÷]: performs division operations.
5. *Sign-change key* [+/−]: changes the sign of the number displayed, positive to negative or negative to positive.

Special Functions

6. *Reciprocal key* [1/x]: divides the displayed number into 1; sometimes a second function key.

$$\frac{1}{R_T} = \frac{1}{20} + \frac{1}{30} + \frac{1}{50}$$

[2] [0] [1/x] [+] [3] [0] [1/x] [+] [5] [0] [1/x] [=] [1/x]

display 9.6774

7. *Square function key* [x^2]: raises the displayed number to the second power; sometimes a second function key.

$$P = 0.02^2 \times 150$$

[.] [0] [2] [x^2] [×] [1] [5] [0] [=] display 0.06

8. *Square-root key* [$\sqrt{x}$]: calculates the square root of the displayed number; sometimes a second function key.

$$Z_T = \sqrt{30^2 + 40^2}$$

[3] [0] [x^2] [+] [4] [0] [x^2] [=] [$\sqrt{x}$] display 50

Logarithm Functions

9. *Logarithm key* [log]: finds the base 10 logarithm of the number displayed.

$$n = \frac{\log 16}{\log 2}$$

[1] [6] [log] [÷] [2] [log] [=]

display 4

10. *Natural log key* [ln]: finds the natural (base *e*) logarithm of the number displayed.

$$36v = 100e^{-t}$$

[3] [6] [÷] [1] [0] [0] [=] [ln]

display −1.0217

11. *Antilogarithm key* [10^x]: calculates the value of 10 raised to the power displayed; usually a second function key.

$$\log a = 2.357$$

[2] [.] [3] [5] [7] [10^x]

display 227.51

12. *Antiln key* [e^x]: calculates the value of e (2.71828) raised to the power displayed; usually a second function key.

$$v = 100e^{-2}$$

[1] [0] [0] [×] [2] [+/−] [e^x] [=]

display 13.534

Trigonometric Functions

13. *Sine key* [sin]: finds the sine of the angle displayed.

$$X_L = 1000 \times \sin 30°$$

[1] [0] [0] [0] [×] [3] [0] [sin] [=]

display 500

14. *Arcsin or inverse sine key* [$\sin^{-1}$]: finds the angle corresponding to the displayed sine; usually a second function key.

$$500 = 1000 \times \sin \theta$$

[5] [0] [0] [÷] [1] [0] [0] [0] [=] [$\sin^{-1}$]

display 30

15. *Cosine key* [cos]: finds the cosine of the angle displayed.

$$R = 1000 \times \cos 30°$$

[1] [0] [0] [0] [×] [3] [0] [cos] [=]

display 866.03

16. *Arc-cosine or inverse cosine key* [$\cos^{-1}$]: finds the angle corresponding to the displayed cosine; usually a second function key.

$$866.03 = 1000 \times \cos \theta$$

[8] [6] [6] [.] [0] [3] [÷] [1] [0] [0] [0] [=] [$\cos^{-1}$]

display 29.999

17. *Tangent key* [tan]: finds the tangent of the angle displayed.

$$X_L = 866.03 \times \tan 30°$$

[8] [6] [6] [.] [0] [3] [×] [3] [0] [tan] [=]

display 500

18. *Arctangent or inverse tangent* [$\tan^{-1}$]: finds the angle corresponding to the tangent displayed; usually a second function key.

$$\frac{500}{866.03} = \tan \theta$$

[5] [0] [0] [÷] [8] [6] [6] [.] [0] [3] [=] [$\tan^{-1}$]

display 29.999

Exponential Functions

19. *Power key* [y^x]: raises the displayed number to the power immediately following.

$$n = 2^4$$

[2] [y^x] [4]

                                                                                                                 display   16

20. *Exponential key* [EE] or [EXP]: allows input of exponential of scientific or engineering notation.

$$V = 2.15 \times 10^3 \times 3.0 \times 10^{-6}$$

[2] [.] [1] [5] [EE] [3] [×] [3] [EE] [6] [+/−] [=]

                                                                      display   6.45   −03

ACCURACY

Modern-day calculators will typically provide eight- to ten-digit accuracy in their results. For the most part, calculations to this accuracy are unnecessary. It is suggested that you work with four-digit accuracy in finding your answers. Only in extreme cases is it necessary to use more than four digits in problem solving. Let me explain what is meant by four-digit accuracy.

Four-digit accuracy refers to the number of digits used in the calculations. Leading zeros are not counted. This is not the same as working to four decimal places. Examine the following numbers.

$$76547210 = \quad 76550000 = 7.655 \times 10^7$$

$$0.00007654321 = 0.00007654 = 7.654 \times 10^{-8}$$

In both examples the original number is rounded to produce a value containing four weighted digits. In simple terms, always work with a coefficient consisting of four digits. The examples in this book are done in this manner.

Engineering and Scientific Notation

When you hear someone mention 3 Million, a picture of the number 3,000,000 forms in your mind. This is the standard way of noting that number. Manufacturers, as well as engineers, note numbers in a manner all their own. Calculators give answers in an entirely different manner. This appendix is devoted to an explanation of these differing ways of noting numbers.

SCIENTIFIC NOTATION

In working with electronics it becomes necessary to deal with very large, as well as very small numbers. Numbers like 10,000,000 ohms, 0.0000011 farads, 0.000050 amps, and 50,000 volts are messy and time consuming to write down. One method of getting around this is to move the decimal point to the first position after the first whole number, while at the same time keeping tract of how many places the decimal point has been moved. So a number like 480,000 becomes 4.8×10^5.

$$480000. \rightarrow 4.\underset{5}{8.}\underset{4}{0.}\underset{3}{0.}\underset{2}{0.}\underset{1}{0.}$$

$$4.8 \times 10^5$$

Notice the decimal was placed after the first number, 4, and in doing this it was moved 5 places to the left.

Vocabulary

Coefficient: The coefficient is a number between 1 and 10. The coefficient represents the numerical value attached to the number in question. In the previous example, 4.8 was the coefficient. The following examples will show how the coefficient is determined.

Exponent: The exponent is a statement of how far the decimal point has been moved from its original position. The exponent is a power of 10. It is helpful to view the number line when examining exponents. The following examples will show how the exponent is determined.

Examples

$$0.0008876 = 8.876 \times 10^{-4}$$

[+6]	[+5]	[+4]	[+3]	[+2]	[+1]	[0].	[−1]	[−2]	[−3]	[−4]	[−5]	[−6]	[−7]	power of 10
[ ]	[ ]	[ ]	[ ]	[ ]	[ ]	[**0**].	[0]	[0]	[0]	[8]	[8]	[7]	[6]	
[ ]	[ ]	[ ]	[ ]	[ ]	[ ]	[ ]	[**0**].	[0]	[0]	[8]	[8]	[7]	[6]	1 place
[ ]	[ ]	[ ]	[ ]	[ ]	[ ]	[ ]	[ ]	[**0**].	[0]	[8]	[8]	[7]	[6]	2 places
[ ]	[ ]	[ ]	[ ]	[ ]	[ ]	[ ]	[ ]	[ ]	[**0**].	[8]	[8]	[7]	[6]	3 places
[ ]	[ ]	[ ]	[ ]	[ ]	[ ]	[ ]	[ ]	[ ]	[ ]	[**8**].	[8]	[7]	[6]	4 places

The decimal place has been moved to the right.   A total of 4 places.

$$2500 = 2.5 \times 10^{3}$$

[+6]	[+5]	[+4]	[+3]	[+2]	[+1]	[0].	[−1]	[−2]	[−3]	[−4]	[−5]	[−6]	[−7]	power of 10
[ ]	[ ]	[ ]	[2]	[5]	[0]	[**0**].	[ ]	[ ]	[ ]	[ ]	[ ]	[ ]	[ ]	
[ ]	[ ]	[ ]	[2]	[5]	[**0**].	[0]	[ ]	[ ]	[ ]	[ ]	[ ]	[ ]	[ ]	1 place
[ ]	[ ]	[ ]	[2]	[**5**].	[0]	[0]	[ ]	[ ]	[ ]	[ ]	[ ]	[ ]	[ ]	2 places
[ ]	[ ]	[ ]	[**2**].	[5]	[0]	[0]	[ ]	[ ]	[ ]	[ ]	[ ]	[ ]	[ ]	3 places

The decimal point has been moved to the left.   A total of 3 places.

$$650 = 6.5 \times 10^{2}$$

[+6]	[+5]	[+4]	[+3]	[+2]	[+1]	[0].	[−1]	[−2]	[−3]	[−4]	[−5]	[−6]	[−7]	power of 10
[ ]	[ ]	[ ]	[ ]	[6]	[5]	[**0**].	[ ]	[ ]	[ ]	[ ]	[ ]	[ ]	[ ]	
[ ]	[ ]	[ ]	[ ]	[6]	[**5**].	[0]	[ ]	[ ]	[ ]	[ ]	[ ]	[ ]	[ ]	1 place
[ ]	[ ]	[ ]	[ ]	[**6**].	[5]	[0]	[ ]	[ ]	[ ]	[ ]	[ ]	[ ]	[ ]	2 places

The decimal point has been moved to the left.   A total of 2 places.

$$0.053 = 5.3 \times 10^{-2}$$

[+6]	[+5]	[+4]	[+3]	[+2]	[+1]	[0].	[−1]	[−2]	[−3]	[−4]	[−5]	[−6]	[−7]	power of 10
[ ]	[ ]	[ ]	[ ]	[ ]	[ ]	[**0**].	[0]	[5]	[3]	[ ]	[ ]	[ ]	[ ]	
[ ]	[ ]	[ ]	[ ]	[ ]	[ ]	[0]	[**0**].	[5]	[3]	[ ]	[ ]	[ ]	[ ]	1 place
[ ]	[ ]	[ ]	[ ]	[ ]	[ ]	[0]	[0]	[**5**].	[3]	[ ]	[ ]	[ ]	[ ]	2 places

The decimal point has been moved to the right.   A total of 2 places.

$$0.0000125 = 1.25 \times 10^{-5}$$

[+6] [+5] [+4] [+3] [+2] [+1] [0].[−1] [−2] [−3] [−4] [−5] [−6] [−7]	power of 10
[ ][ ][ ][ ][ ][ ][**0**].[0][0][0][1][2][5]	
[ ][ ][ ][ ][ ][0][**0**].[0][0][0][1][2][5]	1 place
[ ][ ][ ][ ][ ][0][0][**0**].[0][0][1][2][5]	2 places
[ ][ ][ ][ ][ ][0][0][0][**0**].[0][1][2][5]	3 places
[ ][ ][ ][ ][ ][0][0][0][0][**0**].[1][2][5]	4 places
[ ][ ][ ][ ][ ][0][0][0][0][0][**1**].[2][5]	5 places

The decimal point has been moved to the right. A total of 5 places.

$$1780000 = 1.78 \times 10^6$$

[+6] [+5] [+4] [+3] [+2] [+1] [0].[−1] [−2] [−3] [−4] [−5] [−6] [−7]	power of 10
[1][7][8][0][0][0][**0**].[ ][ ][ ][ ][ ][ ][ ]	
[1][7][8][0][0][**0**].[0][ ][ ][ ][ ][ ][ ][ ]	1 place
[1][7][8][0][**0**].[0][0][ ][ ][ ][ ][ ][ ][ ]	2 places
[1][7][8][**0**].[0][0][0][ ][ ][ ][ ][ ][ ][ ]	3 places
[1][7][**8**].[0][0][0][0][ ][ ][ ][ ][ ][ ][ ]	4 places
[1][**7**].[8][0][0][0][0][ ][ ][ ][ ][ ][ ][ ]	5 places
[**1**].[7][8][0][0][0][0][ ][ ][ ][ ][ ][ ][ ]	6 places

The decimal point has been moved to the left. A total of 6 places.

Notice that when you are dealing with a large number the exponent (power of 10) is positive, while the exponent is negative when the number is small. The following are examples of standard numbers as they would be written in scientific notation.

Number		
Standard Notation	Scientific Notation	
	Coefficient	Exponent
123000.	1.23 ×	10^5
0.00046	4.6 ×	10^{-4}
25000.	2.5 ×	10^4
0.0008876	8.876 ×	10^{-4}
47832189.	4.7832189 ×	10^7
45620.	4.562 ×	10^4
0.0000567	5.67 ×	10^{-5}
7430.	7.43 ×	10^3
0.082	8.2 ×	10^{-2}

ENGINEERING NOTATION

It is time consuming to continually write the power of ten after a number noted in scientific notation. Engineers have simplified the scientific notation method by including symbols that imply the power of ten being used, i.e., 1.5×10^3 ohms is written as 1.5 kohms. You must memorize the symbols in Chart 1-5 to be able to read data supplied by the manufacturer.

Vocabulary

Coefficient: The coefficient is a number between 1 and 1000. The coefficient represents the numerical value attached to the number in question. In the previous example 1.5 was the coefficient.

Prefix: The prefix is a letter or symbol attached to a unit of measure. This letter represents the power of ten associated with the coefficient. See chart 1-5 for examples.

Notice that this system allows the exponent to change by a factor of three each time. This requires some expertise in changing or placing the exponent into a multiple of three. Also understand, these symbols can only correctly be used along with a unit, that is, 87 m is not correct while 87 mAmps would be. Chart 1-5 shows the most used symbols and their corresponding values.

CHART 1-5

Symbol	Name	Value
E	exa	$\times 10^{18}$
P	peta	$\times 10^{15}$
T	tera	$\times 10^{12}$
G	giga	$\times 10^{9}$
M	mega	$\times 10^{6}$
k	kilo	$\times 10^{3}$
m	milli	$\times 10^{-3}$
μ	mu (micro)	$\times 10^{-6}$
n	nano	$\times 10^{-9}$
p	pico	$\times 10^{-12}$
f	femto	$\times 10^{-15}$
a	atto	$\times 10^{-18}$

Scientific to Engineering Conversion

The following are some examples of conversions from scientific notation to engineering notation.

Scientific Notation		Engineering Notation
4.56×10^{10} Hz	$= 45.6 \times 10^{9}$ Hz	$= 45.6$ GHz
6.5×10^{-4} Amps	$= 650 \times 10^{-6}$ Amps	$= 650$ μAmps
4.008×10^{-3} Volts	$=$	4.008 mVolts
9.7×10^{4} Ohms	$= 97 \times 10^{3}$ Ohms	$= 97$ kOhms

Equivalent Engineering Values

Addition and subtraction require that all units have the same prefix. This necessitates the conversion from one engineering symbol to another. This conversion is done in the same manner as any conversion from scientific notation to engineering notation. The following chart shows a few of these conversions.

Mega	Kilo	Units	Milli	Micro
0.00485	4.85	4850	4850000	4850000000
5	5000	5000000	5000000000	5000000000000
0.00000003	0.00003	0.03	30	30000

Notice that these conversions always move the decimal point three places at a time. Again if you are moving from a large exponent to a smaller exponent you must make the coefficient larger, while when making conversions from a small exponent to a larger one, the reverse is true.

C

Glossary

Ampere (amp): standard unit of current. One coulomb of charge past a given point in one seconds time. C2

Amplitude: peak value of the sine wave. C14

Anode: terminal where electrons exit. Negative terminal on power source or positive terminal on components. C18

Apparent power: the amount of power provided by the power source. C19

Atom: basic building block for all things. See nucleus electrons and orbital. C2.

Balanced (neutral): equal number of positive and negative charges in a material. C2

Band stop filter: High and Low pass filter connected in parallel to reject one band of frequencies. C20

Bandwidth: the amount of frequency between cutoff points. measured in Hz. C20, 21

Bessel filter: exhibits as shallow rolloff and a smooth pass band—provides a linear phase shift and transient response with minimal distortion. C20

Bode Plot: A graph of frequency response for filter or other circuit. Frequency vs dB level. C20

Bridge circuit: two series strings in parallel. balanced or unbalanced determined by potential difference between two points. C6

Butterworth filter: exhibits a standard rolloff and a flat pass band. Most commonly used filter. C20

Capacitance: indicates how fast a capacitor will accumulate a charge. C16

Capacitor: constructed of two plates separated by an insulator. Serves as a shock absorber to ac circuits. Stores DC voltage. C16

Cathode: terminal where electrons enter. Positive terminal on power sources or negative terminal on component. C18

Cauer (eliptic) filter: exhibits a rippled steep rolloff with a rippled pass band. C20

Chebychev filter: exhibits a steep rolloff with a rippled pass band. C20

cgs: centimeter-gram-second system of measurement. C13

Conductance: lack of opposition to current flow. Reciprocal of resistance value. Measured in mhos or seimens. C2

Conductors: have a very small amount of resistance. C11

Coulomb: unit of charge equaling 6.25×10^{18} electrons. C2

Current: the rate of electrical flow—measured in amperes (amps). C2

Current dividers: (parallel circuit) divides mainline current between resistors as branch currents. C8

Current meter: inserted into the circuit in series fashion—polarity must be followed—measures the number of amps flowing through point of insertion. C7

Current rating: amount of current fuse can handle with out opening. C11

Damping resistor: decreases voltage output, widens band width and lowers the quality of the circuit. There are three sources; internal resistance of the generator, impedance of load and external addition of a parallel damping resistor. C21

Decibels: the unit used to measure power, voltage and current changes. C20

Dielectric strength: the strength (insulating properties) of the material used to separate the capacitor plates. C16

Differentiator: prime function is to provide immediate output for any change in input. C17

Directly proportional: if there is an increase in A then there must be an increase in B. C3

Doped: having an enhanced electrical charge or characteristic. C16

Eddy currents: produced in the inductor core and result in energy loss. C21

Efficiency: (η) the rating of power transmitted from the primary to secondary windings of a transformer. C15

Electricity: the movement of electrons. C2

Electron: negatively charged portion of atom. C2

Energy: force times distance. C2

Farad: unit for capacitance. C16

Faraday's law: There are three determining factors in the production of voltage in a generator—1) number of turns 2) speed in which the magnet is introduced into the coil 3) the strength of the magnetic field. C13

Flemming's rule: see figure 13-8 for complete explanation. C13

Flux: (ϕ) flow of magnetism, the lines of force between north and south poles. C13

Flux density: (B) strength of the magnetic field. C13

Filters: 1) used to separate ac and dc components of a signal 2) used to allow only certain frequencies to be transmitted. C20
active—requires use of amplifier
passive—are combinations of resistors, capacitors and inductors

Fuse: most commonly used device for protection from circuit overload. C11
slow blow—used on devices that require more current on initial operation than during normal operation
fast blow—used on circuits that require immediate shut down in the event of an over current condition

APPENDIX C / GLOSSARY

Fusible link: weakest portion of the circuit—during over current, link will serve as a fuse. C11

Galvanometer: extremely sensitive current meter measures both positive and negative current, meter points to middle if no current is flowing. C6

Ground: a common connection between all components—commonly has a zero potential—positive voltage has negative ground and visa versa. C7

Heat dissipation: (power) electron movement increases resistance which causes added friction—the heat produced is proportional to the amount of work (# of electrons moved) being done in a particular time period. C3

Impedance blocks: combination of resistors, capacitors and inductors—used to break down large and complicated circuits. C19

Integrator: prime function is to provide a slow changing output for a quick changing input signal. C17

Inversely proportional: if there is an increase in A then there must be a decrease in B and visa versa. C3

Isotopes: molecules which are heavier than the normal due to the addition of extra neutrons. C2

j-(j-factor): represents imaginary number, the square root of one—used in rectangular notation as a prefix to any reactance. C19

Joule: unit of work equivalent to watt-second—see power and work. C2

KCL (Kirchhoff's current law): current entering a point must equal the current leaving the same point. C6

Kilowatt hour: standard unit of electrical work

KVL (Kirchhoff's voltage law): voltage is additive—the sum of the voltage drops throughout any one current path must equal the supply voltage. C2

Lenz' Law: states when a magnet is moved into a coil the current produced will be such that the induced magnetic field of the coil will oppose that of the magnet. C13

Logarithm table: consists of two number systems an arithmetic and geometric system—used to perform exponential functions. C17

Mainline: part of the circuit where all the current flows. C5

Matrix: method of laying out resistance, voltage, currents and power in an organized manner. C4

Mid-scale: 1/2 scale current position on an ohm meter. C12

mil: measurement for wire diameter 1 mil = 0.001 inch. C11

Millman's theory: sum of current provided by each current generator is divided by the sum of the conductance of each generator. C9

MKS: meter-kilogram-second system of measurement. C13

mmf (magneto-motive-force): the product of number of turns (of wire used in coil construction) and the amount of current which flows through the coil. C13

Multiplier: current limiting resistor used in volt meter construction. C12

Negative charge: caused by excess of electrons. C2

Neutral: not positive or negative—also see balanced. C2

Neutron: portion of atom that is neutral in charge. C2

Norton's Rule: Any complex network can be reduced to a single current source—in parallel with a current limiting resistor and a load. C9

Octave: is doubling or halving the reference frequency. C20

ohm: (Ω) unit of resistance 1 ohm = one volt per ampere. C2

Ohm's Law: $E = I \times R$ C3

ohm meter: has its own voltage source so only used on a dead circuit—must be calibrated before each use. Used to determine the amount of resistance in a circuit. C7, 12

Open circuit voltage: maximum voltage which can be generated between two points. C9

Parallel circuit: where each load is fed by the battery separately—mutliple path system. C5

Parallel resonance: magnifies the impedance of the circuit—minimizes the current, occurs when X_L and X_C are equal in a Tank circuit. C19

Period (T): refers to the length of time between sine or other wave cycles. C14

Permeability (μ): the ability of metal to consentrate flux lines. C13

PI (π) configuration: delta configuration—see TEE configuration. C9

Polar notation: two components—relates to the hypotenuse and angle. C19

Positive charge: deficiency in electrons. C2

Power: see heat dissipation—electrical unit is measured as a joules/second or watt. C3

Power factor: the cosine of the phase angle—gives a statement of circuit efficiency. C19

Proton: positively charged portion of atom. C2

Q-quality factor: the ratio between inductive reactance and ac resistance. C21

Quadrant: refers to each of the four (90 degree) sections of a sine or other wave

Ratio: numerical representation of how one thing relates to another. C8

Reactance: is an opposition to current flow exhibited by capacitors and inductors—measured in ohms. C15

Reactive power: considered the amount of power that is wasted in the circuits operation. C19

Real power: measured in watts—considered as the power used by circuit. C19

Resistance: the opposition to current flow—see ohm. C2

Resistivity: the amount of resistance in a solid (wire) specific (ρ) determined by the amount of ohms contained in a cross section of the area of a solid. C11

Resonance: a special RCL circuit combination when X_C and X_L are equal—series or parallel—used in communication circuitry for tuning purposes. C19

Series circuit: has only one path for current. C4

Series resonance: minimizes circuit impedance—input voltage is magnified—no phase shift—maximum current flow and voltage drop across components. C19

Shunt resistors: generally any resistor installed in parallel—used to control current flow through meter movements. C12

Shell: orbital of electrons around an atom. see valance shell. C2

Short: a lower than normal amount of resistance. may cause damage to circuit. C3

SI: international standard of measuring. C13

Square wave: a wave produced by a circuit being either off or on—used in digital circuitry.

Superposition: the sum of the effects produced by each power source. C9

Tank circuit: a capacitor and inductor connected in parallel—provides maximum impedance and minimum mainline current. C21

TEE configuration: *Wye*—a three port filter where the components are connected in a manner resembling a *T* or *Y*. C9, C20

Thevenin's rule: any complex network can be reduced to a single voltage source, a current limiting resistance, and a load. C9

Time constant: predicted value of time a circuit will be transient. R × C or L/R. C17

Transformer: a special type of inductor used to step up or step down voltage or current—offers complete separation of ac and dc voltages. C15, C20

Transients (current or voltage): the time that the current and voltages are constantly changing as the power is turned on or off. C17

Triangle wave: signal commonly called a voltage ramp or sawtooth wave—used in video circuits. C14

Trim pot: a small potentiometer often used in printed circuits. C12

Troubleshooting: an organized attempt to find the fault with a circuit.—six step procedure. C7

Valance shell: outer most orbital of atom—determines the electrical properties of the material. C2

Voltage: electrical pressure also called electromotive force. C2

Voltmeter: connected to circuit in a parallel manner—polarity must be followed—used to measure the amount of voltage present between the points of connection—sensitivity of at least 20 kΩ recommended. C7, C12

Voltage divider: series circuit—divides supply voltage between resistors as IR drops. C8

Voltage rating (of fuse): indicates the maximum supply voltage a fuse can protect.

Voltage source: supply—will transmit voltage to the load better when the source resistance is low compared to the load resistance. C9, C18

Answers to Selected Problems

CHAPTER 2 PRACTICE PROBLEMS

2. 856 coulombs   4. 2.8 amps   6. 2.667 ohms
8. 100000 ohms   10. 16.67 mAmps   12. 4 ohms
14. 36 watts   16. G. volt   17. D. Ohms
18. F. coulomb   19. A. amp   20. B. watt
21. C. joule   22. E. mho   23. C. G 24. A. Q
25. D. I 26. F. P 27. B. E 28. G. R
29. E. W 30. E. A 31. A. w 32. D. J
33. C. Ω 34. B. ℧   35. G. C 36. F. V

CHAPTER 3 PRACTICE PROBLEMS

2. $P = 137.5$ watts, $E = 550$ volts
4. $I = 46.47$ mA, $E = 2.323$ mV
6. $I = 7.44$ mA, $P = 0.2606$ watts
8. $R = 1800$ ohms, $P = 1.125$ watts
10. $I = 0.125$ mA, $P = 1.875$ mwatts
12. $R = 187.5$ KΩ, $P = 108$ μw
14. $R = 320$ Ω, $I = 37.5$ mA
16. $I = 66.67$ μA, $P = 1.467$ mwatt
18. $R = 17.85$ Ω, $I = 1.4$ mA
20. $P = 4$ watts, $I = 0.4$ amps
    $R = 25$ Ω
22. $P = 5.5$ watts, $I = 0.05$ A
    $E = 110$ volts
24. $E = 30$ volts, $R = 600$ Ω
    $W = 4.5$ mjoules, $T = 3$ mseconds
    $Q = 150$ μcoulombs

CHAPTER 4 PRACTICE PROBLEMS

2. 33.33 mAmps   4. 2.5 watts
6. $V_1 = 3.158$ v, $V_2 = 6.947$ v, $V_3 = 9.895$ v
   $R_T = 950$ ohms, $I_T = 21.05$ mA, $P_T = 0.4211$ watts
   $I_1 = 21.05$ mA, $I_2 = 21.05$ mA, $I_3 = 21.05$ mA
   $P_1 = 66.48$ mwatt, $P_2 = 146.3$ mwatt, $P_3 = 208.3$ watt
8. $I_T = 140$ mA, $R_T = 71.43$ Ω, $P_T = 1.4$ watt
   $I_1 = 140$ mA, $R_1 = 15$ Ω, $P_1 = 294$ mwatt
   $I_2 = 140$ mA, $R_2 = 30.71$ Ω, $P_2 = 602$ mwatt
   $V_3 = 3.6$ v, $R_3 = 25.71$ Ω, $P_3 = 504$ mwatt
10. $I_T = 2.5$ mA, $V_T = 6.75$ v, $P_T = 16.88$ mwatt
    $I_1 = 2.5$ mA, $V_1 = 550$ mV, $P_1 = 1.375$ mwatt
    $R_3 = 1.28$ kΩ, $V_2 = 3$ v, $P_2 = 7.5$ mwatt
    $I_3 = 2.5$ mA, $V_3 = 3.2$ v, $P_3 = 8$ mwatt
12. $R_T = 112.5$ kΩ, $V_T = 225$ v
    $R_1 = 50$ kΩ, $V_2 = 54$ v
    $R_3 = 35.5$ kΩ, $V_3 = 71$ v
14. $R_T = 10$ kΩ, $V_1 = 7$ mV
    $I_T = 3.5$ μA, $V_2 = 16.45$ mV
    $P_T = 122.5$ μwatt, $V_3 = 11.55$ mV

CHAPTER 5 PRACTICE PROBLEMS

2. I_{12} kohms $= 8.332$ mA, I_{22} kohms $= 4.545$ mA
   $I_{6.8}$ kohms $= 14.70$ mA, I_{15} kohms $= 6.665$ mA
4. $P_1 = 312.5$ mwatts, $P_2 = 520.8$ mwatts
   $P_3 = 231.5$ mwatts, $P_T = 1.065$ wattts
6. 5600 Ω 8. 38.26 ohms

10. $I_T = 11$ amps, $P_T = 55$ watts
12. $I_T = 7.064$ mA, $E_T = 75.02$ v, $P_T = 529.9$ mwatts
    $R_1 = 18.01$ kΩ, $R_2 = 33$ kΩ, $R_3 = 120$ kΩ
    $E_1 = 75.02$ v, $E_2 = 75.02$ v, $E_3 = 75.02$ v
    $P_1 = 312.5$ mwatts, $P_2 = 170.5$ mwatts,
    $P_3 = 46.89$ mwatts
14. $R_T = 1.667$ kΩ, $I_T = 5.991$ mA, $P_T = 59.91$ mwatts
    $I_1 = 3.03$ mA, $I_2 = 0.833$ mA, $I_3 = 2.128$ mA
    $P_1 = 30.3$ mwatts, $P_2 = 8.33$ mwatts,
    $P_3 = 21.28$ mwatts
    $E_1 = 10$ v, $E_2 = 10$ v, $E_3 = 10$ v
16. $R_T = 1.544$ kΩ, $R_1 = 3.9$ kΩ, $R_3 = 5.586$ kΩ
    $I_T = 77.73$ mA, $I_1 = 30.77$ mA, $I_2 = 25.53$ mA
    $E_1 = 120$ v, $E_2 = 120$ v, $E_3 = 120$ v
    $P_T = 9.327$ watts, $P_1 = 3.692$ watts, $P_3 = 2.564$ watts
18. $R_1 = 560$ Ω, $I_T = 28.25$ mA, $P_T = 141.2$ mwatts
    $I_1 = 8.929$ mA, $I_2 = 4.167$ mA, $I_3 = 15.15$ mA
    $E_1 = 5$ v, $E_2 = 5$ v, $E_3 = 5$ v
    $P_1 = 44.64$ mwatts, $P_2 = 20.83$ mwatts,
    $P_3 = 75.75$ mwatts
20. $R_T = 1.509$ kΩ, $I_T = 13.26$ mA, $P_T = 265.1$ mwatts
    $I_1 = 2.941$ mA, $I_2 = 4.255$ mA, $I_3 = 6.061$ mA
    $E_1 = 20$ v, $E_2 = 20$ v, $E_3 = 20$ v
    $P_1 = 58.82$ mwatts, $P_2 = 85.11$ mwatts,
    $P_3 = 121.2$ mwatts
22. $R_T = 718.0$ Ω, $E_T = 15$ v, $P_T = 313.4$ mwatts
    $I_T = 20.89$ mA, $I_2 = 3.846$ mA, $I_3 = 4.545$ mA
    $E_1 = 15$ v, $E_2 = 15$ v, $E_3 = 15$ v
    $P_1 = 187.5$ mwatts, $P_2 = 57.69$ mwatts,
    $P_3 = 68.18$ mwatts
24. $R_T = 3.535$ Ω, $R_1 = 10$ Ω, $R_3 = 8.602$ Ω
    $I_T = 3.395$ A, $I_2 = 0.8$ A, $I_3 = 1.395$ A
    $E_1 = 12$ v, $E_2 = 12$ v, $E_3 = 12$ v
    $P_T = 40.74$ watts, $P_1 = 14.4$ watts, $P_2 = 9.6$ watts

CHAPTER 6 PRACTICE PROBLEMS

2. $R_t = 23.37$ kΩ, $P_T = 26.38$ mwatt, $E_T = 24.83$ v
   $R_3 = 5.271$ kΩ, $E_1 = 9.031$ v, $I_T = 1.063$ mA
   $P_1 = 9.596$ mwatts, $P_2 = 10.84$ mwatts,
   $P_3 = 5.950$ mwatt
4. Parallel   6. Parallel   8. Series
10. $R_T = 9$ kΩ, $R_1 = 5$ kΩ, $R_3 = 6.667$ kΩ
    $I_1 = 1$ mA, $I_2 = 0.4$ mA, $I_T = 1$ mA
    $P_1 = 5$ mwatts, $P_2 = 1.6$ mwatts, $P_3 = 2.4$ mwatts
    $P_T = 9$ mwatts, $V_1 = 5$ v, $V_3 = 4$ v
12. $R_1 = 9.054$ kΩ, $R_2 = 694.4$ Ω, $R_T = 9.730$ kΩ
    $V_1 = 11.17$ v, $V_2 = 0.8333$ v, $V_3 = 0.8333$ v
    $I_1 = 1.233$ mA, $I_3 = 33.33$ μA, $I_T = 1.233$ mA
    $P_1 = 13.77$ mwatts, $P_3 = 27.77$ μwatts,
    $P_T = 14.80$ mwatts
14. $I_1 = 5$ mA, $I_2 = 5$ mA, $I_3 = 2$ mA, $I_4 = 3$ mA,
    $I_5 = 3$ mA, $I_6 = 3$ mA, $V_1 = 30$ v, $V_2 = 10$ v,
    $V_3 = 60$ v, $V_4 = 18$ v, $V_5 = 12$ v, $V_6 = 30$ v
16. $V_A = 197.5$ v, $I_4 = 13.75$ mA, $V_4 = 82.5$ v
    $I_2 = 3.75$ mA, $V_2 = 60$ v, $V_3 = 60$ v
18. $V_6 = 15$ v

20. $V_1 = 43$ v, $V_2 = 20.12$ v, $V_3 = 26.88$ v
    $V_4 = 14.14$ v, $V_5 = 1.94$ v, $V_6 = 4.04$ v
    $V_7 = 0.51$ v, $V_8 = 0.59$ v, $V_9 = 0.84$ v

CHAPTER 7 PRACTICE PROBLEMS

2. Either R_2 or R_3 is shorted.   4. R_3 is open.
6. The circuit has an open.
8. 1) R_5 or R_6 could be shorted.
   2) R_4 could be shorted.

CHAPTER 8 PRACTICE PROBLEMS

2. $V_1 = 12.17$ v   4. $V_2 = 31.78$ v   6. $V_1 = 13.60$ v
8. $V_2 = 34.4$ v   12. $I_1 = 15.78$ mA
14. $I_1 = 17.65$ mA   16. $I_2 = 976.6$ mA
18. $I_2 = 14.16$ μA   20. $I_1 = 1.141$ A

CHAPTER 9 PRACTICE PROBLEMS

2. 14.93 kΩ   4. 16.67 kΩ   6. 6.05 kΩ
8. $I_N = 8.333$ mA, $R_N = 1.2$ kΩ
   $I_{500} = 5.882$ mA, $V_{500} = 2.941$ v
   $I_{1500} = 3.704$ mA, $V_{1500} = 5.556$ v
   $I_{4700} = 1.695$ mA, $V_{4700} = 7.966$ v
   $I_{15000} = 617.3$ μA, $V_{15000} = 9.259$ v
10. 75.91 kΩ, 157.4 kΩ, 128.8 kΩ

CHAPTER 10 PRACTICE PROBLEMS

2. $I_1 = 6.401$ mA, $I_2 = 4.451$ mA, $I_3 = 1.95$ mA
   $V_1 = 14.08$ v, $V_2 = 20.92$ v, $V_3 = 10.92$ v
4. $I_1 = 142.1$ μA, $I_2 = 138.6$ μA, $I_3 = 3.5$ μA
   $V_1 = 14.21$ v, $V_2 = 20.79$ v, $V_3 = 0.77$ v
6. $I_1 = 4.399$ mA, $I_2 = 4.156$ mA, $I_3 = 0.243$ mA,
   $I_4 = 4.156$ mA
   $V_1 = 24.63$ v, $V_2 = 11.22$ v, $V_3 = 0.3645$ v,
   $V_4 = 9.143$ v
8. $I_1 = 71.43$ μA, $I_2 = 64.94$ μA, $I_3 = 6.49$ μA,
   $I_4 = 64.94$ μA
   $V_1 = 23.57$ v, $V_2 = 9.74$ v, $V_3 = 1.428$ v, $V_4 = 11.69$ v
10. $I_1 = 4.114$ mA, $I_2 = 1.266$ mA, $I_3 = 2.8$ mA
    $V_1 = 9.05$ v, $V_2 = 5.95$ v, $V_3 = 15.96$ v
12. $I_1 = 16.43$ μA, $I_2 = 44.29$ μA, $I_3 = 60.72$ μA
    $V_1 = 1.643$ v, $V_2 = 6.644$ v, $V_3 = 13.36$ v
14. $I_1 = 3.637$ mA, $I_2 = 1.771$ mA, $I_3 = 1.866$ mA
    $I_4 = 1.771$ mA, $I_5 = 1.771$ mA, $I_6 = 3.637$ mA
    $V_1 = 8.001$ v, $V_2 = 0.8855$ v, $V_3 = 10.45$ v
    $V_4 = 4.781$ v, $V_5 = 4.781$ v, $V_6 = 6.547$ v
16. $I_1 = 71.72$ μA, $I_2 = 40.04$ μA, $I_3 = 36.68$ μA
    $I_4 = 40.04$ μA, $I_5 = 40.04$ μA, $I_6 = 71.72$ μA
    $V_1 = 7.172$ v, $V_2 = 1.882$ v, $V_3 = 6.97$ v
    $V_4 = 4.004$ v, $V_5 = 1.081$ v, $V_6 = 0.8606$ v
18. $I_1 = 4.489$ mA, $I_2 = 2.101$ mA, $I_3 = 2.701$ mA,
    $I_4 = 2.224$ mA
    $V_1 = 7.051$ v, $V_2 = 10.22$ v, $V_3 = 17.95$ v, $V_4 = 14.78$ v

20. $I_1 = 46.88\ \mu A$, $I_2 = 50\ \mu A$, $I_3 = 46.88\ \mu A$, $I_4 = 50\ \mu A$ $V_1 = 4.688$ v, $V_2 = 7.5$ v, $V_3 = 10.3$ v, $V_4 = 7.5$ v

22. $I_1 = 3.824$ mA, $I_2 = 0.8824$ mA, $I_3 = 0.2036$ mA
$I_4 = 2.942$ mA, $I_5 = 0.6788$ mA, $I_6 = 0.2036$ mA
$I_7 = 3.824$ mA, $I_8 = 0.8824$ mA, $I_9 = 0.2036$ mA
$V_1 = 8.413$ v, $V_2 = 4.147$ v, $V_3 = 1.14$ v
$V_4 = 9.709$ v, $V_5 = 2.647$ v, $V_6 = 0.5497$ v
$V_7 = 6.883$ v, $V_8 = 2.912$ v, $V_9 = 0.9569$ v

24. $I_1 = 23.22\ \mu A$, $I_2 = 14.27\ \mu A$, $I_3 = 2.19\ \mu A$
$I_4 = 8{,}955\ \mu A$, $I_5 = 12.08\ \mu A$, $I_6 = 2.19\ \mu A$
$I_7 = 23.22\ \mu A$, $I_8 = 14.27\ \mu A$, $I_9 = 2.19\ \mu A$
$V_1 = 2.322$ v, $V_2 = 2.141$ v, $V_3 = 0.4818$ v
$V_4 = 5.015$ v, $V_5 = 1.45$ v, $V_6 = 0.4818$ v
$V_7 = 7.663$ v, $V_8 = 1.427$ v, $V_9 = 0.4818$ v

CHAPTER 11 PRACTICE PROBLEMS

2. #12 @ 20°C = 100.2 v, #14 @ 20°C = 88.44 v
4. 1883 ohms
6. Yes. The 120 v fuse would provide adaquate protection for the 25 v circuit.
8. 2.28 ohms

CHAPTER 12 PRACTICE PROBLEMS

2. $S = 33.33\ k\Omega/v$ **4.** $S = 1\ k\Omega/v$
6. $R_{INT} = 200\ k\Omega$ **8.** $2500\ k\Omega$, $250\ k\Omega/v$

10.
Range	Individual	Cascaded
2.5 v	3 kΩ	3 kΩ
20 v	38 kΩ	35 kΩ
50 v	98 kΩ	60 kΩ

12.
Range	Individual	Cascaded
1 mA	51.02 Ω	45.92 Ω
10 mA	5.01 Ω	4.592 Ω
100 mA	.5001 Ω	.5102 Ω

14.
Range	Individual	Cascaded
1 mA	105.3 Ω	94.74 Ω
10 mA	10.05 Ω	9.474 Ω
100 mA	1.001 Ω	0.8421 Ω
500 mA	0.2 Ω	0.2105 Ω

16. Midscale = 45 on the $R \times 10K$ range
18. Midscale = 12 on the $R \times 10K$ range
$I_1 = 92.3\ \mu A$, $I_5 = 70.6\ \mu A$, $I_{12} = 50\ \mu A$
$I_{20} = 37.5\ \mu A$, $I_{50} = 19.4\ \mu A$ $I_{100} = 10.7\ \mu A$
20. $R_{MT} = 120\ k\Omega$, $V_{battery} = 6$ V, $R_{LIM} = 114\ k\Omega$
$R_{100} = 1.2\ k\Omega$, $R_1 = 12\ \Omega$
22. 15 kΩ **24.** 5 Ω

CHAPTER 13 PRACTICE PROBLEMS

2. 18.6×10^{-3} wb **4.** $\phi = 0.625$ gauss/m²
6. mmf = 90 ampturns **8.** $H = 4554$ ampturns/meter
10. $\mu_r = 267$
12. $H = 15625$ ampturns/meter, $B = 1.821$ teslas
14. mmf = 5.4 ampturns **16.** $d\phi/dt = 6$ mwb/s
18. $I = 1.488$ amps **20.** $B = 2.546$ teslas
22. $H = 200$ ampturns/meter **24.** $\mu_r = 3333000$

CHAPTER 14 PRACTICE PROBLEMS

2. $V_P = 50\ \mu v$, $V_{rms} = 35.36\ \mu v$, $T = 1.818\ \mu s$, $f = 550$ kHz
$\lambda_{meter} = 545.5$ meters, $\lambda_{feet} = 1789$ feet
quadrant = 3rd, $e_{in} = -32.14\ \mu v$

4. $V_{PP} = 550$ mv, $V_P = 275$ mv, $V_{rms} = 194.5$ mv, $t = 32.4$ ps
$T = 111.1$ ps, $\lambda_{meter} = 0.03333$ meters,
$\lambda_{feet} = 0.1093$ feet, quadrant = 2nd

6. $V_{PP} = 4.89$ v, $V_{rms} = 1.729$ v, $T = 9.708$ ns, $f = 103$ kHz
$\lambda_{meter} = 2.913$ meters, $\lambda_{feet} = 9.553$ feet,
quadrant = 4th, $e_{in} = -1.873$ v

8. $V_{PP} = 120$ v, $V_{rms} = 42.43$ v, $f = 425.1$ kHz,
$\lambda_{meter} = 705.8$ m, $T = 2.35\ \mu s$, $\theta = 256°$, $e_{in} = -58.22$ v

10. $V_P = 3.150$ v, $V_{PP} = 6.3$ v, $\theta = 165°$, $f = 365.1$ kHz,
$\lambda_{meter} = 821.6$ m, $T = 2.739\ \mu s$, $t = 1.255\ \mu s$

12. $V_P = 6.25$ v, $V_{PP} = 12.5$ v, $T = 1.575\ \mu s$,
$\lambda_{meter} = 472.4$ m, $\lambda_{feet} = 1550$ feet, quadrant = 1, $t = 0.2843\ \mu s$

14. $V_{PP} = 300$ mv, $V_{rms} = 106.1$ mv, $f = 800$ kHz,
$T = 1.25\ \mu s$, $\lambda_{feet} = 1230$ feet,
$\theta = 229996.8° = 638$ revolutions and 316.8°,
quadrant = 3, $e^{in} = -102.7$ v

16. $V_P = 1.174$ v, $V_{PP} = 2.348$ v, $\theta = 165°$,
$t = 305.6$ ns, $f = 1.5$ MHz, $\lambda_{meter} = 200$ m,
$\lambda_{feet} = 656$ feet

CHAPTER 15 PRACTICE PROBLEMS

2. $V_T = 4.995$ v, $L_1 = 32.98$ mH, $X_{L2} = 1.571$ kohm,
$X_{L3} = 1.257$ kohm, $I_1 = 2.88$ mA, $I_2 = 1.28$ mA,
$I_3 = 1.6$ mA, $L_T = 55.2$ mH, $X_{LT} = 1.734$ kohm

4. $L_1 = 0.6$ H, $L_2 = 0.75$ H, $L_3 = 1$ H,
$X_{L3} = 15710$ ohm, $I_2 = 17.69\ \mu A$, $I_3 = 13.26\ \mu A$,
$I_T = 30.94\ \mu A$

6. $X_L = 518.4$ ohm, $Z = 1.126$ kohm, $\theta = 27.4°$,
$V_R = 13.32$ v, $V_L = 6.903$ v, $I = 13.32$ mA

8. $V_T = 4.497$ v. $V_L = 2.11$ v, $V_R = 3.97$ v,
$X_L = 28.15$ kohm, $L = 80$ mH, $\theta = 28°$

10. $V_R = 12.86$ mv, $V_L = 15.32$ mv, $X_L = 715$ ohm,
$L = 210\ \mu H$, $R = 600$ ohm, $I = 21.44\ \mu A$

12. $I_R = 30\ \mu A$, $I_T = 34\ \mu A$, $Z = 13.24$ K ohms,
$\theta = -61.9°$, $I_L = 16\ \mu A$, $X_L = 28.15$ kohm

14. $V_T = 30$ v, $I_L = 0.3323$ mA, $I_R = 1.584$ mA,
$X_L = 90.22$ kohm, $L = 5.74$ H, $R = 18.93$ kohm

CHAPTER 16 PRACTICE PROBLEMS

2. $0.787\ \mu F$

4. $V_R = 6.38$ v, $V_C = 7.7$ v, $X_C = 145\ \Omega$, $Z = 188\ \Omega$,
$\theta = -50.3°$, $I = 53.2$ mA

6. $f = 92.88$ MHz, $V_R = 37.94$ mv, $R = 20\ \Omega$,
$Z = 26.36\ \Omega$
$\theta = -40.6°$, $I = 1.897$ mA

8. $f = 10.4$ MHz, $V_A = 1.5$ v, $V_R = 1.3$ v, $R = 8\ \Omega$,
$X_C = 4.637\ \Omega$, $I = 162.2$ mA

10. $C = 10\ \mu\text{F}$, $I_C = 452.4$ mA, $I_R = 600$ mA,
    $I_T = 751.4$ mA
    $\theta = 37°$, $Z = 159.7\ \Omega$
12. $f = 2.5$ kHz, $I_C = 22.1\ \mu\text{A}$, $I_R = 37.52\ \mu\text{A}$, $R = 8$ kΩ
    $X_C = 13.57$ kΩ, $Z = 6586\ \Omega$
14. $f = 10.4$ MHz, $V_A = 4$ v, $I_C = 122.8$ mA, $I_T = 254$ mA
    $\theta = 28.9°$, $Z = 15.75\ \Omega$

CHAPTER 17 PRACTICE PROBLEMS

2. $\ln 12 = 2.4849$, $\ln 0.79 = -0.2357$, $\ln 3500 = 8.1605$
   $\ln 2.7 = 0.9933$, $\ln 250 = 5.5215$
4. antiln $3.4563 = 31.70$, antiln $-2.4689 = 0.08468$
   antiln $-0.9765 = 0.3801$, antiln $1.3257 = 3.765$
   antiln $2.3467 = 10.45$
6. $t = 18.67\ \mu\text{s}$ 8. $L = 291.5\ \mu\text{H}$
10. $R = 145.8$ kΩ 12. $t = 33.23\ \mu\text{s}$
14. $t = 54.03\ \mu\text{s}$ 16. $R_{\text{off}} = 10$ MΩ

CHAPTER 18 PRACTICE PROBLEMS

2. 91 AH   4. 216 CCA   6. 108 amps
8. 175 amps   10. 3.3 A   12. 0.75 A
14. 0.068 ohm, 0.044 ohm, 1.5 ohm, 0.39 ohm, 0.9 ohm, 6.4 ohm
16. 73 AH, 220 A, 10.2 v, 0.062 ohm, 2.992 kwatt
20. 35.75 AH   22. 0.156 ohm   24. 2 A
26. 5.64 v, 1.88 ohm   28. 90 ohm   30. 1500 ohm
32. 150 ohm

CHAPTER 19 PRACTICE PROBLEMS

2. $Z = 223.6\ \Omega\ \underline{/26.6°}$, $P_{\text{real}} = 160\ \mu\text{watt}$,
   $P_{\text{apparent}} = 178.9\ \mu\text{watts}$
4. $Z_T = 2495\ \Omega\ \underline{/35.5°}$, $I = 6.012$ mA, $V_{R2} = 1.984$ v
   $V_{XL4} = 1.924$ v, $V_{XC1} = 3$ v
6. $Z_T = 1050\ \Omega\ \underline{/49.6°}$, $I = 19.05$ mA, $V_R = 12.95$ v,
   $V_L = 62.86$ v, $V_C = 46.62$ v
8. $I_L = 6$ mA, $I_C = 3.846$ mA, $I_R = 5.556$ mA,
   $I_T = 5.956$ mA
   $\theta = -21.2°$
10. $Z_T = 2074\ \Omega\ \underline{/0°}$ 12. 756 kHz
14. $V_{\text{out}} = 13.56$ v $\underline{/15.2°}$

CHAPTER 20 PRACTICE PROBLEMS

2. $P_{\text{out}} = 3.733$ mw attenuation   4. $P_{\text{in}} = 6.31$ mw gain
6. $v_{\text{out}} = 408.3$ mv gain   8. $v_{\text{in}} = 221.4$ v attenuation
10. Low pass filter will cutoff at approximately 500 kHz.
12. $R_{\text{TH}} = 1100$ ohm, $X_C = 110$ ohm, $C = 0.4823\ \mu\text{F}$

CHAPTER 21 PRACTICE PROBLEMS

2. $Q = 44.8$, $r_{\text{coil}} = 39.27$ ohm, $C = 80.77$ pF,
   $f_{\text{low}} = 1108$ kHz, $f_{\text{high}} = 1133$ kHz
4. $BW = 2$ kHz, $Q = 250$, $r_{\text{coil}} = 6.28$ ohm
8. $L = 300\ \mu\text{H}$, $C = 2.2$ uF, $f_{\text{res}} = 6.195$ kHz
10. $f_{\text{res}} = 1.73$ MHz, $BW = 363$ kHz, $Z_{\text{tank}} = 9$ Mohm,
    $Q_{\text{effective}} = 4.76$, $I_T = 54.05\ \mu\text{A}$
12. $f_{\text{res}} = 1.29$ MHz, $BW = 246.2$ kHz, $Z_{\text{tank}} = 5$ MHz,
    $v_{\text{out}} = 221$ mv, $Q_{\text{effective}} = 5.24$
14. $L = 0.35$ mH, $Q = 50$, $f_{\text{res}} = 500$ kHz, $BW = 15$ kHz,
    $v_{\text{gen}} = 200$ mv, $R_{\text{source}} = 20$ kohm, $X_L = 110$ ohm,
    $C = 2895$ pF, $r_{\text{coil}} = 2.2$ ohm, $R_{\text{damp}} = 24.44$ kohm

Index

ac voltage, 258–86
   calculating, 263–67
   nonsinusoidal waveforms, 277–81
   period and frequency, 273–77
   sine waves, 259–63
   theory, 267–68
   voltage measurement, 268–73
Algebra, basic, 28–32
Alkaline batteries, 376
All-pass filters, 416
Alnico, 243
Alternating-current generator, 267–68
Aluminum, as a conductor, 200
AM broadcast band, 415
American Wire Gauge, 203
Amperes, 16, 23
Ampere-hour rating, 377–78
Ampere-turns, 247
Ampmeters, 113
Angular velocity, 309, 331
Annealed copper, 203
   as a conductor, 199
   wire table, 204
   See also Copper
Anode, 371
Antilogarithm, 350
Apparent power, 393
Armature windings, insulating cover, 211
Atomic number, elements, 9, 10
Atomic structure, 6–14, 24
Atomic weight, 7
Atoms, 24

   definition, 6–7
   sections of, 7
Ayrton shunt, 225
   design of, 226–28

Balanced (neutral) material, 6
Balanced state, of bridge, 96
Band-pass filters, 415, 426–27
Band stop filters, 416, 426–27
Bandwidth, 415, 435–36
   and resonance, 435–36
Basic algebra, 28–32
Batteries, 370–86
   and capacitors, 325–26
   capacity of, 377–78
      ampere-hour rating, 377–78
      cold cranking amperes (CCA), 378, 382
      increasing, 378
   charging, 378–79
   construction of, 371–72
   definition, 372
   impedance matching, 381
   internal resistance, 380–81, 382
   types of, 372–77
      primary cells, 375–77, 382
      secondary cells, 372–75, 382
   See also Primary cells; Secondary cells
Bessel filters, 416–17, 427
B-H curve, 252–54
Biomedical electronics, 2

Block testing, 126–28
   signal injection, 127
   signal (voltage level) tracing, 126–27
Bode plots, 411–12, 426
Braided wires, 210
Bridge, definition, 96
Btu, 18, 24
Butterworth filters, 417–18, 427
Bypass capacitors, 427
   separating ac/dc voltage using, 425–26

Calculators:
   DRG button, 263
   EE key, 19, 54
   finding parallel resistance using, 86–87
   mem (M) button, 57
   1/x key, 22
   parentheses, use of, 69
   RCL (RM) button, 57
   sum (M+) button, 57
   use of, 3, 17–19
   x^2 key, 40
Candle-feet, 29
Capacitive reactance, 330–35
   and Ohm's law, 334–35
   in parallel, 333–34
   in series, 331–33
   versus inductive reactance, 388–92, 403

479

Capacitors, 324–47
  capacitance, 329–30
    reciprocal of, 330
  capacitive reactance, 330–35
  capacitive voltage divider, 335–36
  compared to batteries, 325
  construction/design, 325–29
  parallel, 330
  quality of, 342
  RC circuits, solution of, 337–42
  series, 330, 413
  troubleshooting, 342–43
    capacitance, 343
    series resistance, 342
    voltage leakage, 342–43
  types of, 328
Carbon-film resistors, 202
Carbon-zinc cell, 375–76
  local action, 375
  polarization, 375–76
Cascaded multiplier voltmeters, design of, 223–25
Cathode, 371
Cauer filters, 419, 427
Ceramic disk capacitors, 328
Cgs system, 247, 249
  flux density, 249
Charge, 23
Charged particles, properties of, 6
Charging, batteries, 378–79
Chebyshev filters, 418–19, 427
Christie, Samuel, 104
Circuit breakers, 213–14
Circuit damping, 451
Circuit identification, 413, 426–27
  parallel circuits, 64–65
  series circuits, 51–52
Circuit protection, 212–14
  circuit breakers, 213–14
  fuses, 212–13
  fusible links, 214
Circuit resistance tests, 130
Circuits:
  load/resistance, 44
  path, 44
  requirements of, 43–46
  schematic symbols, 47
  switches, 211–12
  using, for the matrix, 60
Classic filters, 419–20, 427
Coaxial cables, 210
Coefficient of coupling, 297
Cold cranking amperes (CCA) rating, batteries, 378, 382
Common logarithms, 350
Communications, 2
Complaint verification, 125
Component testing, 128
Conductance, 22–23, 24
Conductors, 9, 199–200
  allowable current carrying capacities of, 208
  resistivity, 200–210

  temperature coefficients of, 209
  wire construction, 210–11
Continuity test, 130
Conventional theory of electron movement, 15
Copper:
  as a conductor, 199
  forms of, 203
  See also Annealed copper
Cosine, 287, 289–90
Coulomb, 15, 23, 24
Coulomb/second, 37
Counterelectromotive force (cemf), 307–9
Coupling capacitors, 427
  separating ac/dc voltage using, 424–25
Cross-multiplication, 137
Current, 15–20, 23, 24, 193
  definition, 16
  flow, 15
  transients in, 348–69
Current division, 143–45
Current flow, 37, 67
  direction of, 51
Current generation, 251
Current levels:
  ELI the ICEman, 351–52
  methods for finding, 353–61
    graphic method, 355–60, 365
    mathematic method, 360–61, 365
    time constants, 352–35
Current-limiting resistance, 379
Current loop, 193
Current meters, 116–17
  cascaded (Ayrton) shunt current meters, 226–28
  construction and design of, 225–28
  individual shunt currents, 225–26
Current rating, fuses, 213
Current source, 164
Cutoff frequency, 414–16, 426
  all-pass filters, 416
  band-pass filters, 415
  band-stop filters, 416
  high-pass filters, 414
  low-pass filters, 414–15

Damping resistance, 447–51
dc circuits, troubleshooting, 112–34
  block testing, 126–28
  complaint verification, 125
  component testing, 128
  ground, 120–25
  meter use, 113–20
  power supply testing, 126
  repair, 128
  special tests, 129–32
  visual inspection, 125–26
Decade box, 105
Decades, 280, 411–12, 426
Decibels, 409–12, 426

Delta and wye circuits, 169–72
  conversion from delta to wye, 169–71
  conversion from wye to delta, 170–71
Diamagnetic materials, 244
Dielectric constant, 326
Dielectric strength, 326
Difference, definition, 20
Differentiator, 362, 365
Digital electronics, 2
Digital meters, internal resistance, 219
Diode, 269, 371
Directly proportional, 140
  definition, 28
Directly proportional relationships, 139
Disk capacitors, 328
Doped insulators, 328
Double pole, double throw switches, 211
Double pole, single throw switches, 211
DRG button, calculators, 263
Dual-element L networks, 421, 427
Dual-trace oscilloscopes, 119

Eddy currents, 292
Edison cell, 375
Efficiency rating, 305
Elasticity, 330
Electrical measurement, fundamental units of, 23
Electrical pressure, 23
Electricity, definition, 14
Electrolyte, 371, 372, 382
Electrolytic capacitors, 328–29
Electromagnetism, 244–46
  concepts of, 250–52
  Faraday's law, 251–52
  Lenz's law, 251
  motor action and Flemming's rule, 250–51
  See also Magnetism
Electromotive force, 20–21, 23, 307
  field intensity, 245–46
  flux density, 246
  magnetomotive force (mmf), 244
  See also Voltage
Electronic fields, 2
  employment areas, 2–3
Electron movement, 14–15
Electron orbitals, 7
Electrons, 7–9, 23, 24, 325–26, 371
  definition, 23
  relations of shells to, 14
Electron theory of current flow, *See* Electron movement
Elements:
  electrons/shells, 12–13
  periodic table of, 11
  and their symbols, 9, 10

ELI the ICEman, 311–17, 319, 337–38, 344, 351–52, 365
Elliptic filter, *See* Cauer filter
Energy, 17–20, 23
    definition, 17
    units, 17
Exponential relationships, 34
Extension wire, 210

Farad, 326
Faraday's law, 251–52
Fast-blow fuse, 213
Ferrites, 243
Ferromagnetic materials, 243
Field intensity, versus flux density, 252–54
Field windings, insulating cover, 211
Filters, 408–30
    construction, 420–22
        L network, 421
        simple filters, 420–21
        T and Π networks, 421–22
    definition, 409
    families, 416–20
        Bessel, 416–17
        Butterworth, 417–18
        Cauer, 419
        Chebyshev, 418–19
        classic, 419–20
    fluctuating dc voltages, 422–26
        ac/dc voltage separation, 422–24
        bypass capacitors, 425–26
        coupling capacitors, 424–25
    types of, 412–16
        all-pass, 416
        band-pass, 415
        band-stop, 416
        high-pass, 414
        low-pass, 414–15
    See also specific types; families
Fixed-ground oscilloscopes, 119
Flemming's rule, 250–51
Flux, 241–42, 252
Flux density, 242–43, 246
    versus field intensity, 252–54
Flywheel effect, 432
FM broadcast band, 415
Foot-pounds, 37
Formulas, memorizing, 3–4
Frequency:
    cutoff, 414–16, 426
    nonsinusoidal waveforms, 277
    and resonance, 435–36
    sine waves, 273–76
Full-wave rectification, 269–70
Fuses, 212–13
    circuit breakers, 213–14
    current rating, 213
    testing, 213
    types of, 212
    voltage rating, 213
Fusible links, 214

Galvanometer, 105
Gauge size, wires, 203
Gilberts, 247
Gold, as a conductor, 199–200
Gram calorie, 18
Graphic method, for finding voltage/current levels, 355–60, 365
Ground, 120–25
    definition, 120
    ground/voltage relationships, 124
    oscilloscopes, 118–19
Ground circuit resistance test, 130–31

Half-step check, 127
Half-wave rectification, 269–70
Harmonic frequencies, nonsinusoidal waveforms, 280
Heat dissipation, *See* Power
Henry, 293
Hertz, 274
High-pass filters, 414, 426, 427
High-pass Π and T filter networks, 422
Hole theory of electron movement, 15
Hookup, oscilloscopes, 118–20
Horizontal input, oscilloscopes, 119
Horizontal position control, oscilloscopes, 118
Horsepower, conversion factor to watts, 37
Hydraulic circuits, 330–31
Hydrometers, 372–74, 382
Hypotenuse, 287, 288, 311, 338
Hysteresis, 253

Impedance, 210
Impedance blocks, 399–402
Impedance matching, 381
Impedance meter, 343
Individual multiplier voltmeters, design of, 222–23
Individual shunt current meter, 225–26
Inductance, 295–301
    mutual, 297–301
    parallel, 296–97
    reciprocal of, 296
    series, 295–96
Inductive kickback, 364
Inductive reactance:
    definition, 309, 319
    and Ohm's law, 310
    in parallel, 310
    ratio between ac resistance and, 435
    reciprocal of, 310
    in series, 310
    symbol for, 309
    versus capacitive reactance, 388–92, 403
Inductors, 286–323
    construction/design, 292–95
    factors limiting efficiency of, 292
    inductive reactance, 306–11
    right triangles, 287–90
    series, 413
    solution of *LR* circuits, 311–17
    transformers, 301–6
    troubleshooting, 317–18
Industrial electronics, 2
Instantaneous voltage, 268
Insulators, 7, 24
Integrator, 362–63, 365
Internal resistance:
    batteries, 380–81, 382
    meters, 219
Inversely proportional, definition, 28
Inverse (reciprocal) of resistance, *See* Conductance
Isotopes, 7

***j*-factor, 397–98, 403**
Joules, 19, 23, 24, 37
Joules/coulomb, 37
Joules/second, 37

***k* filter design, 419, 427**
Kilo, definition, 18
Kilowatt-hour, 24
Kirchhoff's laws, 170, 182–83
    current law (KCL), 83–84, 183, 193
    voltage law (KVL), 53, 82–83, 182–83, 193

***LCR* bridge, 106, 343**
Lead-acid battery, 372
    load testing, 374–75
    specific gravity testing, 372–74
Leads, ohmmeter, 115
Leclanche cell, *See* Carbon-zinc cell
Left-hand rule for pole identification, 244
Lenz's law, 251
L filter networks, 421, 427
Linear algebra, 176–82
    examples of linear equations, 176
    solving linear equations, 177–81
Load, circuits, 44, 45
Load testing, lead-acid batteries, 374–75
Logarithms, 349–51
    antilogarithm, 350
    characteristic, 350–51
    mantissa, 350–51
Low-pass filters, 414–15, 426
Low-pass Π and T filter networks, 422
LR circuits, solution of, 311–17
    ELI the ICEman, 311–17
    parallel-circuit triangles, 312
    phase angle of triangles, 312–13
    RL circuits, 313–17
    series-circuit triangles, 311
    triangles, 311
LR time constants, 364–65

Magnetic dipoles, 241, 253–54
Magnetic domain, 241
Magnetic field, 307
Magnetic strength, See Inductance
Magnetism, 240–57
   B-H curve, 252–54
   definition, 241
   electromagnetism, 244–46, 250–52
   flux, 241–42
   flux density, 242–43
   permeability, 243–44
   pole identification, left-hand rule for, 244
   reluctance, 246
   units and systems, 247–50
   See also Electromagnetism
Magnetometer, 243
Magnetomotive force (mmf), 244, 252
Mainline, 66
Manganese-alkaline cell, 376
Mantissa, 350–51
Mathematic method, for finding voltage/current levels, 360–61, 365
m-derived filter, 419, 427
Mercury cells, 376
Mesh current analysis, 183–93
   applications, 186–93
   procedure, 184–86
Metal-air cells, 376
Meter movements, 219–21
   adjustments, 221
   parts, 220
   resistance, 220
   specifications, 220
Meters:
   ampmeters, 113
   cascaded mutiplier voltmeters, 223–25
   current meters, 116–17
   digital meters, 219
   extending ranges of, 235–36
   galvanometers, 105
   hydrometers, 372–74, 382
   impedance meters, 343
   individual multiplier voltmeters, 222–23
   individual shunt current meters, 225–26
   internal resistance, 219
   magnetometers, 243
   meter movements, 219–21
   moving-coil (D'Arsonval) construction, 219
   multimeters, 113
   ohmmeters, 113, 114–15, 229, 318, 319, 343
   resistance meters, 229–32
   transistorized meters, 219
   use of, 113–20
   vacuum-tube meters, 219
   voltmeters, 113, 114, 221–25, 269
   VOM meters, 113, 217–39, 268

Mho, 22, 23, 67
Microfarads, 326, 343
Microswitches, 212
Mil, 201
Millifarads, 326
Millman's theorem, 165–68
   applied to a dual-voltage circuit, 166–67
   applied to other circuits, 167–68
MKS system, 247–49, 293
   flux density, 249
Moleculoy, as a conductor, 200
Motor action, 250–51
Moving-coil (D'Arsonval) construction, meters, 219
Multimeter, definition, 113
Mutual inductance, 297–301
   in parallel, 298–99
   in series, 298
   See also Inductance; Transformers

Nanofarads, 326
Napier, John, 349
Natural logarithms, 350
Negative charge, 6
Network theorems, 150–74
   delta and wye circuits, 169–72
   Millman's theorem, 165–68
   Norton's rule, 159–63
   source impedance and load matching, 163–65
   superposition, 151–54
   Thevenin's rule, 154–59
Neutron, 7
Nickel-cadmium battery, 375
Nickel-iron battery, 375
Nonconductors, See Insulators
Nonsinusoidal waveforms, 277–81
   frequency, 277
   harmonics, 280
   phase, 277–78
   square wave, 280–81
   triangle waveforms (sawtooth waveforms), 281
Nontempered silver, as a conductor, 199
North-seeking pole, 241–42
Norton's rule, 159–63
   compared to Thevenin's rule, 160–61
   and parallel circuits, 161–62
   procedure, 159–60
   simple Norton analysis, 160
Nucleus, 7

Octaves, 280, 411–12, 426
Ohm, 21, 23
Ohmmeters, 113, 114–15, 229, 318, 319, 343
   design of, 233
   leads, 115

   scaling, 232–33
   shunts, 230–32
Ohm's law, 4, 27–49, 54, 66, 71, 161, 314, 315, 340, 341
   basic algebra, 28–32
   finding primary/secondary winding impedance with, 303
   first, 32–34
      and graphs, 33
   and power, 37–41
   and reactance, 310, 334–35
   second, 34–35
   third, 35–36
   using, 41–43
Open circuit, 35, 44
Open-circuit voltage, 156
Orbits, 7–8
Oscilloscopes, 117–20, 268, 281, 317, 404
   controls, 118
   definition, 117
   ground, 118–19
   hookup, 118–20

Panel meters, 220
Parallel capacitance, 333–34
Parallel circuits, 63–80
   analysis, 71–76
   circuit identification, 64–65
   concepts, 65–71
   definition, 64
   fabricating resistor values, 76–77
   schematic representations of, 65
Parallel-circuit triangles, 312, 338
Parallel inductance, 296–97
   See also Inductance
Parallel resonance, 402, 404, 433–34
Parallel resonant analysis, 442–47
Parallel tank circuit, 413, 427
Paramagnetic materials, 243–44
Passive filters, 409
Peak-to-peak voltage, 268
Peak voltage, 268, 270
Period and frequency, sine waves, 273–76
Permeability, 243–44
Phase, nonsinusoidal waveforms, 277–78
Phase angle, 312–13, 337, 338, 394, 404
Picofarads, 326, 343
Polar notation, 396, 398–99
Pole identification, left-hand rule for, 244
Positive charge, 6
Potassium hydroxide, 376, 382
Potential, 20
Potential difference, 20–21, 23, 124
Power, 23, 37, 409–12, 426
   and Ohm's law, 37–41
   source and load resistance and, 164
Power cord wire, 210

Power factor, 392–94
   apparent power, 393
   reactive power, 394
   real power, 393–94
Power supply testing, 126
Primary cells, 375–77, 382
   carbon-zinc cell, 375–76
   manganese-alkaline cell, 376
   mercury cell, 376
   metal-air cells, 376
Primary winding, transformers, 301–3, 306, 319
Proton, 7
Pulse shaping, 362–63
Pushbutton switches, 212
Pythagorean theorem, 287–88, 313, 315, 316, 318, 340, 342

Ratios:
   definition, 138
   as directly proportional statements, 139
   understanding, 136–39
RC circuits, solution of, 337–42
   ELI the ICEman, 337–38, 344
   parallel-circuit triangles, 338
   series-circuit triangles, 338
   triangles, 337–38
      phase angle of, 338
RCL bridge, 317–18, 319
RCL circuit analysis, 387–407
   applied voltage as a reference, 394–96
   capacitive vs. inductive reactance, 388–92, 403
   impedance blocks, 399–402
   polar notation, 398–99
   power factor, 392–94
      apparent power, 393
      reactive power, 394
      real power, 393–94
   rectangular notation, 396–98
   resonance, 402–3
RC time constants, 363
Reactance, 295, 306–11, 319
   and Ohm's law, 310, 334–35
   See also Capacitive reactance; Inductive reactance
Reactive power, 394
Real power, 393–94
Reciprocal method of calculation, 76–77
Rectangular notation, 396–98
Rectified voltage, 269
Reluctance, 246, 249
Repair, of faulty components, 128
Resistance, 21–22, 23, 24, 200–210
   circuits, 44
   temperature coefficient of, 208–10
   of wire, 201–5
      calculation of, 205–8
Resistance meters:
   construction/design of, 229–32
      basic meter design, 229–30
      complete VOM meter layout, 233–35
      ohmmeter scaling, 232–33
      ohmmeter shunts, 230–32
Resistance wires, 202
Resistors:
   carbon-film resistors, 202
   fabricating resistor values, 76–77
   wirewound resistors, 202
Resonance, 402–4, 431–51
   circuit analysis, 436–47
      parallel resonant analysis, 442–47
      series resonant analysis, 437–42
   damping resistance, 447–51
   frequency/bandwidth, 435–36
   parallel resonance, 402, 433–34
   quality factor, 435
   series resonance, 402, 431
   tank circuit, 432–33
Retentivity, 253
Right triangles, 287–90
   Pythagorean theorem, 287–88, 313, 315, 316, 318, 340, 342
   trigonometric relations, 288–90
   See also Triangles
Ringing, definition, 317
RL circuits, 313–17
RMS voltage, 269–70
Rolloff, 412
   Bessel filters, 416–17
   Butterworth filters, 417–18
   Cauer filters, 419
   Chebyshev filters, 418–19
   classic filters, 419–20
Root-mean-square (rms) voltage, 269
Rotary switches, 211–12

Saturation, 253
Sawtooth waveforms, 281
Schematics, voltage test points, 121
Schematic symbols:
   circuits, 47
   parallel circuits, 65
   switches, 211
Secondary cells, 372–75, 382
   lead-acid battery, 372
   nickel-cadmium battery, 375
   nickel-iron battery, 375
   silver-zinc battery, 375
Secondary winding, transformers, 301–3, 306, 319
Semiconductors, 9, 24
Sensitivity, 221–22
   VOM meters, 218
Series capacitance, 331–33
Series capacitors, 413
Series circuits, 50–62
   analysis, 56–60
   circuit identification, 51–52
   concepts, 52–56
   current flow, direction of, 51

Series-circuit triangles, 311, 338
Series inductance, 295–96
   See also Inductance
Series inductor, 413
Series-parallel circuits, 81–111
   analysis, 89–104
   finding total resistance, 211
   Kirchhoff's law, 82–85
Series resistance, capacitors, 342
Series resonance, 402, 404, 431
Series resonant analysis, 437–42
Series tank circuit, 413, 427
Short, 35
Short-circuit voltage test, 131–32
Shunt resistors, 225
Siemens, 22, 23
Signal injection, 127
Signal (voltage level) tracing, 126–27, 342
Silver, as a conductor, 199
Silver-zinc battery, 375
Simplification, definition, 102
Simplified circuits, 102–3
Sine, 287, 289–90
Sine waves, 259–63, 331
   amplitude, 259
   period, 259
   period and frequency, 273–76
   quadrants, 261
   sine values, 261–63
   wavelength, 276–77
Single-element filters, 421–22, 427
Single pole, double throw switches, 211
Single pole, single throw switches, 211
Single-trace oscilloscopes, 119
SI system, 247
Skin effect, 292
Slide switches, 211–12
Slow-blow fuse, 212
Soh-Cah-Toa, 289–90
Solid-core wires, 210
Source impedance and load matching, 163–65
Square law principle, 242
Square wave, 280–81, 362, 365
Standard voltage divider formula, 142
Step-up/step-down transformers, 301
Storage batteries, *See* Secondary cells
Stored energy, 295
Stranded wires, 210
Suborbitals, 9
Superposition, 151–54
   example with opposing voltages, 152–53
   required steps, 153
Supposition, 76
Susceptance, 310, 333
Switches:
   circuits, 211–12
   microswitches, 212
   pushbutton switches, 212
   rotary, 211–12
   schematic symbols, 211

Switches (*cont.*)
    slide switches, 211–12
Symbols, elements, 10

Tangent, 287, 289–90
Tank circuit, 413, 432–33
    flywheel effect, 432
Tchebycheff filters, *See* Chebyshev filters
Temperature coefficient, of resistance, 208–210
Testing:
    block, 126–28
    component, 128
    fuses, 213
    power supply, 126
Thevenin's rule, 154–59
    procedure, 154
    and resistance in parallel with the battery, 157–59
    and resistance in series with the load, 156–57
    simple Thevenin analysis, 155–56
T and Π filter networks, 421–22
Time constant:
    definition, 353
    and differentiator design, 362
    and integrator design, 363
Time/division control, oscilloscopes, 118
Times-minus formula, 77
Total circuit resistance, 54
Total current, 66
Total power dissipation, 55
Total resistance finding, 85–89
Transformers, 301–6, 319
    conversion factors, 302
    efficiency, 305
    phase relations, 306
    turns ratio, 301–3
    See also Mutual inductance
Transformer windings, insulating cover, 211
Transistorized meters, internal resistance, 219
Transmission points, most favorable, 164
Triangles, 311, 337–38
    parallel-circuit triangles, 312
    phase angle of, 312–13, 338
    right, 287–90
    series-circuit triangles, 311
Triangle wave, 362
Trignometric relations, 288–91
Triple-element filters, 421, 427
Troubleshooting:
    capacitors, 342–43
    dc circuits, 112–34
    inductors, 317–18
Trunk resistances, 87
Turns ratio, 301–3
Twin lead, 210

Unloaded voltage division, 139–43

Vacuum-tube meters, internal resistance, 219
Valence shell, 8–9, 24
VAR, *See* Volt-amperes reactance (VAR)
Vertical input, oscilloscopes, 119
Vertical position control, oscilloscopes, 118
Visual inspection, dc circuits, 125–26
Volt, 23
Voltage, 23, 24, 37, 193, 409–12, 426
    ground/voltage relationships, 124
    Kirchhoff's voltage law (KVL), 53, 82–83, 182–83, 193
    measurement of, 268–76
        average or dc voltage, 269
        conversion factors, 269–70
        instantaneous voltage, 268
        peak-to-peak voltage, 268
        peak voltage, 268, 270
        RMS voltage, 269–70
    open-circuit, 156
    transients in, 348–69
Voltage division, 135–49
    capacitive, 335–36
    loaded, 145–47
    understanding ratios, 136–39
    unloaded, 139–43
Voltage leakage, capacitors, 342–43
Voltage levels:
    ELI the ICEman, use of, 351–52
    methods for finding, 353–61
        graphic method, 355–60, 365
        mathematic method, 360–61, 365
    time constants, 352–35
Voltage level tracing, 126–27
Voltage ramp, 281
Voltage rating, fuses, 213
Voltage source, 163
Voltaic batteries, *See* Primary cells
Volt-amperes, 37, 404
Volt-amperes reactance (VAR), 394, 404
Voltmeters, 113, 114, 269
    cascaded multiplier voltmeters, 223–25
    construction and design of, 221–25
    individual multiplier voltmeters, 222–23
    sensitivity, 221–222
    Volt-ohm-milliampere meters, *See* VOM meters, 113
    Volts/division control, oscilloscopes, 118
VOM meters, 113, 268
    design, 217–39
    internal resistance, 219
    leads, 115, 116
    loading, 218–19
    sensitivity, 218
VTVM, 268, 269

Watt, 23
Watt-second, 18, 23
Wavelength, 276–77
Wave shaping, 362–63
Wheatstone bridge, 104–5
Wire, construction of, 210–11
Wire resistance, 205–8
Wirewound resistors, 202
Work, 23, 37
    definition, 17
Wye and delta networks, filters, 421–22

Zero circuit resistance, 35
Z meter, 343